Two Manuals Combined into One Book

Military Mountaineering

And

Rappelling

Department of Defense

Pentagon
Publishing

PREFACE

Mountains exist in almost every country in the world and almost every war has included some type of mountain operations. This pattern will not change; therefore, soldiers will fight in mountainous terrain in future conflicts. Although mountain operations have not changed, several advancements in equipment and transportation have increased the soldiers' capabilities. The helicopter now allows access to terrain that was once unreachable or could be reached only by slow methodical climbing. Inclement weather, however, may place various restrictions on the capabilities of air assets available to a commander. The unit must then possess the necessary mountaineering skills to overcome adverse terrain to reach an objective.

This field manual details techniques soldiers and leaders must know to cope with mountainous terrain. These techniques are the foundation upon which the mountaineer must build. They must be applied to the various situations encountered to include river crossings, glaciers, snow-covered mountains, ice climbing, rock climbing, and urban vertical environments. The degree to which this training is applied must be varied to conform to known enemy doctrine, tactics, and actions. This FM also discusses basic and advanced techniques to include acclimatization, illness and injury, equipment, anchors, evacuation, movement on glaciers, and training.

This field manual is a training aid for use by *qualified* personnel in conjunction with FM 3-97.6, Mountain Operations, which is used for planning operations in mountainous terrain. Personnel using FM 3-97.61 should attend a recognized Department of Defense Mountain Warfare School for proper training. **Improper use of techniques and procedures by untrained personnel may result in serious injury or death.** Personnel should be certified as Level I, Basic Mountaineer; Level II, Assault Climber; or Level III, Mountain Leader before using FM 3-97.61 for training (see Appendix A).

The measurements in this manual are stated as they are used in training (either metric or standard). Appendix B contains a measurement conversion chart for your convenience.

The proponent of this publication is HQ TRADOC. Submit changes for improving this publication to doctrine@benning.army.mil or on DA Form 2028 (Recommended Changes to Publications and Blank Forms) and forward to the Commander, U.S. Army Infantry School, ATTN: ATSH-RBO, Fort Benning, GA 31905-5593.

Unless otherwise stated, whenever the masculine gender is used, both men and women are included.

FIELD MANUAL
No. 3-97.61

HEADQUARTERS
DEPARTMENT OF THE ARMY
WASHINGTON, DC, 26 August 2002

MILITARY MOUNTAINEERING

CONTENTS

*This publication supersedes TC 90-6-1, dated 26 April 1989.

CHAPTER 1

MOUNTAIN TERRAIN, WEATHER, AND HAZARDS

Commanders must consider the effects terrain and weather will have on their operations, mainly on their troops and logistics efforts. Weather and terrain combine to challenge efforts in moving supplies to forward areas. Spring storms, which may deposit a foot of snow on dry roads, combined with unprepared vehicles create hazardous situations. Helicopters are a valuable asset for use in moving men and supplies, but commanders should not plan to use them as the only means of movement and resupply. Alternate methods must be planned due to the variability of weather. Units scheduled for deployment in mountainous terrain should become self-sufficient and train under various conditions. Commanders must be familiar with the restraints that the terrain can place on a unit.

Section I. MOUNTAIN TERRAIN

Operations in the mountains require soldiers to be physically fit and leaders to be experienced in operations in this terrain. Problems arise in moving men and transporting loads up and down steep and varied terrain in order to accomplish the mission. Chances for success in this environment are greater when a leader has experience operating under the same conditions as his men. Acclimatization, conditioning, and training are important factors in successful military mountaineering.

1-1. DEFINITION

Mountains are land forms that rise more than 500 meters above the surrounding plain and are characterized by steep slopes. Slopes commonly range from 4 to 45 degrees. Cliffs and precipices may be vertical or overhanging. Mountains may consist of an isolated peak, single ridges, glaciers, snowfields, compartments, or complex ranges extending for long distances and obstructing movement. Mountains usually favor the defense; however, attacks can succeed by using detailed planning, rehearsals, surprise, and well-led troops.

1-2. COMPOSITION

All mountains are made up of rocks and all rocks of minerals (compounds that cannot be broken down except by chemical action). Of the approximately 2,000 known minerals, seven rock-forming minerals make up most of the earth's crust: quartz and feldspar make up granite and sandstone; olivene and pyroxene give basalt its dark color; and amphibole and biotite (mica) are the black crystalline specks in granitic rocks. Except for calcite, found in limestone, they all contain silicon and are often referred to as silicates.

1-3. ROCK AND SLOPE TYPES

Different types of rock and different slopes present different hazards. The following paragraphs discuss the characteristics and hazards of the different rocks and slopes.

a. **Granite.** Granite produces fewer rockfalls, but jagged edges make pulling rope and raising equipment more difficult. Granite is abrasive and increases the danger of ropes or accessory cords being cut. Climbers must beware of large loose boulders. After a rain, granite dries quickly. Most climbing holds are found in cracks. Face climbing can be found, however, it cannot be protected.

b. **Chalk and Limestone.** Chalk and limestone are slippery when wet. Limestone is usually solid; however, conglomerate type stones may be loose. Limestone has pockets, face climbing, and cracks.

c. **Slate and Gneiss.** Slate and gneiss can be firm and or brittle in the same area (red coloring indicates brittle areas). Rockfall danger is high, and small rocks may break off when pulled or when pitons are emplaced.

d. **Sandstone.** Sandstone is usually soft causing handholds and footholds to break away under pressure. Chocks placed in sandstone may or may not hold. Sandstone should be allowed to dry for a couple of days after a rain before climbing on it—wet sandstone is extremely soft. Most climbs follow a crack. Face climbing is possible, but any outward pull will break off handholds and footholds, and it is usually difficult to protect.

e. **Grassy Slopes.** Penetrating roots and increased frost cracking cause a continuous loosening process. Grassy slopes are slippery after rain, new snow, and dew. After long, dry spells clumps of the slope tend to break away. Weight should be distributed evenly; for example, use flat hand push holds instead of finger pull holds.

f. **Firm Spring Snow (Firn Snow).** Stopping a slide on small, leftover snow patches in late spring can be difficult. Routes should be planned to avoid these dangers. Self-arrest should be practiced before encountering this situation. Beginning climbers should be secured with rope when climbing on this type surface. Climbers can glissade down firn snow if necessary. Firn snow is easier to ascend than walking up scree or talus.

g. **Talus.** Talus is rocks that are larger than a dinner plate, but smaller than boulders. They can be used as stepping-stones to ascend or descend a slope. However, if a talus rock slips away it can produce more injury than scree because of its size.

h. **Scree.** Scree is small rocks that are from pebble size to dinner plate size. Running down scree is an effective method of descending in a hurry. One can run at full stride without worry—the whole scree field is moving with you. Climbers must beware of larger rocks that may be solidly planted under the scree. Ascending scree is a tedious task. The scree does not provide a solid platform and will only slide under foot. If possible, avoid scree when ascending.

1-4. ROCK CLASSIFICATIONS

Rock is classified by origin and mineral composition.

a. **Igneous Rocks.** Deep within the earth's crust and mantle, internal heat, friction and radioactive decay creates magmas (melts of silicate minerals) that solidify into igneous rocks upon cooling. When the cooling occurs at depth, under pressure, and over time, the minerals in the magma crystallize slowly and develop well, making coarse-grained plutonic rock. The magma may move upward, propelled by its own lower density, either melting and combining with the overlying layers or forcing them aside. This results in an intrusive rock. If the melt erupts onto the surface it cools rapidly and the minerals form little or no crystal matrix, creating a volcanic or extrusive rock.

(1) ***Plutonic or Intrusive Rocks.*** Slow crystallization from deeply buried magmas generally means good climbing, since the minerals formed are relatively large and interwoven into a solid matrix. Weathering develops protrusions of resistant minerals, which makes for either a rough-surfaced rock with excellent friction, or, if the resistant

crystals are much larger than the surrounding matrix, a surface with numerous knobby holds. Pieces of foreign rock included in the plutonic body while it was rising and crystallizing, or clusters of segregated minerals, may weather differently than the main rock mass and form "chicken heads."

(a) Intrusions are named according to location and size. Large (100 square kilometers or larger) masses of plutonic rock are called "batholiths" and small ones "stocks." Most plutonic rock is in the granite family, differing only in the amounts of constituent minerals contained. A core of such batholiths is in every major mountain system in the world. In the Alps, Sierras, North Cascades, Rockies, Adirondacks, and most other ranges this core is at least partly exposed.

(b) Small plutonic intrusions are stocks, forced between sedimentary strata, and dikes, which cut across the strata. Many of these small intrusive bodies are quickly cooled and thus may look like extrusive rock.

(2) ***Volcanic or Extrusive Rocks.*** Explosive eruptions eject molten rock so quickly into the air that it hardens into loose aerated masses of fine crystals and uncrystallized glass (obsidian). When this ash consolidates while molten or after cooling, it is called "tuff," a weak rock that breaks down quickly and erodes easily. Quieter eruptions, where widespread lava flows from large fissures, produce basalt. Basaltic rocks are fine-grained and often sharp-edged.

(3) ***Jointing Rocks.*** In plutonic rocks, joints or cracks are caused by internal stresses such as contraction during cooling or expansion when overlying rock erodes or exfoliates. Some joints tend to follow a consistent pattern throughout an entire mountain and their existence can often be predicted. Therefore, when a ledge suddenly ends, the joint—and thus the ledge—may begin again around the corner. When molten rock extrudes onto the surface as a lava flow or intrudes into a cold surrounding mass as a dike or sill, the contraction from rapid cooling usually causes so much jointing that climbing can be extremely hazardous. Occasionally, this jointing is regular enough to create massed pillars with usable vertical cracks such as Devil's Tower in Wyoming.

b. **Sedimentary Rocks.** Sedimentary rocks are born high in the mountains, where erosion grinds down debris and moves it down to rivers for transportation to its final deposition in valleys, lakes, or oceans. As sediments accumulate, the bottom layers are solidified by pressure and by mineral cements precipitated from percolating groundwater. Gravel and boulders are transformed into conglomerates; sandy beaches into sandstone; beds of mud into mudstone or shale; and shell beds and coral reefs into limestone or dolomite.

(1) Though in general sedimentary rocks are much more friable than those cooled from molten magmas, pressure and cementing often produce solid rocks. In fact, by sealing up internal cracks cementing can result in flawless surfaces, especially in limestone.

(2) Most high mountain ranges have some sedimentary peaks. Ancient seafloor limestone can be found on the summits of the Himalayas and the Alps. The Canadian Rockies are almost exclusively limestone. With the exception of the Dolomites, in general sedimentary rocks do not offer high-angle climbing comparable to that of granite.

c. **Metamorphic Rocks.** These are igneous or sedimentary rocks that have been altered physically and or chemically by the tremendous heat and pressures within the earth. After sediments are solidified, high heat and pressure can cause their minerals to recrystallize. The bedding planes (strata) may also be distorted by folding and squeezing. Shale changes to slate or schist, sandstone and conglomerate into quartzite, and limestone to marble. These changes may be minimal, only slightly altering the sediments, or extensive enough to produce gneiss, which is almost indistinguishable from igneous rock.

(1) Metamorphic rocks may have not only joints and bedding, but cleavage or foliation, a series of thinly spaced cracks caused by the pressures of folding. Because of this cleavage, lower grades of metamorphic rocks may be completely unsuitable for climbing because the rock is too rotten for safe movement.

(2) Higher degrees of metamorphism or metamorphism of the right rocks provide a solid climbing surface. The Shawangunks of New York are an excellent example of high-grade conglomerate quartzite, which offers world class climbing. The center of the Green Mountain anticline contains heavily metamorphosed schist, which also provides solid climbing.

1-5. MOUNTAIN BUILDING

The two primary mechanisms for mountain-building are volcanic and tectonic activity. Volcanoes are constructed from lava and ash, which begin within the earth as magma. Tectonic activity causes plates to collide, heaving up fold mountains, and to pull apart and crack, forming fault-block mountain ranges.

a. **Plate Tectonics.** The massive slabs composing the outer layer are called tectonic plates. These plates are made up of portions of lighter, granitic continental crust, and heavier, basaltic oceanic crust attached to slabs of the rigid upper mantle. Floating slowly over the more malleable asthenosphere, their movement relative to each other creates earthquakes, volcanoes, ocean trenches, and mountain ridge systems.

b. **Mountain Structure.** The different horizontal and vertical stresses that create mountains usually produce complex patterns. Each type of stress produces a typical structure, and most mountains can be described in terms of these structures.

(1) ***Dome Mountains.*** A simple upward bulge of the crust forms dome mountains such as the Ozarks of Arkansas and Missouri, New York's Adirondacks, the Olympics of Washington, and the High Uintahs of Utah. They are usually the result of the upward movement of magma and the folding of the rock layers overhead. Erosion may strip away the overlying layers, exposing the central igneous core.

(2) ***Fault-Block Mountains.*** Faulting, or cracking of the crust into large chunks, often accompanies upwarp, which results in fault-block mountains. Many forms are created by the motion of these chunks along these faults.

(a) The ranges of the desert country of California, Nevada, and Utah provide the clearest display of faulting. The breakage extends to the surface and often during earthquakes—caused by slippage between the blocks—fresh scarps many feet high develop.

(b) Sometimes a block is faulted on both sides and rises or falls as a unit. More often, however, it is faulted on one side only. The Tetons of Wyoming and the Sierra Nevada display this—along the single zone of faults the range throws up impressive steep scarps, while on the other side the block bends but does not break, leaving a gentler slope from the base of the range to the crest. An example of a dropped block is California's Death Valley, which is below sea level and could not have been carved by erosion.

(3) ***Fold Mountains.*** Tectonic forces, in which continental plates collide or ride over each other, have given rise to the most common mountain form—fold mountains. Geologists call folds geosynclines. Upward folded strata are anticlines and downward folds are synclines. When erosion strips down the overburden of rock from folded mountain ranges, the oldest, central core is all that remains. The Alps and the Appalachians are examples of fold mountains. When the squeezing of a range is intense the rocks of the mountain mass first fold but then may break, and parts of the rocks are pushed sideways and override neighboring formations. This explains why older rocks are often found perched on top of younger ones. Isolated blocks of the over thrust mass may form when erosion strips away links connecting them with their place of origin. Almost every range of folded mountains in the world exhibits an over thrust of one sort or another.

(4) ***Volcanic Mountains.*** Along convergent plate boundaries volcanic activity increases. As it is forced underneath an overriding neighbor, continental crust melts and turns to magma within the mantle. Since it is less dense than the surrounding material it rises and erupts to form volcanoes.

(a) These volcanoes are found in belts, which correspond to continental margins around the world. The best known is the "Ring of Fire" encircling the Pacific Ocean from Katmai in Alaska through the Cascades (Mount Rainier and Mount Saint Helens) down through Mexico's Popocatepetl to the smokes of Tierra del Fuego. This belt then runs west down the Aleutian chain to Kamchatka, south to the volcanoes of Japan and the Philippines, and then east through New Guinea into the Pacific. Smaller volcanic belts are found along the Indonesian-SE Asian arc, the Caucasus region, and the Mediterranean.

(b) Volcanic activity also arises at boundaries where two plates are moving away from each other, creating deep rifts and long ridges where the crust has cracked apart and magma wells up to create new surface material. Examples of this are the Mid-Atlantic Ridge, which has created Iceland and the Azores, and the Rift Valley of East Africa with Kilimanjaro's cone.

(5) ***Complex Mountains.*** Most ranges are complex mountains with portions that have been subject to several processes. A block may have been simply pushed upward without tilting with other portions folded, domed, and faulted, often with a sprinkling of volcanoes. In addition, these processes occur both at the macro and the micro level. One massive fold can make an entire mountain peak; however, there are folds measured by a rope length, and tiny folds found within a handhold. A mountain front may be formed from a single fault, but smaller faults that form ledges and gullies may also be present.

1-6. ROUTE CLASSIFICATION

Military mountaineers must be able to assess a vertical obstacle, develop a course of action to overcome the obstacle, and have the skills to accomplish the plan. Assessment of a vertical obstacle requires experience in the classifications of routes and understanding the levels of difficulty they represent. Without a solid understanding of the difficulty of a chosen route, the mountain leader can place his life and the life of other soldiers in extreme danger. Ignorance is the most dangerous hazard in the mountain environment.

a. In North America the Yosemite Decimal System (YDS) is used to rate the difficulty of routes in mountainous terrain. The YDS classes are:

- Class 1—Hiking trail.
- Class 2—Off-trail scramble.
- Class 3—Climbing, use of ropes for beginners (moderate scrambling).
- Class 4—Belayed climbing. (This is moderate to difficult scrambling, which may have some exposure.)
- Class 5—Free climbing. (This class requires climbers to be roped up, belay and emplace intermediate protection.)

b. Class 5 is further subdivided into the following classifications:

(1) Class 5.0-5.4—Little difficulty. This is the simplest form of free climbing. Hands are necessary to support balance. This is sometimes referred to as advanced rock scrambling.

(2) Class 5.5—Moderate difficulty. Three points of contact are necessary.

(3) Class 5.6—Medium difficulty. The climber can experience vertical position or overhangs where good grips can require moderate levels of energy expenditure.

(4) Class 5.7—Great difficulty. Considerable climbing experience is necessary. Longer stretches of climbing requiring several points of intermediate protection. Higher levels of energy expenditure will be experienced.

(5) Class 5.8—Very great difficulty. Increasing amount of intermediate protection is the rule. High physical conditioning, climbing technique, and experience required.

(6) Class 5.9—Extremely great difficulty. Requires well above average ability and excellent condition. Exposed positions, often combined with small belay points. Passages of the difficult sections can often be accomplished under good conditions. Often combined with aid climbing (A0-A4).

(7) Class 5.10—Extraordinary difficulty. Climb only with improved equipment and intense training. Besides acrobatic climbing technique, mastery of refined security technique is indispensable. Often combined with aid climbing (A0-A4).

(8) Class 5.11-5.14—Greater increases of difficulty, requiring more climbing ability, experience, and energy expenditure. Only talented and dedicated climbers reach this level.

c. Additional classifications include the following.

(1) Classes are further divided into a, b, c, and d categories starting from 5.10 to 5.14 (for example, 5.10d).

(2) Classes are also further divided from 5.9 and below with +/- categories (for example, 5.8+).

(3) All class 5 climbs can also be designated with "R" or "X," which indicates a run-out on a climb. This means that placement of intermediate protection is not possible on portions of the route. (For example, in a classification of 5.8R, the "R" indicates periods of run-out where, if a fall was experienced, ground fall would occur.) Always check the local guidebook to find specific designation for your area.

(4) All class 5 climbs can also be designated with "stars." These refer to the popularity of the climb to the local area. Climbs are represented by a single "star" up to five "stars;" a five-star climb is a classic climb and is usually aesthetically pleasing.

d. Aid climb difficulty classification includes:

(1) A0—"French-free." This technique involves using a piece of gear to make progress; for example, clipping a sling into a bolt or piece of protection and then pulling up on it or stepping up in the sling. Usually only needed to get past one or two more difficult moves on advanced free climbs.

(2) A1—Easy aid. The placement of protection is straight forward and reliable. There is usually no high risk of any piece of protection pulling out. This technique requires etriers and is fast and simple.

(3) A2—Moderate aid. The placement of protection is generally straight forward, but placement can be awkward and strenuous. Usually A2 involves one or two moves that are difficult with good protection placement below and above the difficult moves. No serious fall danger.

(4) A3—Hard aid. This technique requires testing your protection. It involves several awkward and strenuous moves in a row. Generally solid placements which will hold a fall and are found within a full rope length. However, long fall potential does exist, with falls of 40 to 60 feet and intermediate protection on the awkward placements failing. These falls, however, are usually clean and with no serious bodily harm.

(5) A4—Serious aid. This technique requires lots of training and practice. More like walking on eggs so none of them break. Leads will usually take extended amounts of time which cause the lead climber to doubt and worry about each placement. Protection placed will usually only hold a climber's weight and falls can be as long as two-thirds the rope length.

(6) A5—Extreme aid. All protection is sketchy at best. Usually no protection placed on the entire route can be trusted to stop a fall.

(7) A6—Extremely severe aid. Continuous A5 climbing with A5 belay stations. If the leader falls, the whole rope team will probably experience ground fall.

(8) Aid climbing classes are also further divided into +/- categories, such as A3+ or A3-, which would simply refer to easy or hard.

e. Grade ratings (commitment grades) inform the climber of the approximate time a climber trained to the level of the climb will take to complete the route.

- I—Several hours.
- II—Half of a day.
- III—About three-fourths of a day.
- IV—Long hard day (usually not less than 5.7).
- V—1 1/2 to 2 1/2 days (usually not less than 5.8).
- VI—Greater than 2 days.

f. Climbing difficulties are rated by different systems. Table 1-1 shows a comparison of these systems.

- YDS (Yosemite Decimal System)—Used in the United States.
- UIAA (Union des International Alpine Association)—Used in Europe.
- British—The British use adjectives and numbers to designate the difficulty of climbs. This system can be confusing if the climber is not familiar with it.
- French—The French use numbers and letters to designate the difficulty of climbs.
- Brazilian—Brazil uses Roman Numerals and adjectives to designate difficulty.
- Australian—Australia uses only numbers to designate difficulty.

YDS	UIAA	BRITISH	FRENCH	BRAZIL	AUSTRALIA
Class 1	I	easy (E)			
Class 2	II	easy (E)			
Class 3	III	easy (E)	1a, b, c		
Class 4	III-	moderate (MOD)	1a, b, c		
5.0	III	moderate (MOD)	2a, b		4
5.1	III+	difficult (DIFF)	2a, b		5
5.2	IV-	hard difficult	2c, 3a		6
5.3	IV	very difficult	3b, c, 4a		7
5.4	IV+	hard very difficult	3b, c, 4a	II	8, 9
5.5	V-	mild severe	3b, c, 4a	IIsup	10, 11
5.6	V	severe, hard severe, 4a	4a, b, c	III	12, 13
5.7	V+	severe, hard severe, 4b	4a, b, c	IIIsup	14
5.8	VI-	hard severe, hard very severe, 4c	5a, b	IV	15
5.9	VI	5a	5b, c	IVsup	16, 17
5.10a	VII-	E1, 5b	5b, c	V	18
5.10b	VII	E1, 5b	5b, c	Vsup	19
5.10c	VII	E1, 5b	5b, c	VI	20
5.10d	VII+	E1/E2, 5b-5c	5b, c	VIsup	21
5.11a	VIII-	E3, 6a	6a, b, c	VII	22
5.11b	VIII	E3/E4, 6a	6a, b, c	VII	23
5.11c	VIII	E4, 6b	6a, b, c	VIIsup	24
5.12a	IX-	E5, E6/7, 6c	7a	VIII	26
5.12b	IX	E5, E6/7, 6c	7a	VIIIsup	27
5.12c	IX	E5, E6/7, 6c	7a		28
5.12d	IX+	E6/7, 7a	7a		29

Table 1-1. Rating systems.

g. Ice climbing ratings can have commitment ratings and technical ratings. The numerical ratings are often prefaced with WI (waterfall ice), AI (alpine ice), or M (mixed rock and ice).

(1) ***Commitment Ratings.*** Commitment ratings are expressed in Roman numerals.

- I—A short, easy climb near the road, with no avalanche hazard and a straightforward descent.
- II—A route of one or two pitches within a short distance of rescue assistance, with little objective hazard.
- III—A multipitch route at low elevation, or a one-pitch climb with an approach that takes about an hour. The route requires anywhere from a few hours to a long day to complete. The descent may require building rappel anchors, and the route might be prone to avalanche.
- IV—A multipitch route at higher elevations; may require several hours of approach on skis or foot. This route is subject to objective hazards, possibly with a hazardous descent.
- V—A long climb in a remote setting, requiring all day to complete the climb itself. Requires many rappels off anchors for the descent. This route has sustained exposure to avalanche or other objective hazards.

- VI—A long ice climb in an alpine setting, with sustained technical climbing. Only elite climbers will complete it in a day. A difficult and involved approach and descent, with objective hazards ever-present, all in a remote area.
- VII—Everything a grade VI has, and more of it. Possibly days to approach the climb, and objective hazards rendering survival as questionable. Difficult physically and mentally.

(2) ***Technical Ratings.*** Technical ratings are expressed as Arabic numerals.

- 1—A frozen lake or stream bed.
- 2—A pitch with short sections of ice up to 80 degrees; lots of opportunity for protection and good anchors.
- 3—Sustained ice up to 80 degrees; the ice is usually good, with places to rest, but it requires skill at placing protection and setting anchors.
- 4—A sustained pitch that is vertical or slightly less than vertical; may have special features such as chandeliers and run-outs between protection.
- 5—A long, strenuous pitch, possibly 50 meters of 85- to 90-degree ice with few if any rests between anchors. The pitch may be shorter, but on featureless ice. Good skills at placing protection are required.
- 6—A full 50-meter pitch of dead vertical ice, possibly of poor quality; requires efficiency of movement and ability to place protection while in awkward stances.
- 7—A full rope length of thin vertical or overhanging ice of dubious adhesion. An extremely tough pitch, physically and mentally, requiring agility and creativity.
- 8—Simply the hardest ice climbing ever done; extremely bold and gymnastic.

1-7. CROSS-COUNTRY MOVEMENT

Soldiers must know the terrain to determine the feasible routes for cross-country movement when no roads or trails are available.

a. A pre-operations intelligence effort should include topographic and photographic map coverage as well as detailed weather data for the area of operations. When planning mountain operations, additional information may be needed about size, location, and characteristics of landforms; drainage; types of rock and soil; and the density and distribution of vegetation. Control must be decentralized to lower levels because of varied terrain, erratic weather, and communication problems inherent to mountainous regions.

b. Movement is often restricted due to terrain and weather. The erratic weather requires that soldiers be prepared for wide variations in temperature, types, and amounts of precipitation.

(1) Movement above the timberline reduces the amount of protective cover available at lower elevations. The logistical problem is important; therefore, each man must be self-sufficient to cope with normal weather changes using materials from his rucksack.

(2) Movement during a storm is difficult due to poor visibility and bad footing on steep terrain. Although the temperature is often higher during a storm than during clear weather, the dampness of rain and snow and the penetration of wind cause soldiers to chill quickly. Although climbers should get off the high ground and seek shelter and

warmth, if possible, during severe mountain storms, capable commanders may use reduced visibility to achieve tactical surprise.

c. When the tactical situation requires continued movement during a storm, the following precautions should be observed:

- Maintain visual contact.
- Keep warm. Maintain energy and body heat by eating and drinking often; carry food that can be eaten quickly and while on the move.
- Keep dry. Wear wet-weather clothing when appropriate, but do not overdress, which can cause excessive perspiration and dampen clothing. As soon as the objective is reached and shelter secured, put on dry clothing.
- Do not rush. Hasty movement during storms leads to breaks in contact and accidents.
- If lost, stay warm, dry, and calm.
- Do not use ravines as routes of approach during a storm as they often fill with water and are prone to flash floods.
- Avoid high pinnacles and ridgelines during electrical storms.
- Avoid areas of potential avalanche or rock-fall danger.

1-8. COVER AND CONCEALMENT

When moving in the mountains, outcroppings, boulders, heavy vegetation, and intermediate terrain can provide cover and concealment. Digging fighting positions and temporary fortifications is difficult because soil is often thin or stony. The selection of dug-in positions requires detailed planning. Some rock types, such as volcanic tuff, are easily excavated. In other areas, boulders and other loose rocks can be used for building hasty fortifications. In alpine environments, snow and ice blocks may be cut and stacked to supplement dug-in positions. As in all operations, positions and routes must be camouflaged to blend in with the surrounding terrain to prevent aerial detection.

1-9. OBSERVATION

Observation in mountains varies because of weather and ground cover. The dominating height of mountainous terrain permits excellent long-range observation. However, rapidly changing weather with frequent periods of high winds, rain, snow, sleet, hail, and fog can limit visibility. The rugged nature of the terrain often produces dead space at midranges.

a. Low cloud cover at higher elevations may neutralize the effectiveness of OPs established on peaks or mountaintops. High wind speeds and sound often mask the noises of troop movement. Several OPs may need to be established laterally, in depth, and at varying altitudes to provide visual coverage of the battle area.

b. Conversely, the nature of the terrain can be used to provide concealment from observation. This concealment can be obtained in the dead space. Mountainous regions are subject to intense shadowing effects when the sun is low in relatively clear skies. The contrast from lighted to shaded areas causes visual acuity in the shaded regions to be considerably reduced. These shadowed areas can provide increased concealment when combined with other camouflage and should be considered in maneuver plans.

1-10. FIELDS OF FIRE

Fields of fire, like observation, are excellent at long ranges. However, dead space is a problem at short ranges. When forces cannot be positioned to cover dead space with direct fires, mines and obstacles or indirect fire must be used. Range determination is deceptive in mountainous terrain. Soldiers must routinely train in range estimation in mountainous regions to maintain their proficiency.

Section II. MOUNTAIN WEATHER

Most people subconsciously "forecast" the weather. If they look outside and see dark clouds they may decide to take rain gear. If an unexpected wind strikes, people glance to the sky for other bad signs. A conscious effort to follow weather changes will ultimately lead to a more accurate forecast. An analysis of mountain weather and how it is affected by mountain terrain shows that such weather is prone to patterns and is usually severe, but patterns are less obvious in mountainous terrain than in other areas. Conditions greatly change with altitude, latitude, and exposure to atmospheric winds and air masses. Mountain weather can be extremely erratic. It varies from stormy winds to calm, and from extreme cold to warmth within a short time or with a minor shift in locality. The severity and variance of the weather causes it to have a major impact on military operations.

1-11. CONSIDERATIONS FOR PLANNING

Mountain weather can be either a dangerous obstacle to operations or a valuable aid, depending on how well it is understood and to what extent advantage is taken of its peculiar characteristics.

a. Weather often determines the success or failure of a mission since it is highly changeable. Military operations plans must be flexible, especially in planning airmobile and airborne operations. The weather must be anticipated to allow enough time for planning so that the leaders of subordinate units can use their initiative in turning an important weather factor in their favor. The clouds that often cover the tops of mountains and the fogs that cover valleys are an excellent means of concealing movements that normally are made during darkness or in smoke. Limited visibility can be used as a combat multiplier.

b. The safety or danger of almost all high mountain regions, especially in winter, depends upon a change of a few degrees of temperature above or below the freezing point. Ease and speed of travel depend mainly on the weather. Terrain that can be crossed swiftly and safely one day may become impassable or highly dangerous the next due to snowfall, rainfall, or a rise in temperature. The reverse can happen just as quickly. The prevalence of avalanches depends on terrain, snow conditions, and weather factors.

c. Some mountains, such as those found in desert regions, are dry and barren with temperatures ranging from extreme heat in the summer to extreme cold in the winter. In tropical regions, lush jungles with heavy seasonal rains and little temperature variation often cover mountains. High rocky crags with glaciated peaks can be found in mountain ranges at most latitudes along the western portion of the Americas and Asia.

d. Severe weather may decrease morale and increase basic survival problems. These problems can be minimized when men have been trained to accept the weather by being self-sufficient. Mountain soldiers properly equipped and trained can use the weather to their advantage in combat operations.

1-12. MOUNTAIN AIR

High mountain air is dry and may be drier in the winter. Cold air has a reduced capacity to hold water vapor. Because of this increased dryness, equipment does not rust as quickly and organic material decomposes slowly. The dry air also requires soldiers to increase consumption of water. The reduced water vapor in the air causes an increase in evaporation of moisture from the skin and in loss of water through transpiration in the respiratory system. Due to the cold, most soldiers do not naturally consume the quantity of fluids they would at higher temperatures and must be encouraged to consciously increase their fluid intake.

a. Pressure is low in mountainous areas due to the altitude. The barometer usually drops 2.5 centimeters for every 300 meters gained in elevation (3 percent).

b. The air at higher altitudes is thinner as atmospheric pressure drops with the increasing altitude. The altitude has a natural filtering effect on the sun's rays. Rays are absorbed or reflected in part by the molecular content of the atmosphere. This effect is greater at lower altitudes. At higher altitudes, the thinner, drier air has a reduced molecular content and, consequently, a reduced filtering effect on the sun's rays. The intensity of both visible and ultraviolet rays is greater with increased altitude. These conditions increase the chance of sunburn, especially when combined with a snow cover that reflects the rays upward.

1-13. WEATHER CHARACTERISTICS

The earth is surrounded by an atmosphere that is divided into several layers. The world's weather systems are in the lower of these layers known as the "troposphere." This layer reaches as high as 40,000 feet. Weather is a result of an atmosphere, oceans, land masses, unequal heating and cooling from the sun, and the earth's rotation. The weather found in any one place depends on many things such as the air temperature, humidity (moisture content), air pressure (barometric pressure), how it is being moved, and if it is being lifted or not.

a. Air pressure is the "weight" of the atmosphere at any given place. The higher the pressure, the better the weather will be. With lower air pressure, the weather will more than likely be worse. In order to understand this, imagine that the air in the atmosphere acts like a liquid. Areas with a high level of this "liquid" exert more pressure on an area and are called high-pressure areas. Areas with a lower level are called low-pressure areas. The average air pressure at sea level is 29.92 inches of mercury (hg) or 1,013 millibars (mb). The higher in altitude, the lower the pressure.

(1) ***High Pressure.*** The characteristics of a high-pressure area are as follows:

- The airflow is clockwise and out.
- Otherwise known as an "anticyclone".
- Associated with clear skies.
- Generally the winds will be mild.
- Depicted as a blue "H" on weather maps.

(2) ***Low Pressure.*** The characteristics of a low-pressure area are as follows:

- The airflow is counterclockwise and in.
- Otherwise known as a "cyclone".
- Associated with bad weather.
- Depicted as a red "L" on weather maps.

b. Air from a high-pressure area is basically trying to flow out and equalize its pressure with the surrounding air. Low pressure, on the other hand, is building up vertically by pulling air in from outside itself, which causes atmospheric instability resulting in bad weather.

c. On a weather map, these differences in pressure are depicted as isobars. Isobars resemble contour lines and are measured in either millibars or inches of mercury. The areas of high pressure are called "ridges" and lows are called "troughs."

1-14. WIND

In high mountains, the ridges and passes are seldom calm; however, strong winds in protected valleys are rare. Normally, wind speed increases with altitude since the earth's frictional drag is strongest near the ground. This effect is intensified by mountainous terrain. Winds are accelerated when they converge through mountain passes and canyons. Because of these funneling effects, the wind may blast with great force on an exposed mountainside or summit. Usually, the local wind direction is controlled by topography.

a. The force exerted by wind quadruples each time the wind speed doubles; that is, wind blowing at 40 knots pushes four times harder than a wind blowing at 20 knots. With increasing wind strength, gusts become more important and may be 50 percent higher than the average wind speed. When wind strength increases to a hurricane force of 64 knots or more, soldiers should lay on the ground during gusts and continue moving during lulls. If a hurricane- force wind blows where there is sand or snow, dense clouds fill the air. The rocky debris or chunks of snow crust are hurled near the surface. During the winter season, or at high altitudes, commanders must be constantly aware of the wind-chill factor and associated cold-weather injuries (see Chapter 2).

b. Winds are formed due to the uneven heating of the air by the sun and rotation of the earth. Much of the world's weather depends on a system of winds that blow in a set direction.

c. Above hot surfaces, air expands and moves to colder areas where it cools and becomes denser, and sinks to the earth's surface. The results are a circulation of air from the poles along the surface of the earth to the equator, where it rises and moves to the poles again.

d. Heating and cooling together with the rotation of the earth causes surface winds. In the Northern Hemisphere, there are three prevailing winds:

(1) ***Polar Easterlies***. These are winds from the polar region moving from the east. This is air that has cooled and settled at the poles.

(2) ***Prevailing Westerlies***. These winds originate from approximately 30 degrees north latitude from the west. This is an area where prematurely cooled air, due to the earth's rotation, has settled to the surface.

(3) ***Northeast Tradewinds***. These are winds that originate from approximately 30° north from the northeast.

e. The jet stream is a long meandering current of high-speed winds often exceeding 250 miles per hour near the transition zone between the troposphere and the stratosphere known as the tropopause. These winds blow from a generally westerly direction dipping down and picking up air masses from the tropical regions and going north and bringing down air masses from the polar regions.

f. The patterns of wind mentioned above move air. This air comes in parcels called "air masses." These air masses can vary from the size of a small town to as large as a country. These air masses are named from where they originate:

- Maritime—over water.
- Continental—over land
- Polar—north of 60° north latitude.
- Tropical—south of 60° north latitude.

Combining these parcels of air provides the names and description of the four types of air masses:

- Continental Polar—cold, dry air mass.
- Maritime Polar—cold, wet air mass.
- Maritime Tropical—warm, wet air mass.
- Continental Tropical—warm, dry air mass.

g. Two types of winds are peculiar to mountain environments, but do not necessarily affect the weather.

(1) ***Anabatic Wind (Valley Winds).*** These winds blow up mountain valleys to replace warm rising air and are usually light winds.

(2) ***Katabatic Wind (Mountain Wind).*** These winds blow down mountain valley slopes caused by the cooling of air and are occasionally strong winds.

1-15. HUMIDITY

Humidity is the amount of moisture in the air. All air holds water vapor even if it cannot be seen. Air can hold only so much water vapor; however, the warmer the air, the more moisture it can hold. When air can hold all that it can the air is "saturated" or has 100 percent relative humidity.

a. If air is cooled beyond its saturation point, the air will release its moisture in one form or another (clouds, fog, dew, rain, snow, and so on). The temperature at which this happens is called the "condensation point". The condensation point varies depending on the amount of water vapor contained in the air and the temperature of the air. If the air contains a great deal of water, condensation can occur at a temperature of 68 degrees Fahrenheit, but if the air is dry and does not hold much moisture, condensation may not form until the temperature drops to 32 degrees Fahrenheit or even below freezing.

b. The adiabatic lapse rate is the rate at which air cools as it rises or warms as it descends. This rate varies depending on the moisture content of the air. Saturated (moist) air will warm and cool approximately 3.2 degrees Fahrenheit per 1,000 feet of elevation gained or lost. Dry air will warm and cool approximately 5.5 degrees Fahrenheit per 1,000 feet of elevation gained or lost.

1-16. CLOUD FORMATION

Clouds are indicators of weather conditions. By reading cloud shapes and patterns, observers can forecast weather with little need for additional equipment such as a barometer, wind meter, and thermometer. Any time air is lifted or cooled beyond its saturation point (100 percent relative humidity), clouds are formed. The four ways air gets lifted and cooled beyond its saturation point are as follows.

a. **Convective Lifting.** This effect happens due to the sun's heat radiating off the Earth's surface causing air currents (thermals) to rise straight up and lift air to a point of saturation.

b. **Frontal Lifting.** A front is formed when two air masses of different moisture content and temperature collide. Since air masses will not mix, warmer air is forced aloft over the colder air mass. From there it is cooled and then reaches its saturation point. Frontal lifting creates the majority of precipitation.

c. **Cyclonic Lifting.** An area of low pressure pulls air into its center from all over in a counterclockwise direction. Once this air reaches the center of the low pressure, it has nowhere to go but up. Air continues to lift until it reaches the saturation point.

d. **Orographic Lifting.** This happens when an air mass is pushed up and over a mass of higher ground such as a mountain. Air is cooled due to the adiabatic lapse rate until the air's saturation point is reached.

1-17. TYPES OF CLOUDS

Clouds are one of the signposts to what is happening with the weather. Clouds can be described in many ways. They can be classified by height or appearance, or even by the amount of area covered vertically or horizontally. Clouds are classified into five categories: low-, mid-, and high-level clouds; vertically-developed clouds; and less common clouds.

a. **Low-Level Clouds.** Low-level clouds (0 to 6,500 feet) are either cumulus or stratus (Figures 1-1 and 1-2, page 1-16). Low-level clouds are mostly composed of water droplets since their bases lie below 6,500 feet. When temperatures are cold enough, these clouds may also contain ice particles and snow.

(1) The two types of precipitating low-level clouds are nimbostratus and stratocumulus (Figures 1-3 and 1-4, page 1-17).

(a) Nimbostratus clouds are dark, low-level clouds accompanied by light to moderately falling precipitation. The sun or moon is not visible through nimbostratus clouds, which distinguishes them from mid-level altostratus clouds. Because of the fog and falling precipitation commonly found beneath and around nimbostratus clouds, the cloud base is typically extremely diffuse and difficult to accurately determine.

(b) Stratocumulus clouds generally appear as a low, lumpy layer of clouds that is sometimes accompanied by weak precipitation. Stratocumulus vary in color from dark gray to light gray and may appear as rounded masses with breaks of clear sky in between. Because the individual elements of stratocumulus are larger than those of altocumulus, deciphering between the two cloud types is easier. With your arm extended toward the sky, altocumulus elements are about the size of a thumbnail while stratocumulus are about the size of a fist.

Figure 1-1. Cumulus clouds.

Figure 1-2. Stratus clouds.

Figure 1-3. Nimbostratus clouds.

Figure 1-4. Stratocumulus clouds.

(2) Low-level clouds may be identified by their height above nearby surrounding relief of known elevation. Most precipitation originates from low-level clouds because rain or snow usually evaporate before reaching the ground from higher clouds. Low-level clouds usually indicate impending precipitation, especially if the cloud is more than 3,000 feet thick. (Clouds that appear dark at their bases are more than 3,000 feet thick.)

b. **Mid-Level Clouds.** Mid-level clouds (between 6,500 to 20,000 feet) have a prefix of alto. Middle clouds appear less distinct than low clouds because of their height. Alto clouds with sharp edges are warmer because they are composed mainly of water droplets. Cold clouds, composed mainly of ice crystals and usually colder than -30 degrees F, have distinct edges that grade gradually into the surrounding sky. Middle clouds usually indicate fair weather, especially if they are rising over time. Lowering middle clouds indicate potential storms, though usually hours away. There are two types of mid-level clouds, altocumulus and altostratus clouds (Figures 1-5 and 1-6).

(1) Altocumulus clouds can appear as parallel bands or rounded masses. Typically a portion of an altocumulus cloud is shaded, a characteristic which makes them distinguishable from high-level cirrocumulus. Altocumulus clouds usually form in advance of a cold front. The presence of altocumulus clouds on a warm humid summer morning is commonly followed by thunderstorms later in the day. Altocumulus clouds that are scattered rather than even, in a blue sky, are called "fair weather" cumulus and suggest arrival of high pressure and clear skies.

(2) Altostratus clouds are often confused with cirrostratus. The one distinguishing feature is that a halo is not observed around the sun or moon. With altostratus, the sun or moon is only vaguely visible and appears as if it were shining through frosted glass.

Figure 1-5. Altocumulus.

Figure 1-6. Altostratus.

c. **High-Level Clouds.** High-level clouds (more than 20,000 feet above ground level) are usually frozen clouds, indicating air temperatures at that elevation below -30 degrees Fahrenheit, with a fibrous structure and blurred outlines. The sky is often covered with a thin veil of cirrus that partly obscures the sun or, at night, produces a ring of light around the moon. The arrival of cirrus indicates moisture aloft and the approach of a traveling storm system. Precipitation is often 24 to 36 hours away. As the storm approaches, the cirrus thickens and lowers, becoming altostratus and eventually stratus. Temperatures are warm, humidity rises, and winds become southerly or south easterly. The two types of high-level clouds are cirrus and cirrostratus (Figure 1-7 and Figure 1-8, page 1-20).

(1) Cirrus clouds are the most common of the high-level clouds. Typically found at altitudes greater than 20,000 feet, cirrus are composed of ice crystals that form when super-cooled water droplets freeze. Cirrus clouds generally occur in fair weather and point in the direction of air movement at their elevation. Cirrus can be observed in a variety of shapes and sizes. They can be nearly straight, shaped like a comma, or seemingly all tangled together. Extensive cirrus clouds are associated with an approaching warm front.

(2) Cirrostratus clouds are sheet-like, high-level clouds composed of ice crystals. They are relatively transparent and can cover the entire sky and be up to several thousand feet thick. The sun or moon can be seen through cirrostratus. Sometimes the only indication of cirrostratus clouds is a halo around the sun or moon. Cirrostratus clouds tend to thicken as a warm front approaches, signifying an increased production of ice crystals. As a result, the halo gradually disappears and the sun or moon becomes less visible.

Figure 1-7. Cirrus.

Figure 1-8. Cirrostratus.

d. **Vertical-Development Clouds.** Clouds with vertical development can grow to heights in excess of 39,000 feet, releasing incredible amounts of energy. The two types of clouds with vertical development are fair weather cumulus and cumulonimbus.

(1) Fair weather cumulus clouds have the appearance of floating cotton balls and have a lifetime of 5 to 40 minutes. Known for their flat bases and distinct outlines, fair weather cumulus exhibit only slight vertical growth, with the cloud tops designating the limit of the rising air. Given suitable conditions, however, these clouds can later develop into towering cumulonimbus clouds associated with powerful thunderstorms. Fair weather cumulus clouds are fueled by buoyant bubbles of air known as thermals that rise up from the earth's surface. As the air rises, the water vapor cools and condenses forming water droplets. Young fair weather cumulus clouds have sharply defined edges and bases while the edges of older clouds appear more ragged, an artifact of erosion. Evaporation along the cloud edges cools the surrounding air, making it heavier and producing sinking motion outside the cloud. This downward motion inhibits further convection and growth of additional thermals from down below, which is why fair weather cumulus typically have expanses of clear sky between them. Without a continued supply of rising air, the cloud begins to erode and eventually disappears.

(2) Cumulonimbus clouds are much larger and more vertically developed than fair weather cumulus (Figure 1-9). They can exist as individual towers or form a line of towers called a squall line. Fueled by vigorous convective updrafts, the tops of cumulonimbus clouds can reach 39,000 feet or higher. Lower levels of cumulonimbus clouds consist mostly of water droplets while at higher elevations, where the temperatures are well below freezing, ice crystals dominate the composition. Under favorable conditions, harmless fair weather cumulus clouds can quickly develop into large cumulonimbus associated with powerful thunderstorms known as super-cells. Super-cells are large thunderstorms with deep rotating updrafts and can have a lifetime of several hours. Super-cells produce frequent lightning, large hail, damaging winds, and

tornadoes. These storms tend to develop during the afternoon and early evening when the effects of heating from the sun are the strongest.

Figure 1-9. Cumulonimbus.

e. **Other Cloud Types.** These clouds are a collection of miscellaneous types that do not fit into the previous four groups. They are orographic clouds, lenticulars, and contrails.

(1) Orographic clouds develop in response to the forced lifting of air by the earth's topography. Air passing over a mountain oscillates up and down as it moves downstream. Initially, stable air encounters a mountain, is lifted upward, and cools. If the air cools to its saturation temperature during this process, the water vapor condenses and becomes visible as a cloud. Upon reaching the mountain top, the air is heavier than the environment and will sink down the other side, warming as it descends. Once the air returns to its original height, it has the same buoyancy as the surrounding air. However, the air does not stop immediately because it still has momentum carrying it downward. With continued descent, the air becomes warmer then the surrounding air and accelerates back upwards towards its original height. Another name for this type of cloud is the lenticular cloud.

(2) Lenticular clouds are cloud caps that often form above pinnacles and peaks, and usually indicate higher winds aloft (Figure 1-10, page 1-22). Cloud caps with a lens shape, similar to a "flying saucer," indicate extremely high winds (over 40 knots). Lenticulars should always be watched for changes. If they grow and descend, bad weather can be expected.

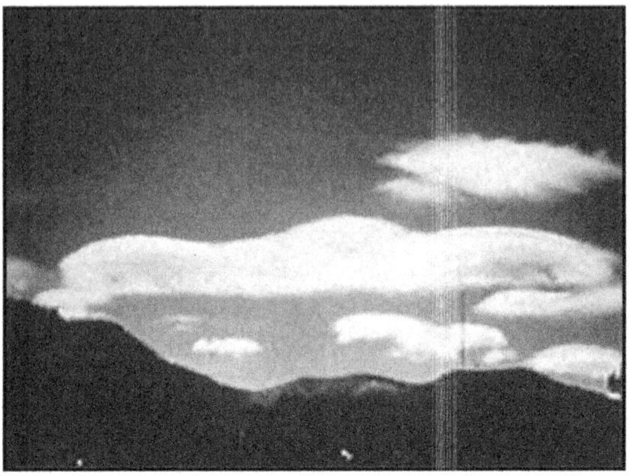

Figure 1-10. Lenticular.

(3) Contrails are clouds that are made by water vapor being inserted into the upper atmosphere by the exhaust of jet engines (Figure 1-11). Contrails evaporate rapidly in fair weather. If it takes longer than two hours for contrails to evaporate, then there is impending bad weather (usually about 24 hours prior to a front).

Figure 1-11. Contrails.

f. **Cloud Interpretation.** Serious errors can occur in interpreting the extent of cloud cover, especially when cloud cover must be reported to another location. Cloud cover always appears greater on or near the horizon, especially if the sky is covered with cumulus clouds, since the observer is looking more at the sides of the clouds rather than between them. Cloud cover estimates should be restricted to sky areas more than 40 degrees above the horizon—that is, to the local sky. Assess the sky by dividing the 360 degrees of sky around you into eighths. Record the coverage in eighths and the types of clouds observed.

1-18. FRONTS

Fronts occur when two air masses of different moisture and temperature contents meet. One of the indicators that a front is approaching is the progression of the clouds. The four types of fronts are warm, cold, occluded, and stationary.

a. **Warm Front.** A warm front occurs when warm air moves into and over a slower or stationary cold air mass. Because warm air is less dense, it will rise up and over the cooler air. The cloud types seen when a warm front approaches are cirrus, cirrostratus, nimbostratus (producing rain), and fog. Occasionally, cumulonimbus clouds will be seen during the summer months.

b. **Cold Front.** A cold front occurs when a cold air mass overtakes a slower or stationary warm air mass. Cold air, being more dense than warm air, will force the warm air up. Clouds observed will be cirrus, cumulus, and then cumulonimbus producing a short period of showers.

c. **Occluded Front.** Cold fronts generally move faster than warm fronts. The cold fronts eventually overtake warm fronts and the warm air becomes progressively lifted from the surface. The zone of division between cold air ahead and cold air behind is called a "cold occlusion." If the air behind the front is warmer than the air ahead, it is a warm occlusion. Most land areas experience more occlusions than other types of fronts. The cloud progression observed will be cirrus, cirrostratus, altostratus, and nimbostratus. Precipitation can be from light to heavy.

d. **Stationary Front.** A stationary front is a zone with no significant air movement. When a warm or cold front stops moving, it becomes a stationary front. Once this boundary begins forward motion, it once again becomes a warm or cold front. When crossing from one side of a stationary front to another, there is typically a noticeable temperature change and shift in wind direction. The weather is usually clear to partly cloudy along the stationary front.

1-19. TEMPERATURE

Normally, a temperature drop of 3 to 5 degrees Fahrenheit for every 1,000 feet gain in altitude is encountered in motionless air. For air moving up a mountain with condensation occurring (clouds, fog, and precipitation), the temperature of the air drops 3.2 degrees Fahrenheit with every 1,000 feet of elevation gain. For air moving up a mountain with no clouds forming, the temperature of the air drops 5.5 degrees Fahrenheit for every 1,000 feet of elevation gain.

a. An expedient to this often occurs on cold, clear, calm mornings. During a troop movement or climb started in a valley, higher temperatures may often be encountered as altitude is gained. This reversal of the normal cooling with elevation is called temperature inversion. Temperature inversions are caused when mountain air is cooled by ice, snow, and heat loss through thermal radiation. This cooler, denser air settles into the valleys and low areas. The inversion continues until the sun warms the surface of the earth or a moderate wind causes a mixing of the warm and cold layers. Temperature inversions are common in the mountainous regions of the arctic, subarctic, and mid-latitudes.

b. At high altitudes, solar heating is responsible for the greatest temperature contrasts. More sunshine and solar heat are received above the clouds than below. The important effect of altitude is that the sun's rays pass through less of the atmosphere and more direct heat is received than at lower levels, where solar radiation is absorbed and reflected by dust and water vapor. Differences of 40 to 50 degrees Fahrenheit may occur between surface temperatures in the shade and surface temperatures in the sun. This is particularly true for dark metallic objects. The difference in temperature felt on the skin between the sun and shade is normally 7 degrees Fahrenheit. Special care must be taken to avoid sunburn and snow blindness. Besides permitting rapid heating, the clear air at high altitudes also favors rapid cooling at night. Consequently, the temperature rises fast after sunrise and drops quickly after sunset. Much of the chilled air drains downward, due to convection currents, so that the differences between day and night temperatures are greater in valleys than on slopes.

c. Local weather patterns force air currents up and over mountaintops. Air is cooled on the windward side of the mountain as it gains altitude, but more slowly (3.2 degrees Fahrenheit per 1,000 feet) if clouds are forming due to heat release when water vapor becomes liquid. On the leeward side of the mountain, this heat gained from the condensation on the windward side is added to the normal heating that occurs as the air descends and air pressure increases. Therefore, air and winds on the leeward slope are considerably warmer than on the windward slope, which is referred to as Chinook winds. The heating and cooling of the air affects planning considerations primarily with regard to the clothing and equipment needed for an operation.

1-20. WEATHER FORECASTING

The use of a portable aneroid barometer, thermometer, wind meter, and hygrometer help in making local weather forecasts. Reports from other localities and from any weather service, including USAF, USN, or the National Weather Bureau, are also helpful. Weather reports should be used in conjunction with the locally observed current weather situation to forecast future weather patterns.

a. Weather at various elevations may be quite different because cloud height, temperature, and barometric pressure will all be different. There may be overcast and rain in a lower area, with mountains rising above the low overcast into warmer clear weather.

b. To be effective, a forecast must reach the small-unit leaders who are expected to utilize weather conditions for assigned missions. Several different methods can be used to create a forecast. The method a forecaster chooses depends upon the forecaster's experience, the amount of data available, the level of difficulty that the forecast situation presents, and the degree of accuracy needed to make the forecast. The five ways to forecast weather are:

(1) ***Persistence Method.*** "Today equals tomorrow" is the simplest way of producing a forecast. This method assumes that the conditions at the time of the forecast will not

change; for example, if today was hot and dry, the persistence method predicts that tomorrow will be the same.

(2) ***Trends Method.*** "Nowcasting" involves determining the speed and direction of fronts, high- and low-pressure centers, and clouds and precipitation. For example, if a cold front moves 300 miles during a 24-hour period, we can predict that it will travel 300 miles in another 24-hours.

(3) ***Climatology Method.*** This method averages weather statistics accumulated over many years. This only works well when the pattern is similar to the following years.

(4) ***Analog Method.*** This method examines a day's forecast and recalls a day in the past when the weather looked similar (an analogy). This method is difficult to use because finding a perfect analogy is difficult.

(5) ***Numerical Weather Prediction.*** This method uses computers to analyze all weather conditions and is the most accurate of the five methods.

1-21. RECORDING DATA

An accurate observation is essential in noting trends in weather patterns. Ideally, under changing conditions, trends will be noted in some weather parameters. However, this may not always be the case. A minor shift in the winds may signal an approaching storm.

a. **Wind Direction.** Assess wind direction as a magnetic direction from which the wind is blowing.

b. **Wind Speed.** Assess wind speed in knots.

(1) If an anemometer is available, assess speed to the nearest knot.

(2) If no anemometer is available, estimate the speed in knots. Judge the wind speed by the way objects, such as trees, bushes, tents, and so forth, are blowing.

c. **Visibility in Meters.** Observe the farthest visible major terrain or man-made feature and determine the distance using any available map.

d. **Present Weather.** Include any precipitation or obscuring weather. The following are examples of present weather:

- Rain—continuous and steady liquid precipitation that will last at least one hour.
- Rain showers—short-term and potentially heavy downpours that rarely last more than one hour.
- Snow—continuous and steady frozen precipitation that will last at least one hour.
- Snow showers—short-term and potentially heavy frozen downpours that rarely last more than one hour.
- Fog, haze—obstructs visibility of ground objects.
- Thunderstorms—a potentially dangerous storm. Thunderstorms will produce lightning, heavy downpours, colder temperatures, tornadoes (not too frequently), hail, and strong gusty winds at the surface and aloft. Winds commonly exceed 35 knots.

e. **Total Cloud Cover.** Assess total cloud cover in eighths. Divide the sky into eight different sections measuring from horizon to horizon. Count the sections with cloud cover, which gives the total cloud cover in eighths. (For example, if half of the sections are covered with clouds, total cloud cover is 4/8.)

f. **Ceiling Height.** Estimate where the cloud base intersects elevated terrain. Note if bases are above all terrain. If clouds are not touching terrain, then estimate to the best of your ability.

g. **Temperature.** Assess temperature with or without a thermometer.

(1) With a thermometer, assess temperature in degrees Celsius (use Fahrenheit only if Celsius conversion is not available). To convert Fahrenheit to Celsius: C = F minus 32 times .55. To convert Celsius to Fahrenheit: F = 1.8 times C plus 32.

Example: 41 degrees F – 32 x .55 = 5 degrees C.
5 degrees C x 1.8 + 32 = 41 degrees F.

(2) Without a thermometer, estimate temperature as above or below freezing ($0^{o}C$), as well as an estimated temperature.

h. **Pressure Trend.** With a barometer or altimeter, assess the pressure trend.

(1) A high pressure moving in will cause altimeters to indicate lower elevation.

(2) A low pressure moving in will cause altimeters to indicate higher elevation.

i. **Observed Weather.** Note changes or trends in observed weather conditions.

(1) Deteriorating trends include:

- Marked wind direction shifts. A high pressure system wind flows clockwise. A low pressure system wind flows counterclockwise. The closer the isometric lines are, the greater the differential of pressure (greater wind speeds).
- Marked wind speed increases.
- Changes in obstructions to visibility.
- Increasing cloud coverage.
- Increase in precipitation. A steady drizzle is usually a long-lasting rain.
- Lowering cloud ceilings.
- Marked cooler temperature changes, which could indicate that a cold front is passing through.
- Marked increase in humidity.
- Decreasing barometric pressure, which indicates a lower pressure system is moving through the area.

(2) Improving trends include:

- Steady wind direction, which indicates no change in weather systems in the area.
- Decreasing wind speeds.
- Clearing of obstructions to visibility.
- Decreasing or ending precipitation.
- Decreasing cloud coverage.
- Increasing height of cloud ceilings.
- Temperature changes slowly warmer.
- Humidity decreases.
- Increasing barometric pressure, which indicates that a higher pressure system is moving through the area.

j. **Update.** Continue to evaluate observed conditions and update the forecast.

Section III. MOUNTAIN HAZARDS

Hazards can be termed natural (caused by natural occurrence), man-made (caused by an individual, such as lack of preparation, carelessness, improper diet, equipment misuse), or as a combination (human trigger). There are two kinds of hazards while in the mountains—subjective and objective. Combinations of objective and subjective hazards are referred to as cumulative hazards.

1-22. SUBJECTIVE HAZARDS

Subjective hazards are created by humans; for example, choice of route, companions, overexertion, dehydration, climbing above one's ability, and poor judgment.

a. **Falling.** Falling can be caused by carelessness, over-fatigue, heavy equipment, bad weather, overestimating ability, a hold breaking away, or other reasons.

b. **Bivouac Site.** Bivouac sites must be protected from rockfall, wind, lightning, avalanche run-out zones, and flooding (especially in gullies). If the possibility of falling exists, rope in, the tent and all equipment may have to be tied down.

c. **Equipment.** Ropes are not total security; they can be cut on a sharp edge or break due to poor maintenance, age, or excessive use. You should always pack emergency and bivouac equipment even if the weather situation, tour, or a short climb is seemingly low of dangers.

1-23. OBJECTIVE HAZARDS

Objective hazards are caused by the mountain and weather and cannot be influenced by man; for example, storms, rockfalls, icefalls, lightning, and so on.

a. **Altitude.** At high altitudes (especially over 6,500 feet), endurance and concentration is reduced. Cut down on smoking and alcohol. Sleep well, acclimatize slowly, stay hydrated, and be aware of signs and symptoms of high-altitude illnesses. Storms can form quickly and lightning can be severe.

b. **Visibility.** Fog, rain, darkness, and or blowing snow can lead to disorientation. Take note of your exact position and plan your route to safety before visibility decreases. Cold combined with fog can cause a thin sheet of ice to form on rocks (verglas). Whiteout conditions can be extremely dangerous. If you must move under these conditions, it is best to rope up. Have the point man move to the end of the rope. The second man will use the first man as an aiming point with the compass. Use a route sketch and march table. If the tactical situation does not require it, plan route so as not to get caught by darkness.

c. **Gullies.** Rock, snow, and debris are channeled down gullies. If ice is in the gully, climbing at night may be better because the warming of the sun will loosen stones and cause rockfalls.

d. **Rockfall.** Blocks and scree at the base of a climb can indicate recurring rockfall. Light colored spots on the wall may indicate impact chips of falling rock. Spring melt or warming by the sun of the rock/ice/snow causes rockfall.

e. **Avalanches.** Avalanches are caused by the weight of the snow overloading the slope. (Refer to paragraph 1-25 for more detailed information on avalanches.)

f. **Hanging Glaciers and Seracs.** Avoid, if at all possible, hanging glaciers and seracs. They will fall without warning regardless of the time of day or time of year. One cubic meter of glacier ice weighs 910 kilograms (about 2,000 pounds). If you must cross these danger areas, do so quickly and keep an interval between each person.

g. **Crevasses.** Crevasses are formed when a glacier flows over a slope and makes a bend, or when a glacier separates from the rock walls that enclose it. A slope of only two to three degrees is enough to form a crevasse. As this slope increases from 25 to 30 degrees, hazardous icefalls can be formed. Likewise, as a glacier makes a bend, it is likely that crevasses will form at the outside of the bend. Therefore, the safest route on a glacier would be to the inside of bends, and away from steep slopes and icefalls. Extreme care must be taken when moving off of or onto the glacier because of the moat that is most likely to be present.

1-24. WEATHER HAZARDS

Weather conditions in the mountains may vary from one location to another as little as 10 kilometers apart. Approaching storms may be hard to spot if masked by local peaks. A clear, sunny day in July could turn into a snowstorm in less than an hour. Always pack some sort of emergency gear.

a. Winds are stronger and more variable in the mountains; as wind doubles in speed, the force quadruples.

b. Precipitation occurs more on the windward side than the leeward side of ranges. This causes more frequent and denser fog on the windward slope.

c. Above approximately 8,000 feet, snow can be expected any time of year in the temperate climates.

d. Air is dryer at higher altitudes, so equipment does not rust as quickly, but dehydration is of greater concern.

e. Lightning is frequent, violent, and normally attracted to high points and prominent features in mountain storms. Signs indicative of thunderstorms are tingling of the skin, hair standing on end, humming of metal objects, crackling, and a bluish light (St. Elmo's fire) on especially prominent metal objects (summit crosses and radio towers).

(1) Avoid peaks, ridges, rock walls, isolated trees, fixed wire installations, cracks that guide water, cracks filled with earth, shallow depressions, shallow overhangs, and rock needles. Seek shelter around dry, clean rock without cracks; in scree fields; or in deep indentations (depressions, caves). Keep at least half a body's length away from a cave wall and opening.

(2) Assume a one-point-of-contact body position. Squat on your haunches or sit on a rucksack or rope. Pull your knees to your chest and keep both feet together. If half way up the rock face, secure yourself with more than one point—lightning can burn through rope. If already rappelling, touch the wall with both feet together and hurry to the next anchor.

f. During and after rain, expect slippery rock and terrain in general and adjust movement accordingly. Expect flash floods in gullies or chimneys. A climber can be washed away or even drowned if caught in a gully during a rainstorm. Be especially alert for falling objects that the rain has loosened.

g. Dangers from impending high winds include frostbite (from increased wind-chill factor), windburn, being blown about (especially while rappelling), and debris being blown about. Wear protective clothing and plan the route to be finished before bad weather arrives.

h. For each 100-meter rise in altitude, the temperature drops approximately one degree Fahrenheit. This can cause hypothermia and frostbite even in summer, especially when combined with wind, rain, and snow. Always wear or pack appropriate clothing.

i. If it is snowing, gullies may contain avalanches or snow sloughs, which may bury the trail. Snowshoes or skis may be needed in autumn or even late spring. Unexpected snowstorms may occur in the summer with accumulations of 12 to 18 inches; however, the snow quickly melts.

j. Higher altitudes provide less filtering effects, which leads to greater ultraviolet (UV) radiation intensity. Cool winds at higher altitudes may mislead one into underestimating the sun's intensity, which can lead to sunburns and other heat injuries. Use sunscreen and wear hat and sunglasses, even if overcast. Drink plenty of fluids.

1-25. AVALANCHE HAZARDS

Avalanches occur when the weight of accumulated snow on a slope exceeds the cohesive forces that hold the snow in place. (Table 1-2, page 1-32, shows an avalanche hazard evaluation checklist.)

a. **Slope Stability.** Slope stability is the key factor in determining the avalanche danger.

(1) ***Slope Angle.*** Slopes as gentle as 15 degrees have avalanched. Most avalanches occur on slopes between 30 and 45 degrees. Slopes above 60 degrees often do not build up significant quantities of snow because they are too steep.

(2) ***Slope Profile.*** Dangerous slab avalanches are more likely to occur on convex slopes, but may occur on concave slopes.

(3) ***Slope Aspect.*** Snow on north facing slopes is more likely to slide in midwinter. South facing slopes are most dangerous in the spring and on sunny, warm days. Slopes on the windward side are generally more stable than leeward slopes.

(4) ***Ground Cover.*** Rough terrain is more stable than smooth terrain. On grassy slopes or scree, the snow pack has little to anchor to.

b. **Triggers.** Various factors trigger avalanches.

(1) ***Temperature.*** When the temperature is extremely low, settlement and adhesion occur slowly. Avalanches that occur during extreme cold weather usually occur during or immediately following a storm. At a temperature just below freezing, the snowpack stabilizes quickly. At temperatures above freezing, especially if temperatures rise quickly, the potential for avalanche is high. Storms with a rise in temperature can deposit dry snow early, which bonds poorly with the heavier snow deposited later. Most avalanches occur during the warmer midday.

(2) ***Precipitation.*** About 90 percent of avalanches occur during or within twenty-four hours after a snowstorm. The rate at which snow falls is important. High rates of snowfall (2.5 centimeters per hour or greater), especially when accompanied by wind, are usually responsible for major periods of avalanche activity. Rain falling on snow will increase its weight and weakens the snowpack.

(3) ***Wind.*** Sustained winds of 15 miles per hour and over transport snow and form wind slabs on the lee side of slopes.

(4) ***Weight.*** Most victims trigger the avalanches that kill them.

(5) ***Vibration.*** Passing helicopters, heavy equipment, explosions, and earth tremors have triggered avalanches.

c. **Snow Pits.** Snow pits can be used to determine slope stability.

(1) Dig the snow pit on the suspect slope or a slope with the same sun and wind conditions. Snow deposits may vary greatly within a few meters due to wind and sun variations. (On at least one occasion, a snow pit dug across the fall line triggered the suspect slope). Dig a 2-meter by 2-meter pit across the fall line, through all the snow, to the ground. Once the pit is complete, smooth the face with a shovel.

(2) Conduct a shovel shear test.

(a) A shovel shear test puts pressure on a representative sample of the snowpack. The core of this test is to isolate a column of the snowpack from three sides. The column should be of similar size to the blade of the shovel. Dig out the sides of the column without pressing against the column with the shovel (this affects the strength). To isolate the rear of the column, use a rope or string to saw from side to side to the base of the column.

(b) If the column remained standing while cutting the rear, place the shovel face down on the top of the column. Tap with varying degrees of strength on the shovel to see what force it takes to create movement on the bed of the column. The surface that eventually slides will be the layer to look at closer. This test provides a better understanding of the snowpack strength. For greater results you will need to do this test in many areas and formulate a scale for the varying methods of tapping the shovel.

(3) Conduct a Rutschblock test. To conduct the test, isolate a column slightly longer than the length of your snowshoes or skis (same method as for the shovel shear test). One person moves on their skis or snowshoes above the block without disturbing the block. Once above, the person carefully places one showshoe or ski onto the block with no body weight for the first stage of the test. The next stage is adding weight to the first leg. Next, place the other foot on the block. If the block is still holding up, squat once, then twice, and so on. The remaining stage is to jump up and land on the block.

d. **Types of Snow Avalanches.** There are two types of snow avalanches: loose snow (point) and slab.

(1) Loose snow avalanches start at one point on the snow cover and grow in the shape of an inverted "V." Although they happen most frequently during the winter snow season, they can occur at any time of the year in the mountains. They often fall as many small sluffs during or shortly after a storm. This process removes snow from steep upper slopes and either stabilizes lower slopes or loads them with additional snow.

(2) Wet loose snow avalanches occur in spring and summer in all mountain ranges. Large avalanches of this type, lubricated and weighed down by meltwater or rain can travel long distances and have tremendous destructive power. Coastal ranges that have high temperatures and frequent rain are the most common areas for this type of avalanche.

(3) Slab avalanches occur when cohesive snow begins to slide on a weak layer. The fracture line where the moving snow breaks away from the snowpack makes this type of avalanche easy to identify. Slab release is rapid. Although any avalanche can kill you, slab avalanches are generally considered more dangerous than loose snow avalanches.

(a) Most slab avalanches occur during or shortly after a storm when slopes are loaded with new snow at a critical rate. The old rule of never travel in avalanche terrain for a few days after a storm still holds true.

(b) As slabs become harder, their behavior becomes more unpredictable; they may allow several people to ski across before releasing. Many experts believe they are susceptible to rapid temperature changes. Packed snow expands and contracts with temperature changes. For normal density, settled snow, a drop in temperature of 10 degrees Celsius (18 degrees Fahrenheit) would cause a snow slope 300 meters wide to contract 2 centimeters. Early ski mountaineers in the Alps noticed that avalanches sometimes occurred when shadows struck a previously sun-warmed slope.

d. **Protective Measures.** Avoiding known or suspected avalanche areas is the easiest method of protection. Other measures include:

(1) ***Personal Safety.*** Remove your hands from ski pole wrist straps. Detach ski runaway cords. Prepare to discard equipment. Put your hood on. Close up your clothing to prepare for hypothermia. Deploy avalanche cord. Make avalanche probes and shovels accessible. Keep your pack on at all times—do not discard. Your pack can act as a flotation device, as well as protect your spine.

(2) ***Group Safety.*** Send one person across the suspect slope at a time with the rest of the group watching. All members of the group should move in the same track from safe zone to safe zone.

e. **Route Selection.** Selecting the correct route will help avoid avalanche prone areas, which is always the best choice. Always allow a wide margin of safety when making your decision.

(1) The safest routes are on ridge tops, slightly on the windward side; the next safest route is out in the valley, far from the bottom of slopes.

(2) Avoid cornices from above or below. Should you encounter a dangerous slope, either climb to the top of the slope or descend to the bottom—well out of the way of the run-out zone. If you must traverse, pick a line where you can traverse downhill as quickly as possible. When you must ascend a dangerous slope, climb to the side of the avalanche path, and not directly up the center.

(3) Take advantage of dense timber, ridges, or rocky outcrops as islands of safety. Use them for lunch and rest stops. Spend as little time as possible on open slopes.

(4) Since most avalanches occur within twenty-four hours of a storm and or at midday, avoid moving during these periods. Moving at night is tactically sound and may be safer.

f. **Stability Analysis.** Look for nature's billboards on slopes similar to the one you are on.

(1) ***Evidence of Avalanching.*** Look for recent avalanches and for signs of wind-loading and wind-slabs.

(2) ***Fracture Lines.*** Avoid any slopes showing cracks.

(3) ***Sounds.*** Beware of hollow sounds—a "whumping" noise. They may suggest a radical settling of the snowpack.

g. **Survival.** People trigger avalanches that bury people. If these people recognized the hazard and chose a different route, they would avoid the avalanche. The following steps should be followed if caught in an avalanche.

(1) Discard equipment. Equipment can injure or burden you; discarded equipment will indicate your position to rescuers.

(2) Swim or roll to stay on tope of the snow. FIGHT FOR YOUR LIFE. Work toward the edge of the avalanche. If you feel your feet touch the ground, give a hard push and try to "pop out" onto the surface.

(3) If your head goes under the snow, shut your mouth, hold your breath, and position your hands and arms to form an air pocket in front of your face. Many avalanche victims suffocate by having their mouths and noses plugged with snow.

(4) When you sense the slowing of the avalanche, you must try your hardest to reach the surface. Several victims have been found quickly because a hand or foot was sticking above the surface.

(5) When the snow comes to rest it sets up like cement and even if you are only partially buried, it may be impossible to dig yourself out. Don't shout unless you hear rescuers immediately above you; in snow, no one can hear you scream. Don't struggle to free yourself—you will only waste energy and oxygen.

(6) Try to relax. If you feel yourself about to pass out, do not fight it. The respiration of an unconscious person is more shallow, their pulse rate declines, and the body temperature is lowered, all of which reduce the amount of oxygen needed. (See Appendix C for information on search and rescue techniques.)

AVALANCHE HAZARD EVALUATION CHECKLIST

Critical Data PARAMETERS:	KEY INFORMATION	Hazard Rating G	Y	R
TERRAIN:	*Is the terrain capable of producing an avalanche?*			
	-Slope angle (steep enough to slide? prime time?)	☐	☐	☐
	-Slope aspect (leeward, shadowed, or extremely sunny?)	☐	☐	☐
	-Slope configuration (anchoring? shape?)	☐	☐	☐
	Overall Terrain Rating:	☐	☐	☐
SNOWPACK:	*Could the snow fail?*			
	-Slab Configuration (slab? depth and distribution?)	☐	☐	☐
	-Bonding Ability (weak layer? tender spots?)	☐	☐	☐
	-Sensitivity (how much force to fail? shear tests? clues?)	☐	☐	☐
	Overall Snowpack Rating:	☐	☐	☐
Weather:	*Is the weather contributing to instability?*			
	-Precipitation (type, amount, intensity? added weight?)	☐	☐	☐
	-Wind (snow transport? amount and rate of deposition?)	☐	☐	☐
	-Temperature (storm trends? effects on snowpack?)	☐	☐	☐
	Overall Weather Rating:	☐	☐	☐
Human:	*What are your alternatives and their possible consequences?*			
	-Attitude (toward life? risk? goals? assumptions?)	☐	☐	☐
	-Technical Skill Level (traveling? evaluating aval. hazard?)	☐	☐	☐
	-Strength/Equipment (strength? prepared for the worst?)	☐	☐	☐
	Overall Human Rating:	☐	☐	☐

Decision/Action:
Overall Hazard Rating/GO or NO Go? **GO ☐ or NOGO ☐**

***HAZARD LEVEL SYMBOLS:**
R = Red light (stop/dangerous)
G = Green light (go/OK)
Y = Yellow light (caution/potentially dangerous).

©Alaska Mountain Safety Center, Inc.

Table 1-2. Avalanche hazard evaluation checklist.

CHAPTER 2

MOUNTAIN LIVING

Units deploying to high elevations must receive advanced training to survive in the harsh mountain environment. Normal activities (navigation, communications, and movement) require specialized techniques. Training should be conducted as realistically as possible, preferably under severe conditions so the soldier gains confidence. Extended training exercises test support facilities and expose the soldier to the isolation common to mountain operations. Training should reflect the harsh mountain environment and should consider the following:

- *Temperature and altitude extremes.*
- *Hygiene and sanitation.*
- *Limited living space (difficulty of bivouac).*
- *Clothing requirements.*

Section I. SURVIVAL

The soldier trained to fight and survive in a mountain environment will have increased confidence in himself. Training should include: psychological preparation, locating water, shelter considerations, fire building, health hazards, and techniques for obtaining food (see FM 21-76).

2-1. WATER SUPPLY

Mountain water should never be assumed safe for consumption. Training in water discipline should be emphasized to ensure soldiers drink water only from approved sources. Fluids lost through respiration, perspiration, and urination must be replaced if the soldier is to operate efficiently.

a. Maintaining fluid balance is a major problem in mountain operations. The sense of thirst may be dulled by high elevations despite the greater threat of dehydration. Hyperventilation and the cool, dry atmosphere bring about a three- to four-fold increase in water loss by evaporation through the lungs. Hard work and overheating increase the perspiration rate. The soldier must make an effort to drink liquids even when he does not feel thirsty. One quart of water, or the equivalent, should be drunk every four hours; more should be drunk if the unit is conducting rigorous physical activity.

b. Three to six quarts of water each day should be consumed. About 75 percent of the human body is liquid. All chemical activities in the body occur in water solution, which assists in removing toxic wastes and in maintaining an even body temperature. A loss of two quarts of body fluid (2.5 percent of body weight) decreases physical efficiency by 25 percent, and a loss of 12 quarts (15 percent of body weight) is usually fatal. Salt lost by sweating should be replaced in meals to avoid a deficiency and subsequent cramping. Consuming the usual military rations (three meals a day) provides sufficient sodium replacement. Salt tablets are not necessary and may contribute to dehydration.

c. Even when water is plentiful, thirst should be satisfied in increments. Quickly drinking a large volume of water may actually slow the soldier. If he is hot and the water is cold, severe cramping may result. A basic rule is to drink small amounts often. Pure water should always be kept in reserve for first aid use. Emphasis must be placed on the three rules of water discipline:

- Drink only treated water.
- Conserve water for drinking. Potable water in the mountains may be in short supply.
- Do not contaminate or pollute water sources.

d. Snow, mountain streams, springs, rain, and lakes provide good sources of water supply. Purification must be accomplished, however, no matter how clear the snow or water appears. Fruits, juices, and powdered beverages may supplement and encourage water intake (do not add these until the water has been treated since the purification tablets may not work). Soldiers cannot adjust permanently to a decreased water intake. If the water supply is insufficient, physical activity must be reduced. Any temporary deficiency should be replaced to maintain maximum performance.

e. All water that is to be consumed must be potable. Drinking water must be taken only from approved sources or purified to avoid disease or the possible use of polluted water. Melting snow into water requires an increased amount of fuel and should be planned accordingly. Nonpotable water must not be mistaken for drinking water. Water that is unfit to drink, but otherwise not dangerous, may be used for other purposes such as bathing. Soldiers must be trained to avoid wasting water. External cooling (pouring water over the head and chest) is a waste of water and an inefficient means of cooling. Drinking water often is the best way to maintain a cool and functioning body.

f. Water is scarce above the timberline. After setting up a perimeter (patrol base, assembly area, defense), a watering party should be employed. After sundown, high mountain areas freeze, and snow and ice may be available for melting to provide water. In areas where water trickles off rocks, a shallow reservoir may be dug to collect water (after the sediment settles). Water should be treated with purification tablets (iodine tablets or calcium hypochlorite), or by boiling at least one to two minutes. Filtering with commercial water purification pumps can also be conducted. Solar stills may be erected if time and sunlight conditions permit (see FM 21-76). Water should be protected from freezing by storing it next to a soldier or by placing it in a sleeping bag at night. Water should be collected at midday when the sun thaw available.

2-2. NUTRITION

Success in mountain operations depends on proper nutrition. Because higher altitudes affect eating habits, precautions must be taken. If possible, at least one hot meal each day should be eaten, which may require personnel to heat their individual rations.

a. The following elements are characteristic of nutritional acclimatization in mountain operations:

- Weight loss during the first two to three days at high elevation.
- A loss of appetite with symptoms of mountain sickness.
- Loss of weight usually stops with acclimatization.
- At progressively higher elevations (greater than 14,000 feet), the tolerance of fatty/high-protein foods rapidly decreases. A high carbohydrate diet may lessen the symptoms of acute mountain sickness and is digested better than fat at high altitudes.

b. Increased fatigue may cause soldiers to become disinterested in eating properly. Decreased consumption may result in malnutrition because of the unpleasant taste of cold rations. Leaders should ensure that fuel tablets and squad stoves are available, or that natural flammable materials are used if possible. Although there is no physiological need for hot food, it does increase morale and a sense of well being. Loss of weight in the first few days occurs because of dehydration, metabolic changes, and loss of appetite. Carbohydrate-containing beverages, such as fruit juices and sports drinks, are an effective means of increasing carbohydrates, energy, and liquid intake when the normal appetite response is blunted at altitude.

c. Three major food components are required to maintain a well-functioning body: proteins, fats, and carbohydrates. These food components provide energy, amino acids, vitamins, fiber, and minerals. All three components must be provided in the correct proportions to maintain a healthy body.

(1) ***Protein***. Proteins consist of a large number of amino acid units that are linked together to form the protein. The amino acids, resulting from digestion of protein, are absorbed through the intestine into the blood, and are used to make or replace body proteins (muscle and body tissue). Sources of readily useable animal proteins include eggs, milk, cheese, poultry, fish, and meats. Other foods such as cereals, vegetables, and legumes also provide amino acids. These proteins are not as balanced in essential amino acid composition as meat, eggs, or milk proteins. The minimum daily protein requirement, regardless of physical activity, is 8 ounces for a 154-pound man. Since amino acids are either oxidized for energy or stored as fats, consuming excess protein is inefficient and may increase the water intake needed for urea nitrogen excretion. Protein requires water for digestion and may facilitate dehydration. Proteins provide the body about four kilocalories of energy per gram and require the most energy for the body to digest.

(2) ***Fats.*** Fats are the most concentrated form of food energy. Of the total daily caloric intake, 25 to 30 percent may be supplied as fats. Main sources of fats are meats, nuts, butter, eggs, milk, and cheese. Fats require more water and oxygen, and are harder to digest at higher altitudes. Fats are the body's natural stored source of energy. Fats provide the body around 9 kilocalories of energy per gram and require less energy for the body to digest than protein but more than carbohydrates.

(3) ***Carbohydrates.*** Carbohydrates are an important source of calories. In the form of glucose, carbohydrates are found in the most important energy-producing cycles in the body's cells. If carbohydrate intake exceeds energy needs, moderate amounts are stored in the muscles and liver. Larger amounts are converted into fats and stored in that form. Carbohydrates should compose up to 50 percent of the total daily caloric intake. Nutritionally, the most useful sources of carbohydrates are foods such as unrefined grains, vegetables, and fruit. Carbohydrates provide the body around four kilocalories of energy per gram and are the easiest to digest.

(4) ***Vitamins.*** Vitamins are classified into two groups on the basis of their ability to dissolve in fat or water. The fat-soluble vitamins include vitamins A, D, E, and K. The water-soluble vitamins include the B vitamins and vitamin C, which are found in cereals, vegetables, fruits, and meats. A well-balanced diet provides all of the required vitamins. Since most water-soluble vitamins are not stored, a proper diet is necessary to ensure adequate levels of these vitamins. If an improper and unbalanced diet is likely to occur

during a deployment, vitamin supplements should be considered, especially if this period is to exceed 10 days.

(5) ***Minerals.*** Mineral elements can be divided into two groups: those needed in the diet in amounts of 100 milligrams or more a day such as calcium, phosphorous, and magnesium; and trace elements needed in amounts of only a few milligrams a day such as iodine, iron, and zinc. Required minerals are contained in a balanced diet (meats, vegetables, fruits).

d. Eating a balanced diet provides the energy needed to conduct daily activities and to maintain the internal body processes. A balanced diet containing adequate amounts of vitamins and minerals ensures an efficient metabolism. Since climbing is a strenuous activity and demands high-energy use, a balanced diet is a necessity.

(1) The efficiency of the body to work above the basal metabolism varies from 20 to 40 percent, depending on the soldier. Over 50 percent of caloric intake is released as heat and is not available when the soldier works. (About 4,500 calories are expended for strenuous work and 3,500 calories for garrison activity.) Heat is a by-product of exertion. Exertion causes excessive bodily heat loss through perspiration and increased radiation. During inactivity in cold weather, the metabolism may not provide enough heat. The "internal thermostat" initiates and causes the muscles to shiver, thus releasing heat. Shivering also requires energy and burns up to 220 calories per hour (estimated for a 100-pound man).

(2) With an abrupt ascent to high altitudes, the soldier experiences physiological acclimatization. The circulatory system labors to provide the needed oxygen to the body. Large meals require the digestive system to work harder than usual to assimilate food. Large meals may be accompanied by indigestion, shortness of breath, cramps, and illness. Therefore, relatively light meals that are high in carbohydrates are best while acclimatizing at higher elevations. Personnel should eat moderately and rest before strenuous physical activity. Since fats and protein are harder to digest, less digestive disturbances may occur if meals are eaten before resting. A diet high in carbohydrates is not as dense in energy and may require eating more often. Carbohydrates, beginning in the morning and continuing through mid-afternoon, are important in maintaining energy levels.

(3) Extra food should be carried in case resupply operations fail. Food should be lightweight and easy to digest, and be eaten hot or cold. Meals-ready-to-eat (MREs) meet these criteria and provide all of the basic food groups. Commanders may consider supplementing MREs with breakfast bars, fruits, juices, candies, cereal bars, and chocolate. Bouillon cubes can replace water and salt as well as warming cold bodies and stimulating the appetite. Hot beverages of soup, juices, powdered milk, and cider should also be considered. Since coffee, tea, and hot chocolate are diuretics, the consumption of these beverages should not be relied upon for hydration.

(4) Warm meals should be provided when possible. When cooking, the heat source must be kept away from equipment and ammunition. At higher elevations, the cooking time may be doubled. To conserve fuel, stoves, fires, and fuel tablets should be protected from the wind. Extra fuel should be stored in tightly sealed, marked, metal containers. Use stoves and heat tabs for warming food and boiling water. Canteen cups and utensils should be cleaned after use. All food items and garbage are carried with the unit. If possible, garbage should be burned or deep buried. Caution must be taken to prevent animals from foraging through rucksacks, ahkios, and burial sites. As all missions are tactical, no trace of a unit should be detected.

(5) Certain drugs, medications, alcohol, and smoking have adverse effects on the circulation, perspiration, hydration, and judgment of soldiers. Therefore, they should be avoided when operating in extremely cold conditions or at high altitudes.

2-3. PERSONAL HYGIENE AND SANITATION

The principles of personal hygiene and sanitation that govern operations on low terrain also apply in the mountains. Commanders must conduct frequent inspections to ensure that personal habits of hygiene are not neglected. Standards must be maintained as a deterrent to disease, and as reinforcement to discipline and morale.

a. **Personal Hygiene.** This is especially important in the high mountains, mainly during periods of cold weather. In freezing weather, the soldier may neglect washing due to the cold temperatures and scarcity of water. This can result in skin infections and vermin infestation. If bathing is difficult for any extended period, the soldier should examine his skin and clean it often. Snow baths in lieu of a water bath are recommended. This helps reduce skin infections and aids the comfort of the soldier.

(1) Snow may be used instead of toilet paper. Soldiers should shave at rest periods in the shelter so that oils stripped in shaving will be replenished. A beard may mask the presence of frostbite or lice. Water-based creams and lotions should be avoided in cold environments since this will further dehydrate tissues and induce frostbite by freezing. The nonwater-based creams can be used for shaving in lieu of soap. Sunscreens and chap sticks should be used on lips, nose, and eyelids. Topical steroid ointments should be carried for rashes. The teeth must also be cleaned to avoid diseases of the teeth and gums. Underwear should be changed when possible, but this should not be considered a substitute for bathing. When operating in areas where resupply is not possible, each soldier should carry a complete change of clothing. If laundering of clothing is difficult, clothes should be shaken and air-dried. Sleeping bags must be regularly cleaned and aired.

(2) The principles of foot hygiene must be followed to protect the feet from cold injuries. The causes of such injuries are present throughout the year in high mountains. Boots should be laced tightly when climbing to provide needed support but not so tight as to constrict circulation. Socks should be worn with no wrinkles since this causes blisters on the feet. Feet should be washed daily, and kept as dry and clean as possible. If regular foot washing is impossible, socks should be changed often (at halts and rest periods or at least once a day) and feet massaged, dried, and sprinkled with foot powder. Talc or antifungal powder should be used when massaging; excess powder is brushed off to avoid clumping, which may cause blisters. Feet can be cleaned with snow, but must be quickly dried. Whenever changing socks, soldiers should closely examine their feet for wrinkles, cracks, blisters, and discoloration. Nails should be trimmed but not too short. Long nails wear out socks; short nails do not provide proper support for the ends of the toes. Medical attention should be sought for any possible problems.

(3) Feet should be sprayed two or three times a day with an aluminum chlorohydrate antiperspirant for a week and then once a day for the rest of the winter. If fissures or cracks occur in the feet, it is best to discontinue spraying until they are healed or to spray less often to control sweating. This process stops about 70 percent of the sweating in the feet.

(4) During periods of extreme cold, there is a tendency for the soldier to become constipated. This condition is brought about by the desire to avoid the inconvenience and

discomfort of defecating. Adequate water intake plus a low protein, high roughage diet can be helpful in preventing constipation.

b. **Sanitation.** In rocky or frozen ground, digging latrines is usually difficult. If latrines are constructed, they should be located downwind from the position and buried after use. In tactical situations, the soldier in a designated, downwind location away from water sources may dig "cat holes." Since waste freezes, it can be covered with snow and ice or pushed down a crevasse. In rocky areas above the timberline, waste may be covered with stones.

Section II. ACCLIMATIZATION AND CONDITIONING

Terrestrial altitude can be classified into five categories. Low altitude is sea level to 5,000 feet. Here, arterial blood is 96 percent saturated with oxygen in most people. Moderate altitude is from 5,000 to 8,000 feet. At these altitudes, arterial blood is greater than 92 percent saturated with oxygen, and effects of altitude are mild and temporary. High altitude extends from 8,000 to 14,000 feet, where arterial blood oxygen saturation ranges from 92 percent down to 80 percent. Altitude illness is common here. Very high altitude is the region from 14,000 to 18,000 feet, where altitude illness is the rule. Areas above 18,000 feet are considered extreme altitudes.

Soldiers deployed to high mountainous elevations require a period of acclimatization before undertaking extensive military operations. The expectation that freshly deployed, unacclimatized troops can go immediately into action is unrealistic, and could be disastrous if the opposing force is acclimatized. Even the physically fit soldier experiences physiological and psychological degradation when thrust into high elevations. Time must be allocated for acclimatization, conditioning, and training of soldiers. Training in mountains of low or medium elevation (5,000 to 8,000 feet) does not require special conditioning and acclimatization procedures. However, some soldiers will have some impairment of operating efficiency at these low altitudes. Above 8,000 feet (high elevation), most unacclimatized soldiers may display some altitude effects. Training should be conducted at progressively higher altitudes, starting at about 8,000 feet and ending at 14,000 feet. Attempts to acclimatize beyond 17,000 feet results in a degradation of the body greater than the benefits gained. The indigenous populations can out-perform even the most acclimatized and physically fit soldier who is brought to this altitude; therefore, employment of the local population may be advantageous.

2-4. SYMPTOMS AND ADJUSTMENTS

A person is said to be acclimatized to high elevations when he can effectively perform physically and mentally. The acclimatization process begins immediately upon arrival at the higher elevation. If the change in elevation is large and abrupt, some soldiers can suffer from acute mountain sickness (AMS), high-altitude pulmonary edema (HAPE), or high-altitude cerebral edema (HACE). Disappearance of the symptoms of acute mountain sickness (from four to seven days) does not indicate complete acclimatization. The process of adjustment continues for weeks or months. The altitude at which complete acclimatization is possible is not a set point but for most soldiers with proper ascent, nutrition and physical activity it is about 14,000 feet.

a. Immediately upon arrival at high elevations, only minimal physical work can be performed because of physiological changes. The incidence and severity of AMS symptoms vary with initial altitude, the rate of ascent, and the level of exertion and individual susceptibility. Ten to twenty percent of soldiers who ascend rapidly (in less than 24 hours) to altitudes up to 6,000 feet experience some mild symptoms. Rapid ascent to 10,000 feet causes mild symptoms in 75 percent of personnel. Rapid ascent to elevations of 12,000 to 14,000 feet will result in moderate symptoms in over 50 percent of the soldiers and 12 to 18 percent may have severe symptoms. Rapid ascent to 17,500 feet causes severe, incapacitating symptoms in almost all individuals. Vigorous activity during ascent or within the first 24 hours after ascent will increase both the incidence and severity of symptoms. Some of the behavioral effects that will be encountered in unacclimatized personnel include:

- Increased errors in performing simple mental tasks.
- Decreased ability for sustained concentration.
- Deterioration of memory.
- Decreased vigilance or lethargy.
- Increased irritability in some individuals.
- Impairment of night vision and some constriction in peripheral vision (up to 30 percent at 6,000 feet).
- Loss of appetite.
- Sleep disturbances.
- Irregular breathing.
- Slurred speech.
- Headache.

b. Judgment and self-evaluation are impaired the same as a person who is intoxicated. During the first few days at a high altitude, leaders have extreme difficulty in maintaining a coordinated, operational unit. The roughness of the terrain and the harshness and variability of the weather add to the problems of unacclimatized personnel. Although strong motivation may succeed in overcoming some of the physical handicaps imposed by the environment, the total impact still results in errors of judgment. When a soldier cannot walk a straight line and has a loss of balance, or he suffers from an incapacitating headache, he should be evacuated to a lower altitude (a descent of at least 1,000 feet for at least 24 hours).

2-5. PHYSICAL AND PSYCHOLOGICAL CONDITIONING

The commander must develop a conditioning/training program to bring his unit to a level where it can operate successfully in mountain conditions. Priorities of training must be established. As with all military operations, training is a major influence on the success of mountain operations.

a. U.S. forces do not routinely train in mountainous terrain. Therefore, extensive preparations are needed to ensure individual and unit effectiveness. Units must be physically and psychologically conditioned and adjusted before undertaking rigorous mountain operations. Units must be conditioned and trained as a team to cope with the terrain, environment, and enemy situation. Certain factors must be considered:

- What are the climatic and terrain conditions of the area of operations?
- How much time is available for conditioning and training?
- Will the unit conduct operations with other U.S. or Allied forces? Are there language barriers? What assistance will be required? Will training and conditioning be required for attached personnel?

- What additional personnel will accompany the unit? Will they be available for training and conditioning?
- What is the current level of physical fitness of the unit?
- What is the current level of individual expertise in mountaineering?
- What type of operations can be expected?
- What is the composition of the advance party? Will they be available to assist in training and acclimatization?
- What areas in the U.S. most closely resemble the area of operations?
- Are predeployment areas and ranges available?
- Does the unit have instructors qualified in mountain warfare?
- What type equipment will be required (to fit the season, mission, terrain)?
- Does the unit have enough of the required equipment? Do personnel know how to use the equipment? Will the equipment go with the advance party, with the unit, or follow after the unit's arrival?
- Does equipment require modification?
- Do weapons and equipment require special maintenance?

b. When the unit arrives in the area of operations, all personnel require a period of conditioning and acclimatization. The time schedule should allow for longer and more frequent periods of rest. The rigors of establishing an assembly area exhaust most unacclimatized personnel. Water, food, and rest must be considered as priorities, ensuring sufficient amounts while individual metabolisms and bodies become accustomed to functioning at higher elevations.

c. Since the acclimatization process cannot be shortened, and the absence of acclimatization hampers the successful execution of operations, deployment to higher elevations must consider the following:

(1) Above 8,000 feet, a unit should ascend at a rate of 1,000 to 2,000 feet per day. Units can leapfrog, taking an extended rest period.

(2) Units should not resort to the use of pharmaceutical pretreatment with carbonic anhydrase inhibitors such as acetazolamide (Diamox). These drugs have side effects that mimic the signs and symptoms of AMS. Inexperienced medics may have difficulty recognizing the differences between the side effects of the drug and a condition that could possibly be life threatening. Additionally, these drugs are diuretics, which results in higher hydration levels (at least 25 percent increase per man per day). These higher hydration levels create a larger logistical demand on the unit by requiring more water, time to acquire water, water purification supplies, and, if in a winter environment, fuels for melting snow and ice for water.

(3) Carbonic anhydrase inhibitors such as acetazolamide are effective in the treatment of mild and severe AMS. These drugs should accompany attached medical personnel because they can treat the soldier suffering the symptoms of AMS and, although rest may be required evacuation may not be needed.

(4) Do not move troops directly to high altitudes even if allowances can be made for inactivity for the first three to five days before mission commitment. Moving troops directly to high altitude can increase the probability of altitude sickness. Even if inactivity follows deployment, the incidence of altitude sickness is more likely than with a gradual ascent.

d. Training on high-altitude effects can prevent psychological preconceptions. Soldiers who have lived on flat terrain may have difficulty when learning to negotiate steep slopes or cliffs, developing a sense of insecurity and fear. They must be slowly introduced to the new terrain and encouraged to develop the confidence required to negotiate obstacles with assurance and ease. They must be taught the many climbing techniques and principles of mountain movement. They overcome their fear of heights by becoming familiar with the problem. The soldier cannot be forced to disregard this fear.

e. Regardless of previous training and the amount of flat cross-country movement practice, the untrained soldier finds mountain movement hard and tiring. A different group of muscles are used, which must be developed and hardened. A new technique of rhythmic movement must be learned. Such conditioning is attained through frequent marches and climbs, while carrying TOE and special equipment loads. This conditions the back and legs, which results in increased ability and endurance. At the same time, the men acquire confidence and ability to safely negotiate the terrain. The better the physical condition of the soldier, the better the chance of avoiding exhaustion. Proper physical conditioning ensures the soldier is an asset and not a liability. The body improves its capacity for exercise, the metabolism becomes more efficient, and blood and oxygen flow quickly and effectively.

f. A physical fitness training program that gradually increases in difficulty should include marches, climbing, and calisthenics. This increases the soldier's endurance. Through a sustained high level of muscular exertion, the soldier's capacity for exertion is increased. Physical conditioning should include long-distance running for aerobic conditioning; calisthenics and weight training to strengthen the heart, lungs, abdomen, legs, back, arms, and hands; a swimming program to increase lung efficiency; and road marches over mountainous terrain with all combat equipment. Upon deploying to high elevations, caution must be exercised by units that are in superior physical condition. The heart rate, metabolism, and lungs must become accustomed to the elevation and thinner air. A conditioning program must be set up on site and integrated in gradual stages where acclimatization, conditioning, and mountaineering skills are realized.

g. Conditioning should begin with basic climbing. It is equally important to instill the will to climb. Confidence goes hand in hand with physical conditioning and skill development. Repetitive practice, to the point of instinctive reaction, is key to learning and maintaining climbing proficiency and technical skills. There are no quick and easy methods to becoming acclimatized and conditioned. Training should gradually challenge the soldier over an extended period and reinforce learning skills.

Section III. MEDICAL CONSIDERATIONS

Improper acclimatization poses many problems for medical personnel. Facilities and supplies may be inadequate to treat all victims. After acclimatization, personnel can still become injured (sprains, strains, fractures, frostbite, hypothermia, and trench foot). Mountain sickness and other illnesses may also occur. Evacuation of the sick and wounded is compounded by the terrain and weather.

2-6. ILLNESS AND INJURY

Units operating in mountainous regions are exposed to varied types of injuries and illnesses not associated with other areas. Medical considerations are like those for other environments; however, there are some unique aspects of mountain operations to be considered if effective support is to be provided. Most injuries in the mountain environment are soft tissue injuries. These include sprains, strains, abrasions, contusions and fractures. As with any other injuries, the most life threatening are treated first with the emphasis on airway control, breathing management, and circulatory support. Skills in basic first aid are essential to the mountain leader and should be reinforced with regular sustainment training.

2-7. TREATMENT AND EVACUATION

In harsh mountain weather, the most important course of action is to provide injured soldiers with medical aid as soon as possible. Immediate first aid is given on site. Due to rough terrain, medical units can seldom reach unit aid stations by vehicle to evacuate casualties. Litter bearers are required to move casualties to the rear where they can be evacuated by ground or air to clearing stations. The victim is protected from the weather and shock during transportation. Rendezvous points are coordinated with medical units as far forward as possible. Training must be accomplished with all litter bearers on evacuation techniques and first aid. Lightly wounded personnel may need assistance to move over rough terrain.

2-8. SOLAR INJURIES

Solar injuries can happen in warm weather or in cold weather. These types of injuries can be just as incapacitating as most other injuries but usually are not fatal. The peak hours of ultraviolet (UV) radiation are between the hours of 1100 and 1500. Due to the long wavelengths of ultraviolet light, cloudy days can be more dangerous than sunny days. On sunny days the soldier takes more care due to the bright conditions. On cloudy days the soldier tends not to wear sunglasses or sunscreen.

a. **Sunburn.** Sunburn is the burning of exposed skin surfaces by ultraviolet radiation.

(1) Contributing factors include fair skin, improper use of para-amino benzoic acid (PABA)-based sunscreens, and exposure to intense ultraviolet rays for extended periods.

(2) Symptoms of sunburn are painful, burning, red or blistered skin with a slight swelling. The skin may be warm to the touch. In severe cases chills, fever, and headaches may occur.

(3) To treat sunburn, apply cool saline dressings to alleviate pain and swelling. Do not pop blisters. If blisters do break, wash thoroughly, bandage, and seek medical attention. A solution of vinegar (acetic) and water can be lightly applied with sterile gauze to alleviate burning. The tannic acid in used tea bags can also be applied to alleviate burning. Administer pain medication if needed.

(4) To prevent sunburn, skin should be covered with clothing or PABA-based sunscreens (at least sun protection factor [SPF] 15) should be applied liberally to exposed skin during the peak hours of UV exposure. The SPF means that you can stay exposed to the suns UV rays that many times longer than without it. (For example, an SPF of 15 means that skin can be exposed to UV rays 15 times longer than without sunscreen.) During sustained activity, the sunscreen should be regularly reapplied to maintain the SPF.

b. **Snowblindness.** Snowblindness is sunburn of the cornea of the eye caused by exposure to ultraviolet radiation.

(1) A contributing factor is the reflection of sunlight from all directions off the snow, ice, and water. Ultraviolet rays can cause vision problems even on cloudy days. They are less filtered at high altitudes than at low altitudes.

(2) Symptoms of snowblindness are painful, red, watery eyes; a gritty feeling; blurred vision; and a headache.

(3) To treat snowblindness, patch both eyes with cold compresses for 24 hours. Topical anesthetics such as Tetracaine Ophthalmic can be used to relieve pain. Avoid rubbing the eyes. If still painful, keep the victim's eyes patched and administer oral pain medication. Snowblindness will usually resolve in about 24 hours for mild to moderate cases. Victims are rarely in need of evacuation unless the case is unusually severe.

(4) To prevent snow blindness, use quality sunglasses even on cloudy days in snow-covered terrain. Proper sunglasses should provide 100 percent UVA and UVB protection and have hoods on the sides to prevent reflected light from entering the eye. (Currently, the U.S. Army does not have these types of "glacier" sunglasses in their inventory and they must be acquired from nonmilitary sources.) In an emergency, improvise slit glasses from materials such as cardboard or birch bark.

2-9. COLD-WEATHER INJURIES

Cold-weather injuries can occur during any season of the year. Death has resulted in temperatures as high as 10 degrees Celsius (50 degrees Fahrenheit). A loss of body heat combined with shock produces devastating results. However, most of these accidents can be prevented by proper planning to include: timely requisition and receipt of supplies and proper clothing; thorough training of personnel with respect to the hazards of cold weather; effective methods for the receipt, dissemination, and use of cold-weather data; periodic inspections of clothing, personnel, and equipment; and personnel receiving a balance of water, rest, and nutrition.

a. Soldiers must be prepared to survive, move, and fight in winter conditions. Intense cold affects the mind as well as the body. Simple tasks take longer to perform, and they take more effort than in a temperate climate. When weather conditions become extreme the problems of survival become more significant. Warmth and comfort become the top priorities. The effects of extreme cold and the probability of injury are magnified due to the lack of proper diet and sleep. The most important measure in the prevention of cold-weather injuries is the education of personnel and their leaders.

b. Cold injuries may be divided into two types: freezing and nonfreezing. The freezing type is known as frostbite. The nonfreezing type includes hypothermia, dehydration, and immersion foot. Cold injuries result from impaired circulation and the action of ice formation and cold upon the tissues of the body. Temperature alone is not a reliable guide as to whether a cold injury can occur. Low temperatures are needed for cold injuries to occur, but freezing temperatures are not. Wind speed can accelerate body heat loss under both wet and cold conditions. All commanders and subordinate leaders/instructors must be familiar with and carry GTA 5-8-12, which includes a wind chill equivalent temperature chart (Figure 2-1, page 2-12).

WIND CHILL FACTOR CHART												
COOLING POWER OF WIND EXPRESSED AS AN EQUIVALENT CHILL TEMPERATURE (UNDER CALM CONDITIONS)												
ESTIMATED WIND SPEED (IN MPH)	ACTUAL THERMOMETER READING (F)											
	50	40	30	20	10	0	-10	-20	-30	-40	-50	-60
	EQUIVALENT TEMPERATURES (F)											
Calm	50	40	30	20	10	0	-10	-20	-30	-40	-50	-60
5	48	37	27	16	6	-5	-15	-26	-36	-47	-57	-68
10	40	28	16	4	-9	-24	-33	-46	-58	-70	-83	-95
15	36	22	9	-5	-18	-32	-45	-58	-72	-85	-99	-112
20	32	18	4	-10	-25	-39	-53	-67	-82	-96	-110	-124
25	30	16	0	-15	-29	-44	-59	-74	-88	-104	-118	-133
30	28	13	-2	-18	-33	-48	-63	-79	-94	-109	-125	-140
35	27	11	-4	-21	-35	-51	-67	-82	-98	-113	-129	-145
40	26	10	-6	-21	-37	-53	-69	-85	-100	-116	-132	-148
Winds greater than 40 MPH have little additional effect.	LITTLE DANGER (For properly clothed person) Maximum danger of false sense of security.					INCREASING DANGER Danger from freezing of exposed flesh.		GREAT DANGER				
	Trench foot and immersion foot may occur at any point on this chart.											

Figure 2-1. Wind chill chart.

c. Many other factors in various combinations determine if cold injuries will occur.

(1) ***Previous Cold Injuries.*** If a soldier has had a cold injury before, he is at higher risk for subsequent cold injuries.

(2) ***Race.*** Blacks are more susceptible to cold-weather injuries than Caucasians.

(3) ***Geographic Origin.*** Personnel from warmer climates are more susceptible to cold injury than those from colder climates.

(4) ***Ambient Temperature.*** The temperature of the air (or water) surrounding the body is critical to heat regulation. For example, the body uses more heat to maintain the temperature of the skin when the temperature of the surrounding air is 37 degrees Fahrenheit than when it is 50 degrees Fahrenheit.

(5) ***Wind Chill Factor.*** The commander should know the wind chill factor. When the forecast gives a figure that falls within the increased danger zone or beyond, caution must be taken to minimize cold injury. The equivalent wind chill temperature is especially important when the ambient temperature is 0 degrees Celsius (32 degrees Fahrenheit) or less. Tissue can freeze if exposed for a prolonged period and if frequent warming is not practiced. The lower the wind chill, the faster tissue freezing can occur. Wind chill is the rate of cooling. Wind does not lower the ambient temperature. The ambient temperature alone determines freezing or nonfreezing injuries. Frostbite Wind chill may cause faster cooling due to increased convection, but not below the ambient temperature.

(6) ***Type of Mission.*** Combat action requiring prolonged immobility and long hours of exposure to low temperatures, or not having an opportunity to warm up increases the possibility of cold injuries.

(7) ***Terrain.*** Minimal cover and wet conditions increase the potential for cold injury.

(8) ***Clothing.*** Clothing for cold weather should be worn with the acronym **C.O.L.D.** in mind.

- **C**—Clothing should be clean since prolonged wear reduces its air-trapping abilities and clogs air spaces with dirt and body oils.
- **O**—Overheating. Avoid overheating. Appropriate measures should be taken when a change in weather or activity alters the amount of clothing needed to prevent overheating and, therefore, accumulation of perspiration.
- **L**—Loose and in layers (to trap air and to conserve body heat). The uniform should be worn completely and correctly to avoid injury to exposed body surfaces. The cold-weather uniform is complete when worn with gloves and inserts.
- **D**—Dry. Keep dry. Wet clothing loses insulation value.

(9) ***Moisture.*** Water conducts heat more rapidly than air (25 percent). When the skin or clothing becomes damp or wet, the risk of cold injury is greatly increased.

(10) ***Dehydration.*** The most overlooked factor causing cold injuries is dehydration. Individuals must retain their body fluids. In cold weather the human body needs special care, and the consumption of water is important to retain proper hydration.

(11) ***Age.*** Within the usual age range of combat personnel, age is not a significant factor.

(12) ***Fatigue.*** Mental weariness may cause apathy leading to neglect of duties vital to survival.

(13) ***Concomitant Injury.*** Injuries resulting in shock or blood loss reduce blood flow to extremities and may cause the injured individual to be susceptible to cold injury, which in turn can accelerate shock.

(14) ***Discipline, Training, and Experience.*** Well-trained and disciplined soldiers suffer less than others from the cold.

(15) ***Nutrition.*** Good nutrition is essential for providing the body with fuel to produce heat in cold weather. The number of calories consumed normally increases as the temperature becomes colder.

(16) ***Excess Activity.*** Excess activity (overheating) results in loss of large amounts of body heat by perspiration. This loss of body heat combined with the loss of insulation value provided by the clothing (due to perspiration dampening the clothing) can subject a soldier to cold injuries.

(17) ***Radical Changes in the Weather.*** Weather conditions in mountainous terrain are known to change considerably throughout the day. Weather can quickly change to extremely cold and wet conditions, especially in higher elevations.

d. Commanders should ensure that the following measures are taken.

(1) Soldiers' uniforms are kept as dry as possible and are protected from the elements.

(2) Soldiers are educated on proper use of clothing systems to avoid the effects of overheating and perspiration (layer dressing and ventilate).

(3) The buddy system is used to watch for early signs of cold-weather injuries.

(4) All soldiers waterproof their equipment.

(5) The rate of movement should be slow, deliberate, and careful. Soldiers should not move out at a force march pace and then be stationary after they have perspired heavily. Soldiers should not wear excessive cold-weather clothing while moving.

e. Medical procedures are needed when sickness and injuries occur. Leaders should—

- Assess the situation (tactical and environmental).
- Approach the victim safely (avoid rock or snow slide).
- Perform emergency first aid.
- Treat for shock (always assume that shock is present).
- Check for other injuries/cold injuries.
- Develop a course of action (decide on a means of evacuation).
- Execute the plan and monitor the victim's condition.

f. Body heat may be lost through radiation, conduction, convection, or evaporation.

(1) ***Radiation.*** The direct heat loss from the body to its surrounding atmosphere is called radiation heat loss. The head can radiate up to 80 percent of the total body heat output. On cold days, personnel must keep all extremities covered to retain heat. This accounts for the largest amount of heat lost from the body.

(2) ***Conduction.*** Conduction is the direct transfer of heat from one object in contact with another (being rained on or sitting in snow).

(3) ***Convection.*** Convection is the loss of heat due to moving air or water in contact with the skin. Wind chill is convection cooling. Clothing that ventilates, insulates, and protects must control the layer of warm air next to the skin.

(4) ***Evaporation.*** The evaporation of perspiration causes heat loss. Wet clothing can cause heat loss by conduction and evaporation. Dressing in layers allows soldiers to remove or add clothing as needed.

g. Some of the most common cold-weather injuries are described in the following paragraphs.

(1) ***Shock.*** Shock is the depressed state of vital organs due to the cardiovascular (heart) system not providing enough blood. Although shock is not a cold-weather injury, it is a symptom or a result of other injuries. Any illness or injury can produce shock, which increases the instance and severity of a cold-weather injury. Shock should be assumed in all injuries and treated accordingly. Even minor injuries can produce shock due to cold, pain, fear, and loss of blood.

(a) *Symptoms.* Initial symptoms of shock include apprehension, shortness of breath, sweating, cold skin, rapid and faint pulse, and excessive thirst. If the victim is not given adequate first aid immediately, his condition may digress into incoherence, slower heart beat, unconsciousness, and possibly death.

(b) *Treatment.* To treat shock, restore breathing and heart rate through artificial respiration or cardiopulmonary resuscitation. Treat the injury and control hemorrhaging. Make the victim as comfortable as possible and try to relieve the pain. Keep the victim warm but do not overheat him. Elevate the back and head, or feet. If the victim is conscious and has no abdominal injuries, administer water. The victim should receive proper medical attention as soon as possible.

(2) ***Dehydration.*** Dehydration is the loss of body fluids to the point that normal body functions are prevented or slowed. This is usually caused by overexertion and improper water intake. Dehydration precedes all cold-weather injuries and is a major symptom in acute mountain sickness. It contributes to poor performance in all physical activities—even more so than lack of food. Cold weather requirements for water are no different than in the desert. They may, in fact, exceed desert requirements because of the increased difficulty in moving with extra clothing and through the snow. At high altitudes, the air is dry. Combined

with a rapid rate of breathing, as much as two liters of liquid may be lost each day through respiration. A soldier needs about three to six quarts of water each day to prevent dehydration when living and performing physical labor in a cold or mountainous environment. Coffee and tea are diuretics and cause excessive urination and should be avoided. The adequacy of liquid intake can best be judged by the urine color and volume. Dark amber colored urine instead of light yellow or the absence of a need to urinate upon awakening from a night's sleep are indicators of dehydration. Thirst is not a good indicator of hydration.

(a) *Contributing Factors.* Factors that contribute to dehydration in cold weather are:

- The thirst mechanism does not function properly in cold weather.
- Water is often inconvenient to obtain and purify.
- The air in cold climates and at high altitudes lacks moisture.
- Cold causes frequent urination.

(b) *Symptoms.* Symptoms of dehydration include darkening urine, decreased amounts of urine being produced, dry mouth, tiredness, mental sluggishness, lack of appetite, headache, fainting, rapid heartbeat, dizziness, higher temperature, upset stomach, and unconsciousness. The symptoms of dehydration are similar to those of hypothermia. To distinguish between them, open the victim's clothes and feel the stomach. If the stomach is cold, the victim is probably hypothermic; if it is warm, he is probably dehydrated. However, this test is not conclusive since cold-weather dehydrating can also lead to total body cooling. The cold environment may act as a diuretic and impair the body's ability to conserve fluid (cold-induced diuresis and increased rate of urination).

(c) *Treatment.* Prevent dehydration by consuming three to six quarts of fluids each day (forced drinking in the absence of thirst is mandatory) and avoid caffeine and alcohol, which may chemically contribute to dehydration. Keep the victim warm and treat for shock. In advanced cases, administer fluids by mouth if the victim is conscious. Do not let him eat snow; eating snow uses body heat. Allow the victim to rest. If he fails to improve within one hour or is unconscious, evacuate him to a medical facility immediately.

(3) ***Hypothermia.*** Hypothermia is the lowering of the body core temperature at a rate faster than the body can produce heat. Hypothermia may be caused by exposure or by sudden wetting of the body such as falling into a lake or being sprayed with fuel or other liquid. Hypothermia can occur even on moderate days with temperatures of 40 to 50 degrees Fahrenheit with little precipitation if heat loss exceeds heat gain and the condition of the soldier is allowed to deteriorate. Hypothermia is classified as mild (core temperature above 90 degrees Fahrenheit or 32 degrees Celsius) or severe (core temperature below 90 degrees Fahrenheit or 32 degrees Celsius). An individual is considered to be "clinically hypothermic" when the core temperature is less than or equal to 95 degrees Fahrenheit.

(a) *Contributing Factors.* Factors that contribute to hypothermia are:

- Dehydration.
- Poor nutrition.
- Diarrhea.
- Decreased physical activity.
- Accidental immersion in water.
- Change in weather.
- High winds.
- Inadequate types or amounts of clothing.

(b) *Symptoms* The first symptom of hypothermia is when the body core (rectal) temperature falls to about 96 degrees Fahrenheit. Other symptoms include:

- Shivering, which may progress to an uncontrollable point making it hard for an individual to care for himself. Shivering begins after a drop in body temperature of one to two degrees. This is followed by clumsiness (stumbling or falling), slow reactions, mental confusion, and difficulty in speaking.
- Body temperature drop from 95 degrees Fahrenheit to 90 degrees Fahrenheit, which can cause sluggish thinking, irrational thought, apathy, and a false sense of warmth. The victim becomes cold and pale; cannot perform simple tasks; experiences amnesia and hallucinations; develops blueness of skin and decreased heart and respiratory rate with a weak pulse; pupils of the eyes dilate; speech becomes slurred; and visual disturbance occurs.
- Body temperature drop from 90 degrees Fahrenheit to 85 degrees Fahrenheit, which causes irrationality, incoherence, loss of contact with the environment, muscular rigidity, disorientation, and exhaustion. The soldier might stop shivering after his core temperature drops below 90 degrees Fahrenheit.
- Body temperature drop from 85 degrees Fahrenheit and below, which causes muscle rigidity, unconsciousness, comatose state, and faint vital signs. The pulse may be faint or impalpable, and breathing is too shallow to observe.

(c) *Prevention.* Prevent hypothermia by using the buddy system to watch each other for symptoms; consume adequate amounts of liquids daily; rest; and eat properly.

(d) *Avoidance.* Hypothermia can be avoided by dressing in layers, which permits easy additions or deletions to prevent overheating, becoming too cold, or getting wet or windblown. If the soldier is in a situation that precludes staying warm and dry, he should seek shelter. Sweets and physical activity help to produce body heat.

(e) *Treatment.* Treatment methods vary based on the severity of the hypothermia.

- Mild cases: If a soldier shows symptoms of hypothermia, prevent additional heat loss by getting the victim into a shelter; removing wet clothing and replacing it with dry, insulated clothing; insulating the victim from the ground; and sharing a sleeping bag (cover head) to transfer body heat. Make a diagnosis (rectal temperature). Rehydrate the victim with warm liquids, sweets, and food. If the tactical situation allows, build a fire. Above all else, keep the victim conscious until his vital signs are normal, and seek medical assistance. If possible, keep the victim physically active to produce body heat.
- Severe cases: If the victim is unconscious or appears dead without any obvious injury, prevent further heat loss. Rapid rewarming of an unconscious victim may create problems and should not be attempted. It is best to evacuate as soon as possible. At all times, the victim should be handled gently so as not to cause the cold blood from the extremities to rush to the heart. Do not allow the victim to perform ANY physical activity. Immediately transport the victim to the nearest medical facility. Field reheating is not effective and may be hazardous. Provide artificial respiration if breathing stops. If no pulse is detectable, be aware that in hypothermia there is often effective circulation for the victim's hypothermic state. In such a case, cardiac compression (such as CPR) may be fatal. The exception is acute hypothermia with near drowning.

- Breathing warm, moist air is the fastest way to warm the inside of the body. If breathing steam is not possible, place tubing under the rescuer's shirt so the victim will still breathe warm, moist air. This process can be done while on the move. In addition to breathing moist, warm air the victim must be gradually rewarmed using external heat sources. Padded hot water bottles or heated stones should be placed in the armpits.
- If conscious, the victim can be given warm, sweet drinks.
- The Hibler Pack is an improvised method of rewarming hypothermic victims in the field. This is used to heat the body core first so the vital organs are warmed and not the extremities. As the body warms up the warm blood will eventually warm all parts of the body. First lay out a blanket or sleeping bag and place a poncho or space blanket inside of it. The poncho or space blanket should go from the base of the skull to the base of the butt. This keeps the sleeping bag/blanket dry and acts like a vapor barrier. Lay the hypothermic patient inside the sleeping bag/blanket. Using a stove, warm water until it is hot to the touch (but not hot enough to burn the patient) and completely dampen any absorbable materials (such as T-shirt, towel, BDU top, and so on). Place the warm, wet items inside a plastic bag or directly in the armpits and chest of the patient. After the warm, wet item has been placed on the patient, wrap the patient tightly inside the poncho/space blanket and the blanket/sleeping bag. Continually check the temperature of the wet material and keep it warm.
- All bodily systems in hypothermia are brittle so treat the victim gently. As these attempts are being made, try to evacuate the victim. Severe complications may arise as the body temperature rises, which may result in cardiac arrest even though the victim seems to be doing well.

(4) ***Immersion or Trench Foot.*** This is damage to the circulatory and nervous systems of the feet that occurs from prolonged exposure to cold and wet at above freezing temperatures. This can happen wearing boots or not. A soldier may not feel uncomfortable until the injury has already begun.

(a) *Contributing Factors.* Factors that contribute to immersion or trench foot are:

- Stepping into water over the boot tops.
- Not changing socks often enough.
- Improper hygiene.
- Prolonged exposure (three to five days).

(b) *Symptoms.* Symptoms of immersion or trench foot include the sensation of tingling, numbness, and then pain. The toes are pale, and feel cold and stiff. The skin is wet and soggy with the color turning from red to bright red, progressing to pale and mottled, and then grayish blue. As symptoms progress and damage appears, the skin becomes red and then bluish or black. Swelling may occur. Because the early stages of trench foot are not painful, soldiers must be constantly aware to prevent it.

(c) *Treatment.* To prevent this condition, keep the feet dry and clean. Change socks often, drying the insides of boots, massaging the feet, and using foot powder. Drying the feet for 24 hours usually heals mild cases. Moderate cases usually heal within three to five days. The feet should be handled gently—NOT rubbed or massaged. They should be cleaned with soap and water, dried, elevated, and exposed to room temperature. The victim must stay off his feet and seek medical attention. Severe cases, when feet are not allowed to dry, are evacuated as a litter casualty.

(5) ***Blisters.*** When first noticed and before the formation of a blister, cover a hotspot with moleskin (over the area and beyond it). Use tincture benzoin to help the moleskin adhere to and toughen the skin. Once a blister has formed, cover it with a dressing large enough to fit over the blister, and then tape it. Never drain blisters unless they are surrounded by redness, or draining pus indicates infection. If this occurs, drain the blister from the side with a clean sterile needle. After cleaning with soap and water, gently press out the fluid leaving the skin intact. Make a doughnut of moleskin to go around the blister and apply to the skin. For toe blisters, wrap the entire toe with adhesive tape over the moleskin. (Toenails should be trimmed straight across the top, leaving a 90-degree angle on the sides. This provides an arch so that the corners do not irritate the skin.)

(6) ***Frostbite.*** Frostbite is the freezing or crystallization of living tissues due to heat being lost faster than it can be replaced by blood circulation, or from direct exposure to extreme cold or high winds. Exposure time can be minutes or instantaneous. The extremities are usually the first to be affected. Damp hands and feet may freeze quickly since moisture conducts heat away from the body and destroys the insulating value of clothing. Heat loss is compounded with intense cold and inactivity. With proper clothing and equipment, properly maintained and used, frostbite can be prevented. The extent of frostbite depends on temperature and duration of exposure. Frostbite is one of the major nonfatal cold-weather injuries encountered in military operations, but docs not occur above an ambient temperature of 32 degrees Fahrenheit.

(a) *Categories of Frostbite.* Superficial (mild) frostbite involves only the skin (Figure 2-2). The layer immediately below usually appears white to grayish with the surface feeling hard, but the underlying tissue is soft. Deep (severe) frostbite extends beyond the first layer of skin and may include the bone (Figure 2-3). Discoloration continues from gray to black, and the texture becomes hard as the tissue freezes deeper. This condition requires immediate evacuation to a medical facility.

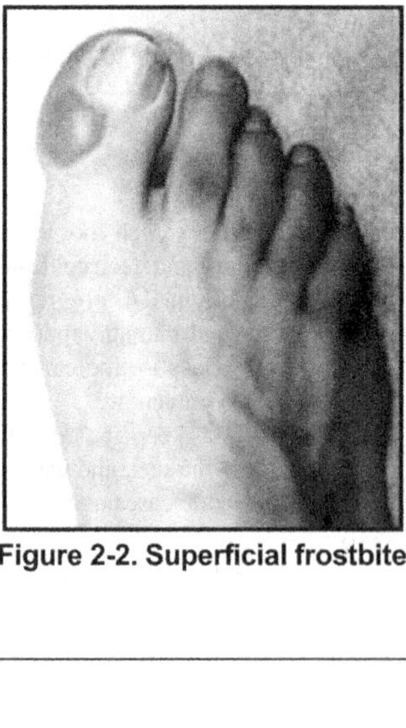

Figure 2-2. Superficial frostbite.

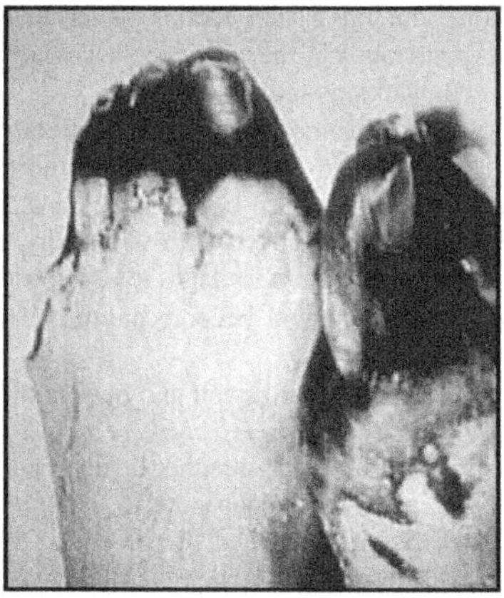

Figure 2-3. Deep frostbite.

(b) *Contributing Factors.* Factors that contribute to frostbite are:

- Dehydration.
- Below-freezing temperatures.
- Skin contact with super cooled metals or liquids.
- Use of caffeine, tobacco, or alcohol.
- Neglect.

(c) *Symptoms.* Symptoms of frostbite vary and may include a cold feeling, pain, burning, numbness, and, in the final stages, a false sense of warmth. The skin first turns red, then pale. It may be bluish in color and then may appear frosty or waxy white. The skin may feel hard, may not be movable over the joints and bony prominences, or may be frozen. Identification of deep versus superficial frostbite is difficult to determine and often requires three to seven days after rewarming for medical personnel to diagnose. Blisters, swelling, and pain may occur after thawing.

(d) *Treatment.* Using the buddy system is one of the primary ways to prevent frostbite. Buddies must watch each other for symptoms of frostbite and provide mutual aid if frostbite occurs. Frostbite should be identified early with prompt first-aid care applied to prevent further damage.

- Treat early signs of frostbite by rewarming with skin-to-skin contact or by sheltering the body part under the clothing next to the body. *Do this immediately.* If tissues have frozen, evacuate the victim before they thaw. If the feet are involved, evacuate the victim as a litter patient.
- Thawing of a frostbitten victim is a hospital procedure. If the victim has frostbite with frozen extremities, protect the frozen parts and evacuate as a litter patient.
- If frostbite is not recognized before it thaws, do not let the area refreeze since this causes more damage. The most often-affected body parts are the hands, fingers, toes, feet, ears, chin, and nose. If evacuation of the victim as a litter case

is not possible and the body part has not yet thawed, have the victim walk out on his own. Walking out on frozen feet is better than having them thaw and refreeze. Self-evacuation may be tactically necessary. Walking on frozen feet does less harm than walking on thawed feet.

- If reheating is inevitable, do not overheat the affected body parts near flame; the warming temperature should not be greater than normal body temperature. Do not rub the parts—the crystallized tissues may break internally and cause more damage. Do not pop blisters; cover them with a dry, sterile dressing. Keep the victim warm (apply loose, bulky bandages to separate toes and fingers.)
- Once a part is rewarmed it will become painful. Pain may be managed with narcotic analgesics.
- Once the foot is rewarmed it will swell and putting the boot back on will not be possible.

(7) ***Constipation.*** Constipation is the infrequent or difficult passage of stools.

(a) *Contributing Factors.* Factors that contribute to constipation are a lack of fluids, improper nutrition, and not defecating when needed.

(b) *Symptoms.* Symptoms include headache, cramping, lack of bowel movement, painful bowel movement, and loss of appetite.

(c) *Treatment.* Constipation is prevented by consuming adequate amounts and varieties of food, drinking from four to six liters of liquid each day, and defecating regularly. If allowed to progress beyond self-care stages, victims will need medical aid.

(8) ***Carbon Monoxide Poisoning.*** This is the replacement of oxygen in the blood with carbon monoxide.

(a) *Contributing Factor.* A contributing factor is inhaling fumes from burning fuel, such as fires, stoves, heaters, and running engines, without proper ventilation.

(b) *Symptoms.* Symptoms are similar to other common illnesses and include headaches, fatigue, excessive yawning, nausea, dizziness, drowsiness, confusion, and unconsciousness. Death may occur. The one visible symptom is bright red lips, mouth, and inside of the eyelids.

(c) *Treatment.* Remove the victim from the source of contamination; administer oxygen, if available; and evacuate to a medical facility. Severe complications may develop even in casualties who appear to have recovered. If the victim is unconscious, administer rescue breathing and CPR as needed.

2-10. HEAT INJURIES

Heat injuries, although associated with hot weather, can occur in cold-weather environments. Most heat injuries can be avoided by planning, periodic inspections of personnel clothing (ventilation) and equipment, a balance of water and food intake, and rest.

a. **Heat Cramps.** Heat cramps are caused by an accumulation of lactic acid in the muscles and a loss of salt through perspiration.

(1) ***Contributing Factor.*** Strenuous exertion causes the body to heat up and to produce heavy perspiration.

(2) ***Symptoms.*** Symptoms of heat cramps include pain and cramping in the arms, legs, back, and stomach. The victim sweats profusely and cannot quench his thirst.

(3) ***Treatment.*** Have the victim rest in a cool, shady area, breath deeply, and stretch the cramped muscle as soon as possible to obtain relief. Loosen the victim's clothing and have him drink cool water. Monitor his condition and seek medical attention if pain and cramps continue.

b. **Heat Exhaustion.** Heat exhaustion may occur when a soldier exerts himself in any environment and he overheats. The blood vessels in the skin become so dilated that the blood flow to the brain and other organs is reduced.

(1) ***Contributing Factors.*** Factors that contribute to heat exhaustion are strenuous activity in hot areas, unacclimatized troops, inappropriate diet, and not enough water or rest.

(2) ***Symptoms.*** Symptoms of heat exhaustion may be similar to fainting but may also include weakness; dizziness; confusion; headache; cold, clammy skin; and nausea. The victim may also have a rapid but weak pulse.

(3) ***Treatment.*** Move the victim to a cool, shady area and loosen his clothes and boots. Have the victim drink water and, if possible, immerse him in water to aid in cooling. Elevate the victim's legs to help restore proper circulation. Monitor his condition and seek medical attention if the symptoms persist.

c. **Heat Stroke.** Heat stroke is a life-threatening situation caused by overexposure to the sun. The body is so depleted of liquids that its internal cooling mechanisms fail to function.

(1) ***Contributing Factors.*** Factors that contribute to heat stroke are prolonged exposure to direct sunlight, overexertion, dehydration, and depletion of electrolytes.

(2) ***Symptoms.*** Symptoms of heat stroke include hot, dry skin; dizziness; confusion and incoherency; headache; nausea; seizures; breathing difficulty; a slow pulse; and loss of consciousness.

(3) ***Treatment.*** Cool the victim at once, and restore breathing and circulation. If the victim is conscious, administer water. If possible, submerge the victim in water to reduce his temperature, treat for shock, and prepare for immediate evacuation.

2-11. ACUTE MOUNTAIN SICKNESS

Acute mountain sickness is a temporary illness that may affect both the beginner and experienced climber. Soldiers are subject to this sickness in altitudes as low as 5,000 feet. Incidence and severity increases with altitude, and when quickly transported to high altitudes. Disability and ineffectiveness can occur in 50 to 80 percent of the troops who are rapidly brought to altitudes above 10,000 feet. At lower altitudes, or where ascent to altitudes is gradual, most personnel can complete assignments with moderate effectiveness and little discomfort.

a. Personnel arriving at moderate elevations (5,000 to 8,000 feet) usually feel well for the first few hours; a feeling of exhilaration or well-being is not unusual. There may be an initial awareness of breathlessness upon exertion and a need for frequent pauses to rest. Irregular breathing can occur, mainly during sleep; these changes may cause apprehension. Severe symptoms may begin 4 to 12 hours after arrival at higher altitudes with symptoms of nausea, sluggishness, fatigue, headache, dizziness, insomnia, depression, uncaring attitude, rapid and labored breathing, weakness, and loss of appetite.

b. A headache is the most noticeable symptom and may be severe. Even when a headache is not present, some loss of appetite and a decrease in tolerance for food occurs. Nausea, even without food intake, occurs and leads to less food intake. Vomiting may occur and contribute to dehydration. Despite fatigue, personnel are unable to sleep. The symptoms usually develop and increase to a peak by the second day. They gradually subside over the next several days so that the total course of AMS may extend from five to seven days. In some instances, the headache may become incapacitating and the soldier should be evacuated to a lower elevation.

c. Treatment for AMS includes the following:

- Oral pain medications such as ibuprofen or aspirin.
- Rest.
- Frequent consumption of liquids and light foods in small amounts.
- Movement to lower altitudes (at least 1,000 feet) to alleviate symptoms, which provides for a more gradual acclimatization.
- Realization of physical limitations and slow progression.
- Practice of deep-breathing exercises.
- Use of acetazolamide in the first 24 hours for mild to moderate cases.

d. AMS is nonfatal, although if left untreated or further ascent is attempted, development of high-altitude pulmonary edema (HAPE) and or high-altitude cerebral edema (HACE) can be seen. A severe persistence of symptoms may identify soldiers who acclimatize poorly and, thus, are more prone to other types of mountain sickness.

2-12. CHRONIC MOUNTAIN SICKNESS

Although not commonly seen in mountaineers, chronic mountain sickness (CMS) (or Monge's disease) can been seen in people who live at sufficiently high altitudes (usually at or above 10,000 feet) over a period of several years. CMS is a right-sided heart failure characterized by chronic pulmonary edema that is caused by years of strain on the right ventricle.

2-13. UNDERSTANDING HIGH-ALTITUDE ILLNESSES

As altitude increases, the overall atmospheric pressure decreases. Decreased pressure is the underlying source of altitude illnesses. Whether at sea level or 20,000 feet the surrounding atmosphere has the same percentage of oxygen. As pressure decreases the body has a much more difficult time passing oxygen from the lungs to the red blood cells and thus to the tissues of the body. This lower pressure means lower oxygen levels in the blood and increased carbon dioxide levels. Increased carbon dioxide levels in the blood cause a systemic vasodilatation, or expansion of blood vessels. This increased vascular size stretches the vessel walls causing leakage of the fluid portions of the blood into the interstitial spaces, which leads to cerebral edema or HACE. Unless treated, HACE will continue to progress due to the decreased atmospheric pressure of oxygen. Further ascent will hasten the progression of HACE and could possibly cause death.

While the body has an overall systemic vasodilatation, the lungs initially experience pulmonary vasoconstriction. This constricting of the vessels in the lungs causes increased workload on the right ventricle, the chamber of the heart that receives de-oxygenated blood from the right atrium and pushes it to the lungs to be re-oxygenated. As the right ventricle works harder to force blood to the lungs, its overall output is decreased thus decreasing the overall pulmonary perfusion. Decreased pulmonary perfusion causes decreased cellular respiration—the transfer of oxygen from the alveoli to the red blood cells. The body is now experiencing increased carbon dioxide levels due to the decreased oxygen levels, which now causes pulmonary vasodilatation. Just as in HACE, this expanding of the vascular structure causes leakage into interstitial space resulting in pulmonary edema or HAPE. As the edema or fluid in the lungs increases, the capability to pass oxygen to the red blood cells decreases thus creating a vicious cycle, which can quickly become fatal if left untreated.

2-14. HIGH-ALTITUDE PULMONARY EDEMA

HAPE is a swelling and filling of the lungs with fluid, caused by rapid ascent. It occurs at high altitudes and limits the oxygen supply to the body.

a. HAPE occurs under conditions of low oxygen pressure, is encountered at high elevations (over 8,000 feet), and can occur in healthy soldiers. HAPE may be considered a form of, or manifestation of, AMS since it occurs during the period of susceptibility to this disorder.

b. HAPE can cause death. Incidence and severity increase with altitude. Except for acclimatization to altitude, no known factors indicate resistance or immunity. Few cases have been reported after 10 days at high altitudes. When remaining at the same altitude, the incidence of HAPE is less frequent than that of AMS. No common indicator dictates how a soldier will react from one exposure to another. Contributing factors are:

- A history of HAPE.
- A rapid or abrupt transition to high altitudes.
- Strenuous physical exertion.
- Exposure to cold.
- Anxiety.

c. Symptoms of AMS can mask early pulmonary difficulties. Symptoms of HAPE include:

- Progressive dry coughing with frothy white or pink sputum (this is usually a later sign) and then coughing up of blood.
- Cyanosis—a blue color to the face, hands, and feet.
- An increased ill feeling, labored breathing, dizziness, fainting, repeated clearing of the throat, and development of a cough.
- Respiratory difficulty, which may be sudden, accompanied by choking and rapid deterioration.
- Progressive shortness of breath, rapid heartbeat (pulse 120 to 160), and coughing (out of contrast to others who arrived at the same time to that altitude).
- Crackling, cellophane-like noises (rales) in the lungs caused by fluid buildup (a stethoscope is usually needed to hear them).
- Unconsciousness, if left untreated. Bubbles form in the nose and mouth, and death results.

d. HAPE is prevented by good nutrition, hydration, and gradual ascent to altitude (no more than 1,000 to 2,000 feet per day to an area of sleep). A rest day, with no gain in altitude or heavy physical exertion, is planned for every 3,000 feet of altitude gained. If a soldier develops symptoms despite precautions, immediate descent is mandatory where he receives prompt treatment, rest, warmth, and oxygen. He is quickly evacuated to lower altitudes as a litter patient. A descent of 300 meters may help; manual descent is not delayed to await air evacuation. If untreated, HAPE may become irreversible and cause death. Cases that are recognized early and treated promptly may expect to recover with no aftereffects. Soldiers who have had previous attacks of HAPE are prone to second attacks.

e. Treatment of HAPE includes:

- Immediate descent (2,000 to 3,000 feet minimum) if possible; if not, then treatment in a monoplace hyperbaric chamber.
- Rest (litter evacuation)
- Supplemental oxygen if available.
- Morphine for the systemic vasodilatation and reduction of preload. This should be carefully considered due to the respiratory depressive properties of the drug.
- Furosemide (Lasix), which is a diuretic, given orally can also be effective.
- The use of mannitol should not be considered due to the fact that it crystallizes at low temperatures. Since almost all high-altitude environments are cold, using mannitol could be fatal.
- Nifidipine (Procardia), which inhibits calcium ion flux across cardiac and smooth muscle cells, decreasing contractility and oxygen demand. It may also dilate coronary arteries and arterioles.
- Diphenhydramine (Benadryl), which can help alleviate the histamine response that increases mucosal secretions.

2-15. HIGH-ALTITUDE CEREBRAL EDEMA

HACE is the accumulation of fluid in the brain, which results in swelling and a depression of brain function that may result in death. It is caused by a rapid ascent to altitude without progressive acclimatization. Prevention of HACE is the same as for HAPE. HAPE and HACE may occur in experienced, well-acclimated mountaineers without warning or obvious predisposing conditions. They can be fatal; when the first symptoms occur, immediate descent is mandatory.

a. Contributing factors include rapid ascent to heights over 8,000 feet and aggravation by overexertion.

b. Symptoms of HACE include mild personality changes, paralysis, stupor, convulsions, coma, inability to concentrate, headaches, vomiting, decrease in urination, and lack of coordination. The main symptom of HACE is a severe headache. A headache combined with any other physical or psychological disturbances should be assumed to be manifestations of HACE. Headaches may be accompanied by a loss of coordination, confusion, hallucinations, and unconsciousness. These may be combined with symptoms of HAPE. The victim is often mistakenly left alone since others may think he is only irritable or temperamental; no one should ever be ignored. The symptoms may rapidly progress to death. Prompt descent to a lower altitude is vital.

c. Preventive measures include good eating habits, maintaining hydration, and using a gradual ascent to altitude. Rest, warmth, and oxygen at lower elevations enhance recovery. Left untreated, HACE can cause death.

d. Treatment for HACE includes:

- Dexamethasone injection immediately followed by oral dexamethasone.
- Supplemental oxygen.
- Rapid descent and medical attention.
- Use of a hyberbaric chamber if descent is delayed.

2-16. HYDRATION IN HAPE AND HACE

HAPE and HACE cause increased proteins in the plasma, or the fluid portion of the blood, which in turn increases blood viscosity. Increased viscosity increases vascular pressure. Vascular leakage caused by stretching of the vessel walls is made worse because of this increased vascular pressure. From this, edema, both cerebral and pulmonary, occurs. Hydration simply decreases viscosity.

CHAPTER 3

MOUNTAINEERING EQUIPMENT

Commanders at every level must understand the complexity of operations in a mountainous environment where every aspect of combat operations becomes more difficult. Leaders must understand that each individual has a different metabolism and, therefore, cools down and heats up differently, which requires soldiers to dress-up and dress-down at different intervals. Provided all tactical concerns are met, the concept of uniformity is outdated and only reduces the unit's ability to fight and function at an optimum level. The extreme cold weather clothing system (ECWCS) is specifically designed to allow for rapid moisture transfer and optimum heat retention while protecting the individual from the elements. Every leader is responsible for ensuring that the ECWCS is worn in accordance with the manufacturers' recommendations. Commanders at all levels must also understand that skills learned at an Army mountaineering school are perishable and soldiers need constant practice to remain proficient. The properly trained mountain soldier of today can live better, move faster, and fight harder in an environment that is every bit as hostile as the enemy.

Section I. EQUIPMENT DESCRIPTION AND MAINTENANCE

With mountainous terrain encompassing a large portion of the world's land mass, the proper use of mountaineering equipment will enhance a unit's combat capability and provide a combat multiplier. The equipment described in this chapter is produced by many different manufacturers; however, each item is produced and tested to extremely high standards to ensure safety when being used correctly. The weak link in the safety chain is the user. Great care in performing preventative maintenance checks and services and proper training in the use of the equipment is paramount to ensuring safe operations. The manufacturers of each and every piece of equipment provide recommendations on how to use and care for its product. It is imperative to follow these instructions explicitly.

3-1. FOOTWEAR

Currently, CTA 50-900 provides adequate footwear for most operations in mountainous terrain. In temperate climates a combination of footwear is most appropriate to accomplish all tasks.

a. The hot weather boot provides an excellent all-round platform for movement and climbing techniques and should be the boot of choice when the weather permits. The intermediate cold weather boot provides an acceptable platform for operations when the weather is less than ideal. These two types of boots issued together will provide the unit with the footwear necessary to accomplish the majority of basic mountain missions.

b. Mountain operations are encumbered by extreme cold, and the extreme cold weather boot (with vapor barrier) provides an adequate platform for many basic mountain missions. However, plastic mountaineering boots should be incorporated into training as soon as possible. These boots provide a more versatile platform for any condition that would be encountered in the mountains, while keeping the foot dryer and warmer.

c. Level 2 and level 3 mountaineers will need mission-specific footwear that is not currently available in the military supply system. The two types of footwear they will need are climbing shoes and plastic mountaineering boots.

(1) Climbing shoes are made specifically for climbing vertical or near vertical rock faces. These shoes are made with a soft leather upper, a lace-up configuration, and a smooth "sticky rubber" sole (Figure 3-1). The smooth "sticky rubber" sole is the key to the climbing shoe, providing greater friction on the surface of the rock, allowing the climber access to more difficult terrain.

(2) The plastic mountaineering boot is a double boot system (Figure 3-1). The inner boot provides support, as well as insulation against the cold. The inner boot may or may not come with a breathable membrane. The outer boot is a molded plastic (usually with a lace-up configuration) with a lug sole. The welt of the boot is molded in such a way that crampons, ski bindings, and snowshoes are easily attached and detached.

Note: Maintenance of all types of footwear must closely follow the manufacturers' recommendations.

Figure 3-1. Climbing shoes and plastic mountaineering boots.

3-2. CLOTHING

Clothing is perhaps the most underestimated and misunderstood equipment in the military inventory. The clothing system refers to every piece of clothing placed against the skin, the insulation layers, and the outer most garments, which protect the soldier from the elements. When clothing is worn properly, the soldier is better able to accomplish his tasks. When worn improperly, he is, at best, uncomfortable and, at worst, develops hypothermia or frostbite.

a. **Socks.** Socks are one of the most under-appreciated part of the entire clothing system. Socks are extremely valuable in many respects, if worn correctly. As a system, socks provide cushioning for the foot, remove excess moisture, and provide insulation from cold temperatures. Improper wear and excess moisture are the biggest causes of hot spots and blisters. Regardless of climatic conditions, socks should always be worn in layers.

(1) The first layer should be a hydrophobic material that moves moisture from the foot surface to the outer sock.

(2) The outer sock should also be made of hydrophobic materials, but should be complimented with materials that provide cushioning and abrasion resistance.

(3) A third layer can be added depending upon the climatic conditions.

(a) In severe wet conditions, a waterproof type sock can be added to reduce the amount of water that would saturate the foot. This layer would be worn over the first two layers if conditions were extremely wet.

(b) In extremely cold conditions a vapor barrier sock can be worn either over both of the original pairs of socks or between the hydrophobic layer and the insulating layer. If the user is wearing VB boots, the vapor barrier sock is not recommended.

b. **Underwear.** Underwear should also be made of materials that move moisture from the body. Many civilian companies manufacture this type of underwear. The primary material in this product is polyester, which moves moisture from the body to the outer layers keeping the user drier and more comfortable in all climatic conditions. In colder environments, several pairs of long underwear of different thickness should be made available. A lightweight set coupled with a heavyweight set will provide a multitude of layering combinations.

c. **Insulating Layers.** Insulating layers are those layers that are worn over the underwear and under the outer layers of clothing. Insulating layers provide additional warmth when the weather turns bad. For the most part, today's insulating layers will provide for easy moisture movement as well as trap air to increase the insulating factor. The insulating layers that are presently available are referred to as pile or fleece. The ECWCS (Figure 3-2, page 3-4) also incorporates the field jacket and field pants liner as additional insulating layers. However, these two components do not move moisture as effectively as the pile or fleece.

d. **Outer Layers.** The ECWCS provides a jacket and pants made of a durable waterproof fabric. Both are constructed with a nylon shell with a laminated breathable membrane attached. This membrane allows the garment to release moisture to the environment while the nylon shell provides a degree of water resistance during rain and snow. The nylon also acts as a barrier to wind, which helps the garment retain the warm air trapped by the insulating layers. Leaders at all levels must understand the importance of wearing the ECWCS correctly.

Note: Cotton layers must not be included in any layer during operations in a cold environment.

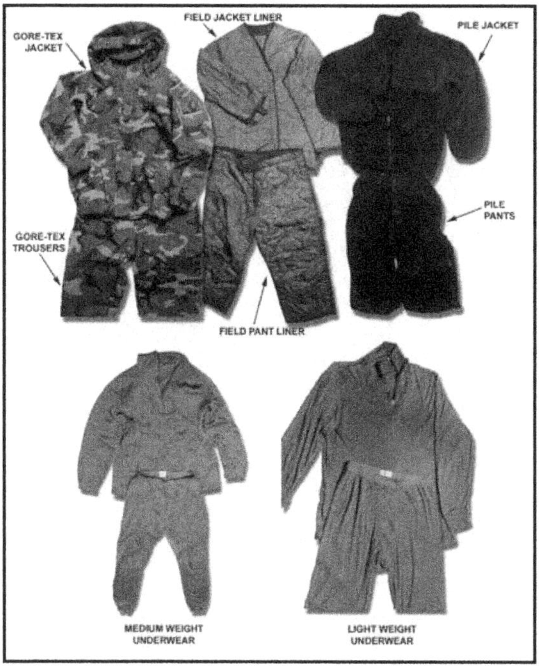

Figure 3-2. Extreme cold weather clothing system.

e. **Gaiters.** Gaiters are used to protect the lower leg from snow and ice, as well as mud, twigs, and stones. The use of waterproof fabrics or other breathable materials laminated to the nylon makes the gaiter an integral component of the cold weather clothing system. Gaiters are not presently fielded in the standard ECWCS and, in most cases, will need to be locally purchased. Gaiters are available in three styles (Figure 3-3).

(1) The most common style of gaiter is the open-toed variety, which is a nylon shell that may or may not have a breathable material laminated to it. The open front allows the boot to slip easily into it and is closed with a combination of zipper, hook-pile tape, and snaps. It will have an adjustable neoprene strap that goes under the boot to keep it snug to the boot. The length should reach to just below the knee and will be kept snug with a drawstring and cord lock.

(2) The second type of gaiter is referred to as a full or randed gaiter. This gaiter completely covers the boot down to the welt. It can be laminated with a breathable material and can also be insulated if necessary. This gaiter is used with plastic mountaineering boots and should be glued in place and not removed.

(3) The third type of gaiter is specific to high-altitude mountaineering or extremely cold temperatures and is referred to as an overboot. It is worn completely over the boot and must be worn with crampons because it has no traction sole.

Figure 3-3. Three types of gaiters.

f. **Hand Wear.** During operations in mountainous terrain the use of hand wear is extremely important. Even during the best climatic conditions, temperatures in the mountains will dip below the freezing point. While mittens are always warmer than gloves, the finger dexterity needed to do most tasks makes gloves the primary cold weather hand wear (Figure 3-4, page 3-6).

(1) The principals that apply to clothing also apply to gloves and mittens. They should provide moisture transfer from the skin to the outer layers—the insulating layer must insulate the hand from the cold and move moisture to the outer layer. The outer layer must be weather resistant and breathable. Both gloves and mittens should be required for all soldiers during mountain operations, as well as replacement liners for both. This will provide enough flexibility to accomplish all tasks and keep the users' hands warm and dry.

(2) Just as the clothing system is worn in layers, gloves and mittens work best using the same principle. Retention cords that loop over the wrist work extremely well when the wearer needs to remove the outer layer to accomplish a task that requires fine finger dexterity. Leaving the glove or mitten dangling from the wrist ensures the wearer knows where it is at all times.

Figure 3-4. Hand wear.

g. **Headwear.** A large majority of heat loss (25 percent) occurs through the head and neck area. The most effective way to counter heat loss is to wear a hat. The best hat available to the individual soldier through the military supply system is the black watch cap. Natural fibers, predominately wool, are acceptable but can be bulky and difficult to fit under a helmet. As with clothes and hand wear, man-made fibers are preferred. For colder climates a neck gaiter can be added. The neck gaiter is a tube of man-made material that fits around the neck and can reach up over the ears and nose (Figure 3-5). For extreme cold, a balaclava can be added. This covers the head, neck, and face leaving only a slot for the eyes (Figure 3-5). Worn together the combination is warm and provides for moisture movement, keeping the wearer drier and warmer.

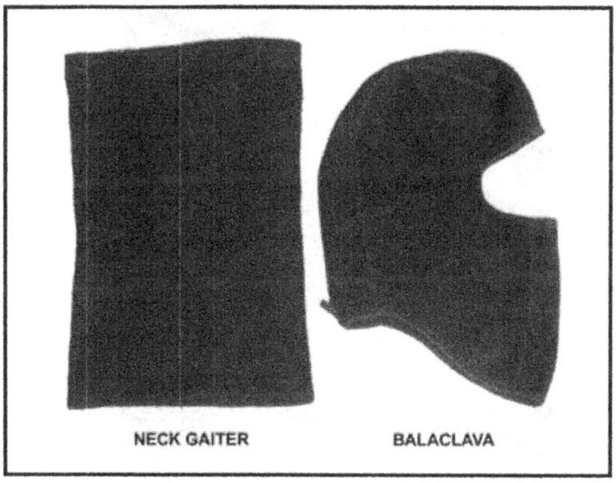

Figure 3-5. Neck gaiter and balaclava.

h. **Helmets**. The Kevlar ballistic helmet can be used for most basic mountaineering tasks. It must be fitted with parachute retention straps and the foam impact pad (Figure 3-6). The level 2 and 3 mountaineer will need a lighter weight helmet for specific climbing scenerios. Several civilian manufacturers produce an effective helmet. Whichever helmet is selected, it should be designed specifically for mountaineering and adjustable so the user can add a hat under it when needed.

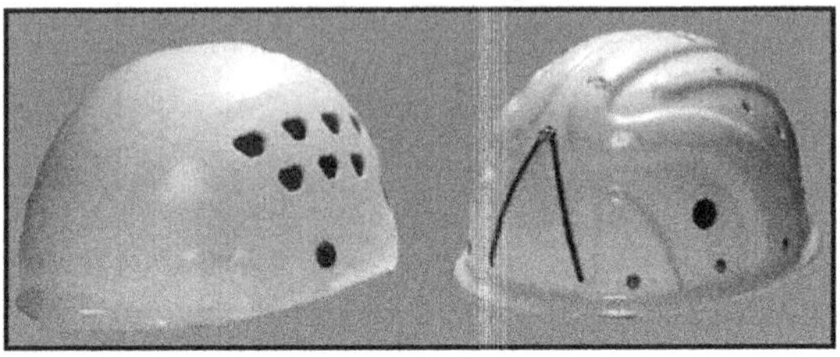

Figure 3-6. Helmets.

i. **Eyewear.** The military supply system does not currently provide adequate eyewear for mountaineering. Eyewear is divided into two catagories: glacier glasses and goggles (Figure 3-7). Glacier glasses are sunglasses that cover the entire eye socket. Many operations in the mountains occur above the tree line or on ice and snow surfaces where the harmful UV rays of the sun can bombard the eyes from every angle increasing the likelihood of snowblindness. Goggles for mountain operations should be antifogging. Double or triple lenses work best. UV rays penetrate clouds so the goggles should be UV protected. Both glacier glasses and goggles are required equipment in the mountains. The lack of either one can lead to severe eye injury or blindness.

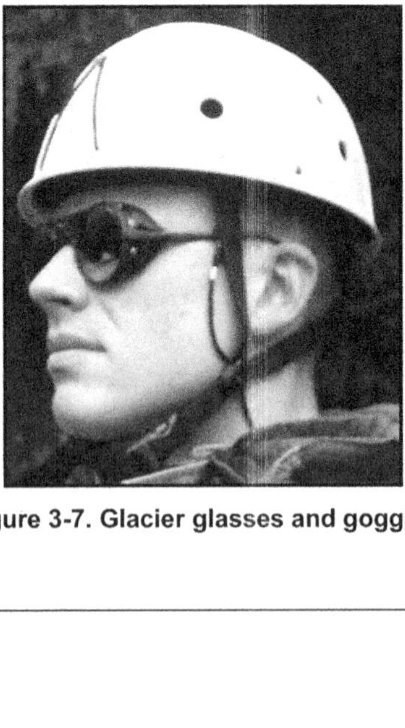

Figure 3-7. Glacier glasses and goggles.

j. **Maintenance of Clothing.** Clothing and equipment manufacturers provide specific instructions for proper care. Following these instructions is necessary to ensure the equipment works as intended.

3-3. CLIMBING SOFTWARE

Climbing software refers to rope, cord, webbing, and harnesses. All mountaineering specific equipment, to include hardware (see paragraph 3-4), should only be used if it has the UIAA certificate of safety. UIAA is the organization that oversees the testing of mountaineering equipment. It is based in Paris, France, and comprises several commissions. The safety commission has established standards for mountaineering and climbing equipment that have become well recognized throughout the world. Their work continues as new equipment develops and is brought into common use. Community Europe (CE) recognizes UIAA testing standards and, as the broader-based testing facility for the combined European economy, meets or exceeds the UIAA standards for all climbing and mountaineering equipment produced in Europe. European norm (EN) and CE have been combined to make combined European norm (CEN). While the United States has no specific standards, American manufacturers have their equipment tested by UIAA to ensure safe operating tolerances.

a. **Ropes and Cord.** Ropes and cords are the most important pieces of mountaineering equipment and proper selection deserves careful thought. These items are your lifeline in the mountains, so selecting the right type and size is of the utmost importance. All ropes and cord used in mountaineering and climbing today are constructed with the same basic configuration. The construction technique is referred to as Kernmantle, which is, essentially, a core of nylon fibers protected by a woven sheath, similar to parachute or 550 cord (Figure 3-8).

Figure 3-8. Kernmantle construction.

(1) Ropes come in two types: static and dynamic. This refers to their ability to stretch under tension. A static rope has very little stretch, perhaps as little as one to two percent, and is best used in rope installations. A dynamic rope is most useful for climbing and general mountaineering. Its ability to stretch up to 1/3 of its overall length makes it the right choice any time the user might take a fall. Dynamic and static ropes come in various

diameters and lengths. For most military applications, a standard 10.5- or 11-millimeter by 50-meter dynamic rope and 11-millimeter by 45-meter static rope will be sufficient.

(2) When choosing dynamic rope, factors affecting rope selection include intended use, impact force, abrasion resistance, and elongation. Regardless of the rope chosen, it should be UIAA certified.

(3) Cord or small diameter rope is indispensable to the mountaineer. Its many uses make it a valuable piece of equipment. All cord is static and constructed in the same manner as larger rope. If used for Prusik knots, the cord's diameter should be 5 to 7 millimeters when used on an 11-mm rope.

b. **Webbing and Slings.** Loops of tubular webbing or cord, called slings or runners, are the simplest pieces of equipment and some of the most useful. The uses for these simple pieces are endless, and they are a critical link between the climber, the rope, carabiners, and anchors. Runners are predominately made from either 9/16-inch or 1-inch tubular webbing and are either tied or sewn by a manufacturer (Figure 3-9). Runners can also be made from a high-performance fiber known as spectra, which is stronger, more durable, and less susceptible to ultraviolet deterioration. Runners should be retired regularly following the same considerations used to retire a rope. For most military applications, a combination of different lengths of runners is adequate.

(1) Tied runners have certain advantages over sewn runners—they are inexpensive to make, can be untied and threaded around natural anchors, and can be untied and retied to other pieces of webbing to create extra long runners.

(2) Sewn runners have their own advantages—they tend to be stronger, are usually lighter, and have less bulk than the tied version. They also eliminate a major concern with the homemade knotted runner—the possibility of the knot untying. Sewn runners come in four standard lengths: 2 inches, 4 inches, 12 inches, and 24 inches. They also come in three standard widths: 9/16 inch, 11/16 inch, and 1 inch.

Figure 3-9. Tied or sewn runners.

c. **Harnesses.** Years ago climbers secured themselves to the rope by wrapping the rope around their bodies and tying a bowline-on-a-coil. While this technique is still a viable way of attaching to a rope, the practice is no longer encouraged because of the increased possibility of injury from a fall. The bowline-on-a-coil is best left for low-angle climbing or an emergency situation where harness material is unavailable. Climbers today can select from a wide range of manufactured harnesses. Fitted properly, the harness should ride high on the hips and have snug leg loops to better distribute the force of a fall to the entire pelvis. This type of harness, referred to as a seat harness, provides a comfortable seat for rappelling (Figure 3-10).

(1) Any harness selected should have one very important feature—a double-passed buckle. This is a safety standard that requires the waist belt to be passed over and back through the main buckle a second time. At least 2 inches of the strap should remain after double-passing the buckle.

(2) Another desirable feature on a harness is adjustable leg loops, which allows a snug fit regardless of the number of layers of clothing worn. Adjustable leg loops allow the soldier to make a latrine call without removing the harness or untying the rope.

(3) Equipment loops are desirable for carrying pieces of climbing equipment. For safety purposes always follow the manufacturer's directions for tying-in.

(4) A field-expedient version of the seat harness can be constructed by using 22 feet of either 1-inch or 2-inch (preferred) tubular webbing (Figure 3-10). Two double-overhand knots form the leg loops, leaving 4 to 5 feet of webbing coming from one of the leg loops. The leg loops should just fit over the clothing. Wrap the remaining webbing around the waist ensuring the first wrap is routed through the 6- to 10-inch long strap between the double-overhand knots. Finish the waist wrap with a water knot tied as tightly as possible. With the remaining webbing, tie a square knot without safeties over the water knot ensuring a minimum of 4 inches remains from each strand of webbing.

(5) The full body harness incorporates a chest harness with a seat harness (Figure 3-10). This type of harness has a higher tie-in point and greatly reduces the chance of flipping backward during a fall. This is the only type of harness that is approved by the UIAA. While these harnesses are safer, they do present several disadvantages—they are more expensive, are more restrictive, and increase the difficulty of adding or removing clothing. Most mountaineers prefer to incorporate a separate chest harness with their seat harness when warranted.

(6) A separate chest harness can be purchased from a manufacturer, or a field-expedient version can be made from either two runners or a long piece of webbing. Either chest harness is then attached to the seat harness with a carabiner and a length of webbing or cord.

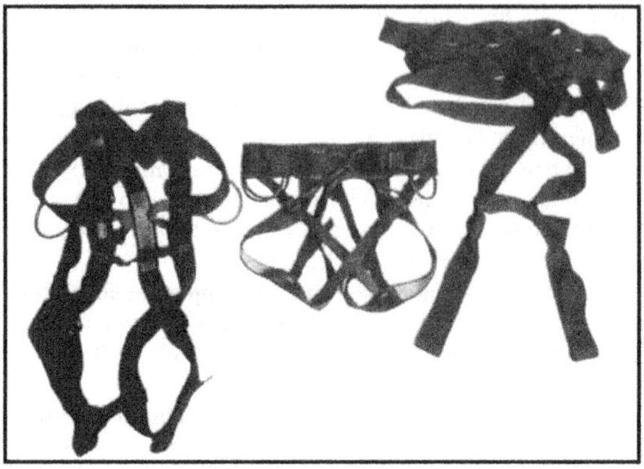

Figure 3-10. Seat harness, field-expedient harness, and full body harness.

3-4. CLIMBING HARDWARE

Climbing hardware refers to all the parts and pieces that allow the trained mountain soldier to accomplish many tasks in the mountains. The importance of this gear to the mountaineer is no less than that of the rifle to the infantryman.

a. **Carabiners.** One of the most versatile pieces of equipment available to the mountaineer is the carabiner. This simple piece of gear is the critical connection between the climber, his rope, and the protection attaching him to the mountain. Carabiners must be strong enough to hold hard falls, yet light enough for the climber to easily carry a quantity of them. Today's high tech metal alloys allow carabiners to meet both of these requirements. Steel is still widely used, but is not preferred for general mountaineering, given other options. Basic carabiner construction affords the user several different shapes. The oval, the D-shaped, and the pear-shaped carabiner are just some of the types currently available. Most models can be made with or without a locking mechanism for the gate opening (Figure 3-11, page 3-12). If the carabiner does have a locking mechanism, it is usually referred to as a locking carabiner. When using a carabiner, great care should be taken to avoid loading the carabiner on its minor axis and to avoid three-way loading (Figure 3-12, page 3-12).

Note: Great care should be used to ensure all carabiner gates are closed and locked during use.

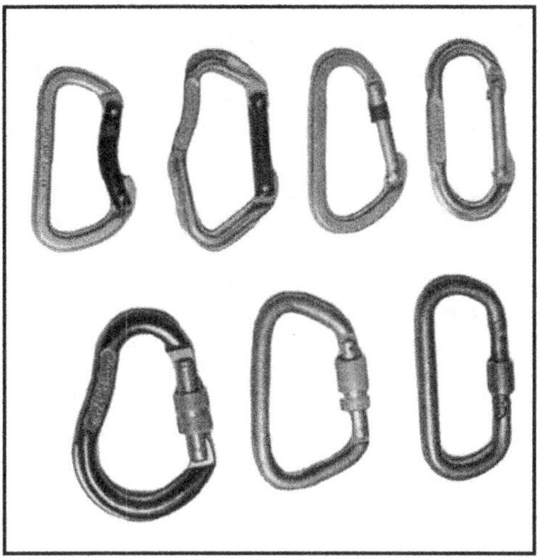

Figure 3-11. Nonlocking and locking carabiners.

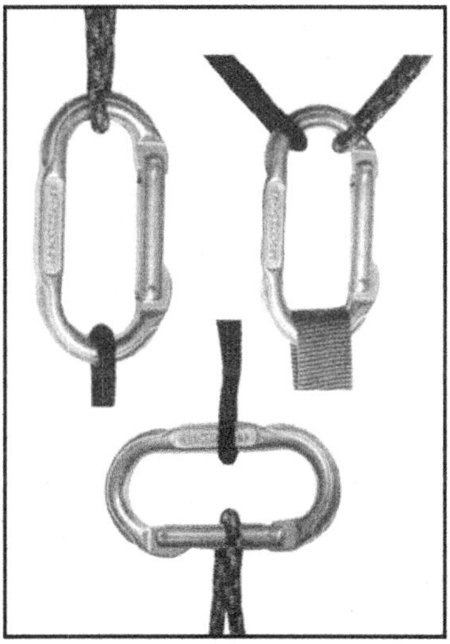

Figure 3-12. Major and minor axes and three-way loading.

(1) The major difference between the oval and the D-shaped carabiner is strength. Because of the design of the D-shaped carabiner, the load is angled onto the spine of the

carabiner thus keeping it off the gate. The down side is that racking any gear or protection on the D-shaped carabiner is difficult because the angle of the carabiner forces all the gear together making it impossible to separate quickly.

(2) The pear-shaped carabiner, specifically the locking version, is excellent for clipping a descender or belay device to the harness. They work well with the munter hitch belaying knot.

(3) Regardless of the type chosen, all carabiners should be UIAA tested. This testing is extensive and tests the carabiner in three ways—along its major axis, along its minor axis, and with the gate open.

b. **Pitons.** A piton is a metal pin that is hammered into a crack in the rock. They are described by their thickness, design, and length (Figure 3-13, page 3-14). Pitons provide a secure anchor for a rope attached by a carabiner. The many different kinds of pitons include: vertical, horizontal, wafer, and angle. They are made of malleable steel, hardened steel, or other alloys. The strength of the piton is determined by its placement rather than its rated tensile strength. The two most common types of pitons are: blades, which hold when wedged into tight-fitting cracks, and angles, which hold blade compression when wedged into a crack.

(1) ***Vertical Pitons***. On vertical pitons, the blade and eye are aligned. These pitons are used in flush, vertical cracks.

(2) ***Horizontal Pitons***. On horizontal pitons, the eye of the piton is at right angles to the blade. These pitons are used in flush, horizontal cracks and in offset or open-book type vertical or horizontal cracks. They are recommended for use in vertical cracks instead of vertical pitons because the torque on the eye tends to wedge the piton into place. This provides more holding power than the vertical piton under the same circumstances.

(3) ***Wafer Pitons***. These pitons are used in shallow, flush cracks. They have little holding power and their weakest points are in the rings provided for the carabiner.

(4) ***Knife Blade Pitons***. These are used in direct-aid climbing. They are small and fit into thin, shallow cracks. They have a tapered blade that is optimum for both strength and holding power.

(5) ***Realized Ultimate Reality Pitons***. Realized ultimate reality pitons (RURPs) are hatchet-shaped pitons about 1-inch square. They are designed to bite into thin, shallow cracks.

(6) ***Angle Pitons***. These are used in wide cracks that are flush or offset. Maximum strength is attained only when the legs of the piton are in contact with the opposite sides of the crack.

(7) ***Bong Pitons***. These are angle pitons that are more than 3.8 centimeters wide. Bongs are commonly made of steel or aluminum alloy and usually contain holes to reduce weight and accommodate carabiners. They have a high holding power and require less hammering than other pitons.

(8) ***Skyhook (Cliffhangers).*** These are small hooks that cling to tiny rock protrusions, ledges, or flakes. Skyhooks require constant tension and are used in a downward pull direction. The curved end will not straighten under body weight. The base is designed to prevent rotation and aid stability.

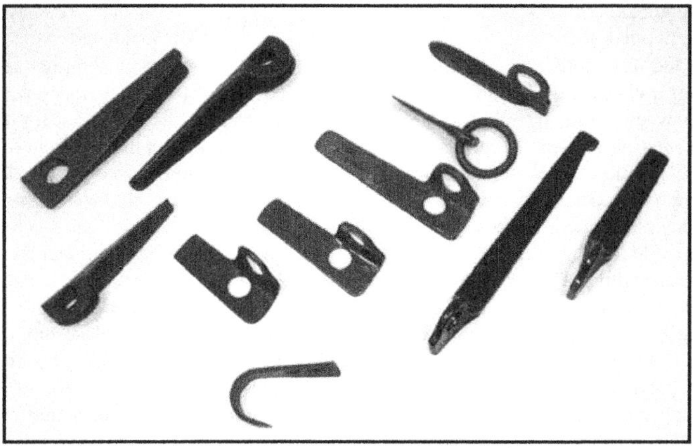

Figure 3-13. Various pitons.

c. **Piton Hammers.** A piton hammer has a flat metal head; a handle made of wood, metal, or fiberglass; and a blunt pick on the opposite side of the hammer (Figure 3-14). A safety lanyard of nylon cord, webbing, or leather is used to attach it to the climber The lanyard should be long enough to allow for full range of motion. Most hammers are approximately 25.5 centimeters long and weigh 12 to 25 ounces. The primary use for a piton hammer is to drive pitons, to be used as anchors, into the rock. The piton hammer can also be used to assist in removing pitons, and in cleaning cracks and rock surfaces to prepare for inserting the piton. The type selected should suit individual preference and the intended use.

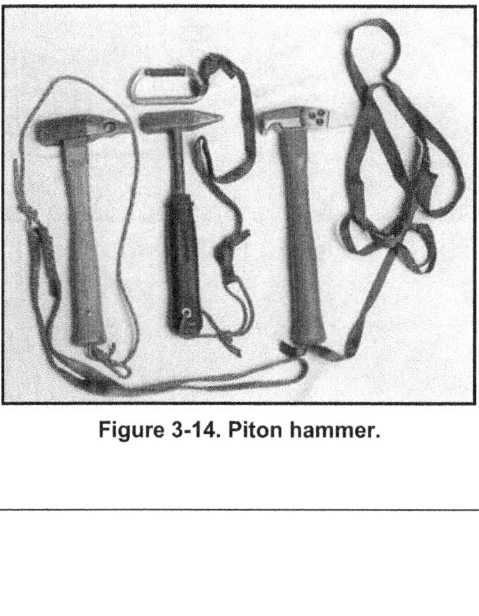

Figure 3-14. Piton hammer.

d. **Chocks.** "Chocks" is a generic term used to describe the various types of artificial protection other than bolts or pitons. Chocks are essentially a tapered metal wedge constructed in various sizes to fit different sized openings in the rock (Figure 3-15). The design of a chock will determine whether it fits into one of two categories—wedges or cams. A wedge holds by wedging into a constricting crack in the rock. A cam holds by slightly rotating in a crack, creating a camming action that lodges the chock in the crack or pocket. Some chocks are manufactured to perform either in the wedging mode or the camming mode. One of the chocks that falls into the category of both a wedge and cam is the hexagonal-shaped or "hex" chock. This type of chock is versatile and comes with either a cable loop or is tied with cord or webbing. All chocks come in different sizes to fit varying widths of cracks. Most chocks come with a wired loop that is stronger than cord and allows for easier placement. Bigger chocks can be threaded with cord or webbing if the user ties the chock himself. Care should be taken to place tubing in the chock before threading the cord. The cord used with chocks is designed to be stiffer and stronger than regular cord and is typically made of Kevlar. The advantage of using a chock rather than a piton is that a climber can carry many different sizes and use them repeatedly.

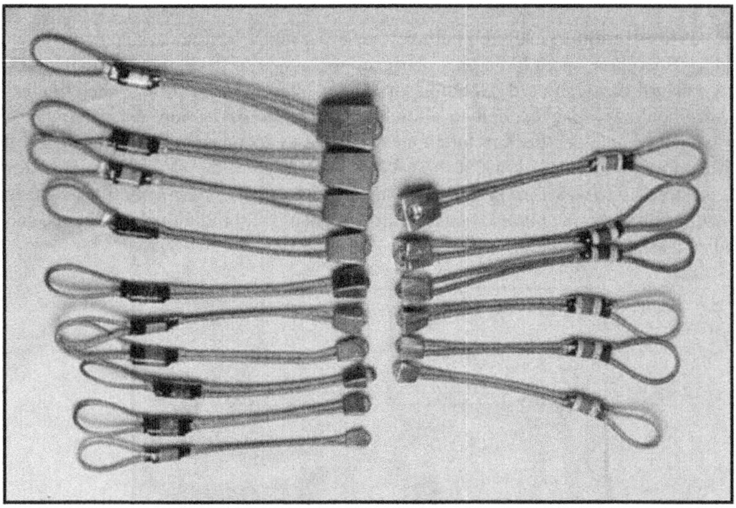

Figure 3-15. Chocks.

e. **Three-Point Camming Device.** The three-point camming device's unique design allows it to be used both as a camming piece and a wedging piece (Figure 3-16). Because of this design it is extremely versatile and, when used in the camming mode, will fit a wide range of cracks. The three-point camming device comes in several different sizes with the smaller sizes working in pockets that no other piece of gear would fit in.

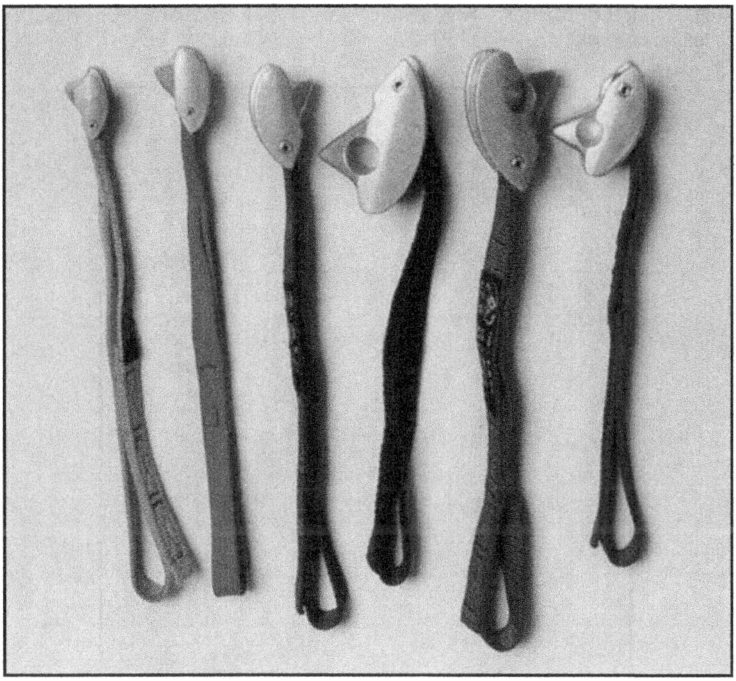

Figure 3-16. Three-point camming device.

f. **Spring-Loaded Camming Devices.** Spring-loaded camming devices (SLCDs) (Figure 3-17) provide convenient, reliable placement in cracks where standard chocks are not practical (parallel or flaring cracks or cracks under roofs). SLCDs have three or four cams rotating around a single or double axis with a rigid or semi-rigid point of attachment. These are placed quickly and easily, saving time and effort. SLCDs are available in many sizes to accommodate different size cracks. Each fits a wide range of crack widths due to the rotating cam heads. The shafts may be rigid metal or semi-rigid cable loops. The flexible cable reduces the risk of stem breakage over an edge in horizontal placements.

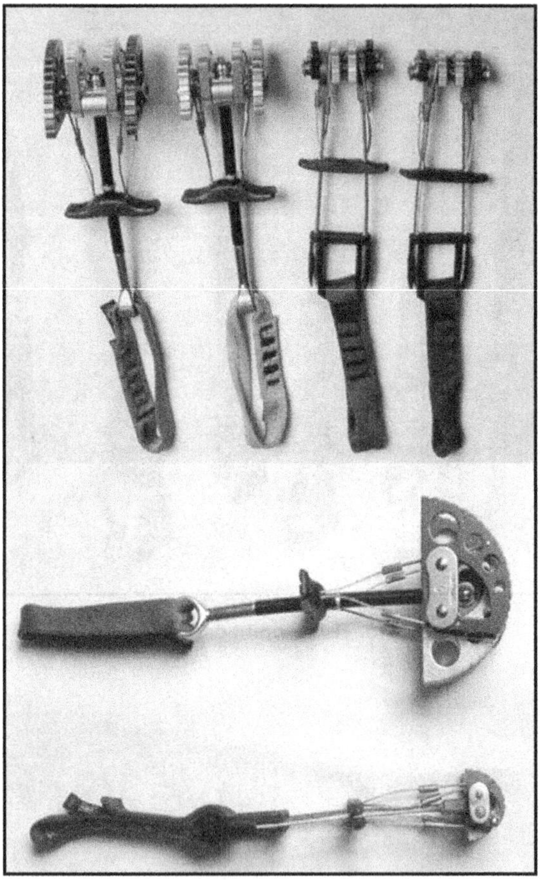

Figure 3-17. Spring-loaded camming devices.

g. **Chock Picks.** Chock picks are primarily used to extract chocks from rock when the they become severely wedged (Figure 3-18). They are also handy to clean cracks with. Made from thin metal, they can be purchased or homemade. When using a chock pick to extract a chock be sure no force is applied directly to the cable juncture. One end of the chock pick should have a hook to use on jammed SLCDs.

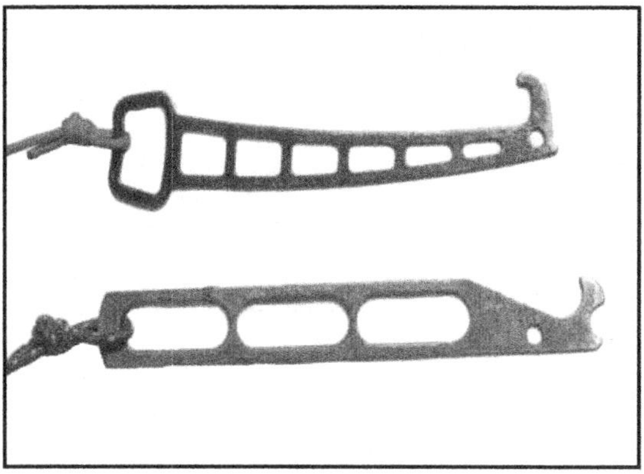

Figure18. Chock picks.

h. **Bolts.** Bolts are screw-like shafts made from metal that are drilled into rock to provide protection (Figure 3-19). The two types are contraction bolts and expansion bolts. Contraction bolts are squeezed together when driven into a rock. Expansion bolts press around a surrounding sleeve to form a snug fit into a rock. Bolts require drilling a hole into a rock, which is time-consuming, exhausting, and extremely noisy. Once emplaced, bolts are the most secure protection for a multidirectional pull. Bolts should be used only when chocks and pitons cannot be emplaced. A bolt is hammered only when it is the nail or self-driving type.

(1) A hanger (for carabiner attachment) and nut are placed on the bolt. The bolt is then inserted and driven into the hole. Because of this requirement, a hand drill must be carried in addition to a piton hammer. Hand drills (also called star drills) are available in different sizes, brands, and weights. A hand drill should have a lanyard to prevent loss.

(2) Self-driving bolts are quicker and easier to emplace. These require a hammer, bolt driver, and drilling anchor, which is driven into the rock. A bolt and carrier are then secured to the emplaced drilling anchor. All metal surfaces should be smooth and free of rust, corrosion, dirt, and moisture. Burrs, chips, and rough spots should be filed smooth and wire-brushed or rubbed clean with steel wool. Items that are cracked or warped indicate excessive wear and should be discarded.

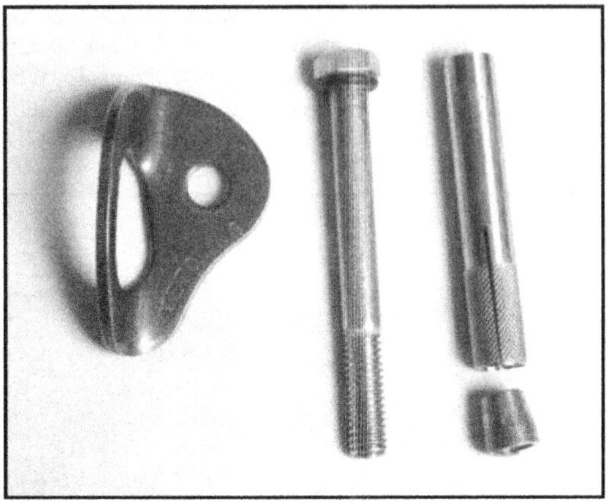

Figure 3-19. Bolts and hangers.

i. **Belay Devices.** Belay devices range from the least equipment intensive (the body belay) to high-tech metal alloy pieces of equipment. Regardless of the belay device choosen, the basic principal remains the same—friction around or through the belay device controls the ropes' movement. Belay devices are divided into three categories: the slot, the tuber, and the mechanical camming device (Figure 3-20).

(1) The slot is a piece of equipment that attaches to a locking carabiner in the harness; a bight of rope slides through the slot and into the carabiner for the belay. The most common slot type belay device is the Sticht plate.

(2) The tuber is used exactly like the slot but its shape is more like a cone or tube.

(3) The mechanical camming device is a manufactured piece of equipment that attaches to the harness with a locking carabiner. The rope is routed through this device so that when force is applied the rope is cammed into a highly frictioned position.

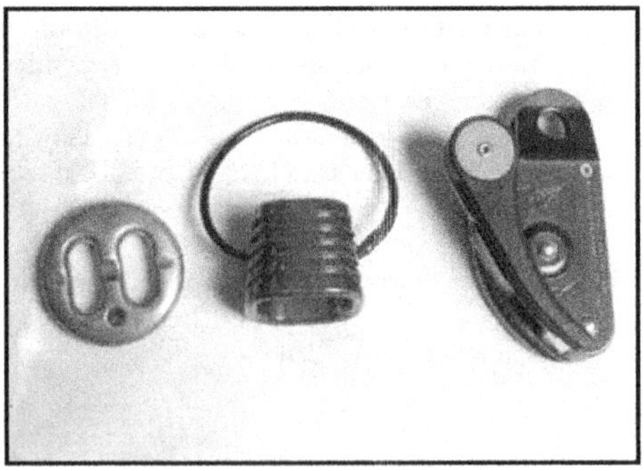

Figure 3-20. Slot, tuber, mechanical camming device.

j. **Descenders.** One piece of equipment used for generations as a descender is the carabiner. A figure-eight is another useful piece of equipment and can be used in conjunction with the carabiner for descending (Figure 3-21).

Note: All belay devices can also be used as descending devices.

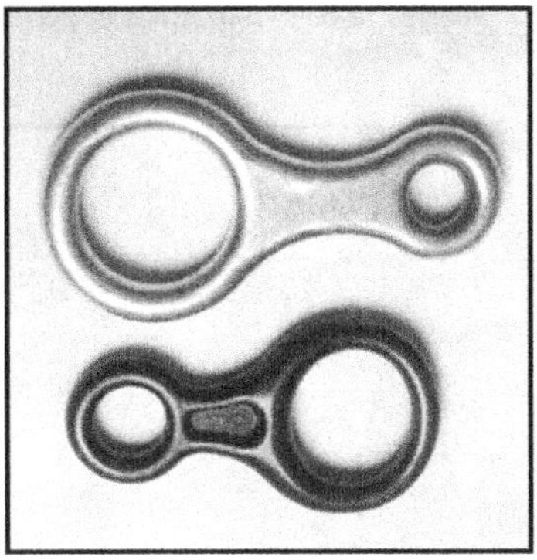

Figure 3-21. Figure-eights.

k. **Ascenders.** Ascenders may be used in other applications such as a personal safety or hauling line cam. All modern ascenders work on the principle of using a cam-like device to allow movement in one direction. Ascenders are primarily made of metal alloys and come in a variety of sizes (Figure 3-22). For difficult vertical terrain, two ascenders work best. For lower angle movement, one ascender is sufficient. Most manufacturers make ascenders as a right and left-handed pair.

Figure 3-22. Ascenders.

l. **Pulleys.** Pulleys are used to change direction in rope systems and to create mechanical advantage in hauling systems. A pulley should be small, lightweight, and strong. They should accommodate the largest diameter of rope being used. Pulleys are made with several bearings, different-sized sheaves (wheel), and metal alloy sideplates (Figure 3-23). Plastic pulleys should always be avoided. The sideplate should rotate on the pulley axle to allow the pulley to be attached at any point along the rope. For best results, the sheave diameter must be at least four times larger than the rope's diameter to maintain high rope strength.

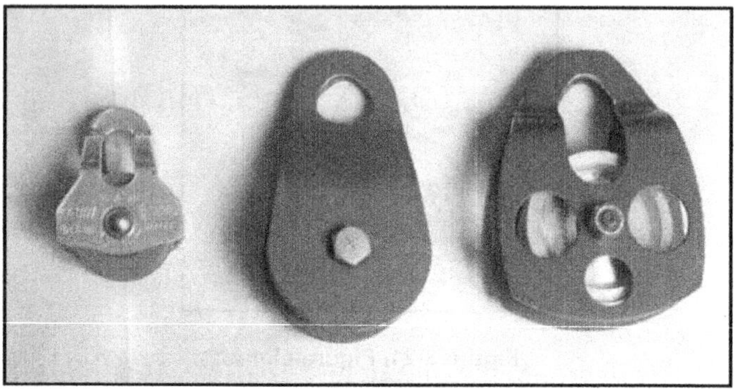

Figure 3-23. Pulley.

3-5. SNOW AND ICE CLIMBING HARDWARE

Snow and ice climbing hardware is the equipment that is particular to operations in some mountainous terrain. Specific training on this type of equipment is essential for safe use. Terrain that would otherwise be inaccessible—snowfields, glaciers, frozen waterfalls—can now be considered avenues of approach using the snow and ice climbing gear listed in this paragraph.

a. **Ice Ax.** The ice ax is one of the most important tools for the mountaineer operating on snow or ice. The climber must become proficient in its use and handling. The versatility of the ax lends itself to balance, step cutting, probing, self-arrest, belays, anchors, direct-aid climbing, and ascending and descending snow and ice covered routes.

(1) Several specific parts comprise an ice ax: the shaft, head (pick and adze), and spike (Figure 3-24, page 3-22).

(a) The shaft (handle) of the ax comes in varying lengths (the primary length of the standard mountaineering ax is 70 centimeters). It can be made of fiberglass, hollow aluminum, or wood; the first two are stronger, therefore safer for mountaineering.

(b) The head of the ax, which combines the pick and the adze, can have different configurations. The pick should be curved slightly and have teeth at least one-fourth of its length. The adze, used for chopping, is perpendicular to the shaft. It can be flat or curved along its length and straight or rounded from side to side. The head can be of one-piece construction or have replaceable picks and adzes. The head should have a hole directly above the shaft to allow for a leash to be attached.

(c) The spike at the bottom of the ax is made of the same material as the head and comes in a variety of shapes.

(2) As climbing becomes more technical, a shorter ax is much more appropriate, and adding a second tool is a must when the terrain becomes vertical. The shorter ax has all the attributes of the longer ax, but it is anywhere from 40 to 55 centimeters long and can have a straight or bent shaft depending on the preference of the user.

b. **Ice Hammer.** The ice hammer is as short or shorter than the technical ax (Figure 3-24). It is used for pounding protection into the ice or pitons into the rock. The only difference between the ice ax and the ice hammer is the ice hammer has a hammerhead instead of an adze. Most of the shorter ice tools have a hole in the shaft to which a leash is secured, which provides a more secure purchase in the ice.

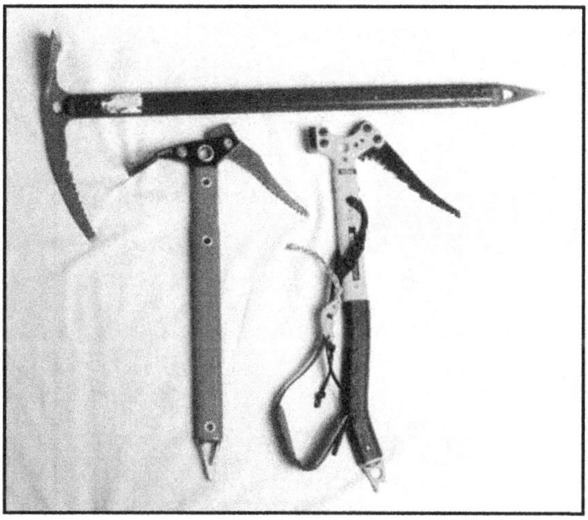

Figure 3-24. Ice ax and ice hammers.

c. **Crampons.** Crampons are used when the footing becomes treacherous. They have multiple spikes on the bottom and spikes protruding from the front (Figure 3-25). Two types of crampons are available: flexible and rigid. Regardless of the type of crampon chosen, fit is the most important factor associated with crampon wear. The crampon should fit snugly on the boot with a minimum of 1 inch of front point protruding. Straps should fit snugly around the foot and any long, loose ends should be trimmed. Both flexible and rigid crampons come in pairs, and any tools needed for adjustment will be provided by the manufacturer.

(1) The hinged or flexible crampon is best used when no technical ice climbing will be done. It is designed to be used with soft, flexible boots, but can be attached to plastic

mountaineering boots. The flexible crampon gets its name from the flexible hinge on the crampon itself. All flexible crampons are adjustable for length while some allow for width adjustment. Most flexible crampons will attach to the boot by means of a strap system. The flexible crampon can be worn with a variety of boot types.

(2) The rigid crampon, as its name implies, is rigid and does not flex. This type of crampon is designed for technical ice climbing, but can be used on less vertical terrain. The rigid crampon can only be worn with plastic mountaineering boots. Rigid crampons will have a toe and heel bail attachment with a strap that wraps around the ankle.

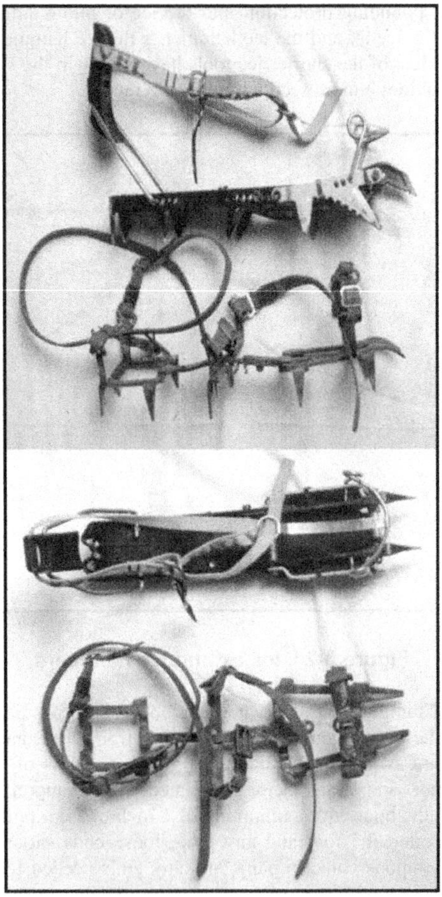

Figure 3-25. Crampons.

d. **Ice Screws.** Ice screws provide artificial protection for climbers and equipment for operations in icy terrain. They are screwed into ice formations. Ice screws are made of chrome-molybdenum steel and vary in lengths from 11 centimeters to 40 centimeters (Figure 3-26). The eye is permanently affixed to the top of the ice screw. The tip consists of milled or hand-ground teeth, which create sharp points to grab the ice when being emplaced. The ice screw has right-hand threads to penetrate the ice when turned clockwise.

(1) When selecting ice screws, choose a screw with a large thread count and large hollow opening. The close threads will allow for ease in turning and better strength. The large hollow opening will allow snow and ice to slide through when turning.

- Type I is 17 centimeters in length with a hollow inner tube.
- Type II is 22 centimeters in length with a hollow inner tube.
- Other variations are hollow alloy screws that have a tapered shank with external threads, which are driven into ice and removed by rotation.

(2) Ice screws should be inspected for cracks, bends, and other deformities that may impair strength or function. If any cracks or bends are noticed, the screw should be turned in. A file may be used to sharpen the ice screw points. Steel wool should be rubbed on rusted surfaces and a thin coat of oil applied when storing steel ice screws.

Note: Ice screws should always be kept clean and dry. The threads and teeth should be protected and kept sharp for ease of application.

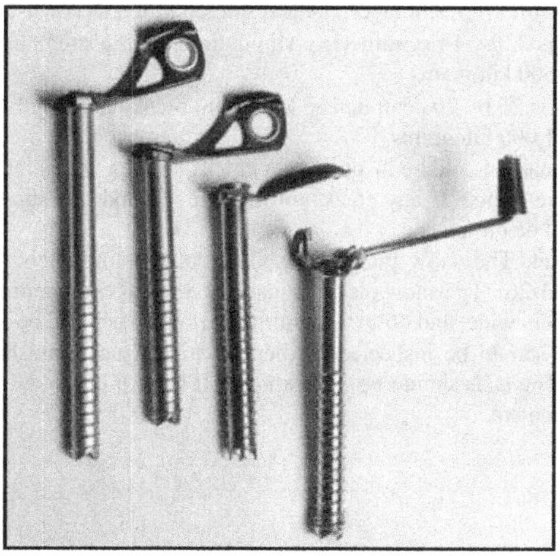

Figure 3-26. Ice screws.

e. **Ice Pitons.** Ice pitons are used to establish anchor points for climbers and equipment when conducting operations on ice. They are made of steel or steel alloys (chrome-molybdenum), and are available in various lengths and diameters (Figure 3-27). They are tubular with a hollow core and are hammered into ice with an ice hammer. The eye is permanently fixed to the top of the ice piton. The tip may be beveled to help grab the ice to facilitate insertion. Ice pitons are extremely strong when placed properly in hard ice. They can, however, pull out easily on warm days and require a considerable amount of effort to extract in cold temperatures.

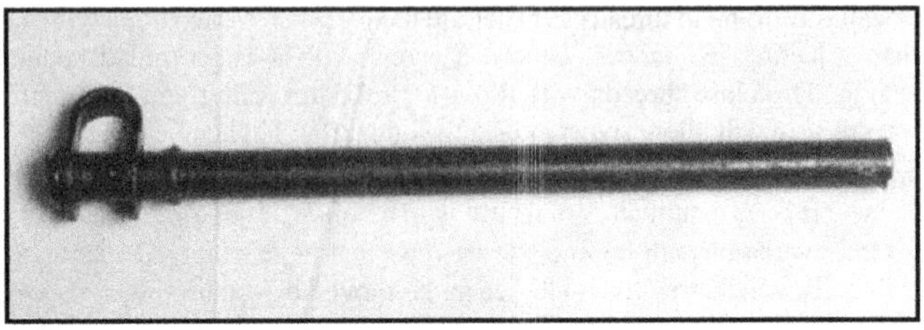

Figure 3-27. Ice piton.

f. **Wired Snow Anchors.** The wired snow anchor (or fluke) provides security for climbers and equipment in operations involving steep ascents by burying the snow anchor into deep snow (Figure 3-28, page 3-26). The fluted anchor portion of the snow anchor is made of aluminum. The wired portion is made of either galvanized steel or stainless steel. Fluke anchors are available in various sizes—their holding ability generally increases with size. They are available with bent faces, flanged sides, and fixed cables. Common types are:

- Type I is 22 by 14 centimeters. Minimum breaking strength of the swaged wire loop is 600 kilograms.
- Type II is 25 by 20 centimeters. Minimum breaking strength of the swaged wire loop is 1,000 kilograms.

The wired snow anchor should be inspected for cracks, broken wire strands, and slippage of the wire through the swage. If any cracks, broken wire strands, or slippage is noticed, the snow anchor should be turned in.

g. **Snow Picket.** The snow picket is used in constructing anchors in snow and ice (Figure 3-28, page 3-26). The snow picket is made of a strong aluminum alloy 3 millimeters thick by 4 centimeters wide, and 45 to 90 centimeters long. They can be angled or T–section stakes. The picket should be inspected for bends, chips, cracks, mushrooming ends, and other deformities. The ends should be filed smooth. If bent or cracked, the picket should be turned in for replacement.

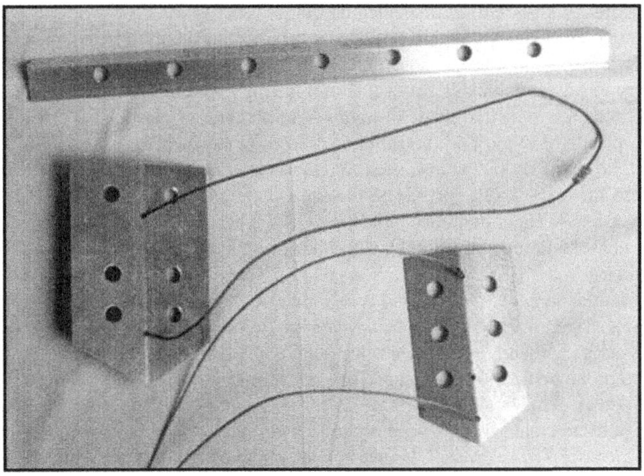

Figure 3-28. Snow anchors, flukes, and pickets.

3-6. SUSTAINABILITY EQUIPMENT

This paragraph describes all additional equipment not directly involved with climbing. This equipment is used for safety (avalanche equipment, wands), bivouacs, movement, and carrying gear. While not all of it will need to be carried on all missions, having the equipment available and knowing how to use it correctly will enhance the unit's capability in mountainous terrain.

a. **Snow Saw**. The snow saw is used to cut into ice and snow. It can be used in step cutting, in shelter construction, for removing frozen obstacles, and for cutting snow stability test pits. The special tooth design of the snow saw easily cuts into frozen snow and ice. The blade is a rigid aluminum alloy of high strength about 3 millimeters thick and 38 centimeters long with a pointed end to facilitate entry on the forward stroke. The handle is either wooden or plastic and is riveted to the blade for a length of about 50 centimeters. The blade should be inspected for rust, cracks, warping, burrs, and missing or dull teeth. A file can repair most defects, and steel wool can be rubbed on rusted areas. The handle should be inspected for cracks, bends, and stability. On folding models, the hinge and nuts should be secure. If the saw is beyond repair, it should not be used.

b. **Snow Shovel.** The snow shovel is used to cut and remove ice and snow. It can be used for avalanche rescue, shelter construction, step cutting, and removing obstacles. The snow shovel is made of a special, lightweight aluminum alloy. The handle should be telescopic, folding, or removable to be compact when not in use. The shovel should have a flat or rounded bottom and be of strong construction. The shovel should be inspected for cracks, bends, rust, and burrs. A file and steel wool can remove rust and put an edge on the blade of the shovel. The handle should be inspected for cracks, bends, and stability. If the shovel is beyond repair, it should be turned in.

c. **Wands.** Wands are used to identify routes, crevasses, snow-bridges, caches, and turns on snow and glaciers. Spacing of wands depends on the number of turns, number of hazards identified, weather conditions (and visibility), and number of teams in the climbing party. Carry too many wands is better than not having enough if they become lost. Wands are 1 to 1.25 meters long and made of lightweight bamboo or plastic shafts pointed on one end with a plastic or nylon flag (bright enough in color to see at a distance) attached to the other end. The shafts should be inspected for cracks, bends, and deformities. The flag should be inspected for tears, frays, security to the shaft, fading, and discoloration. If any defects are discovered, the wands should be replaced.

d. **Avalanche rescue equipment.** Avalanche rescue equipment (Figure 3-29) includes the following:

(1) ***Avalanche Probe.*** Although ski poles may be used as an emergency probe when searching for a victim in an avalanche, commercially manufactured probes are better for a thorough search. They are 9-millimeter thick shafts made of an aluminum alloy, which can be joined to probe up to 360 centimeters. The shafts must be strong enough to probe through avalanche debris. Some manufacturers of ski poles design poles that are telescopic and mate with other poles to create an avalanche probe.

(2) ***Avalanche Transceivers.*** These are small, compact radios used to identify avalanche burial sites. They transmit electromagnetic signals that are picked up by another transceiver on the receive mode.

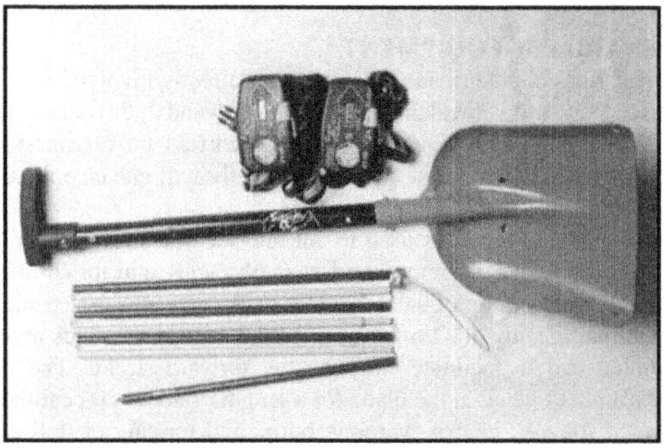

Figure 3-29. Avalanche rescue equipment.

e. **Packs.** Many types and brands of packs are used for mountaineering. The two most common types are internal and external framed packs.

(1) Internal framed packs have a rigid frame within the pack that help it maintain its shape and hug the back. This assists the climber in keeping their balance as they climb or ski. The weight in an internal framed pack is carried low on the body assisting with balance. The body-hugging nature of this type pack also makes it uncomfortable in warm weather.

(2) External framed packs suspend the load away from the back with a ladder-like frame. The frame helps transfer the weight to the hips and shoulders easier, but can be cumbersome when balance is needed for climbing and skiing.

(3) Packs come in many sizes and should be sized appropriately for the individual according to manufacturer's specifications. Packs often come with many unneeded features. A good rule of thumb is: The simpler the pack, the better it will be.

f. **Stoves.** When selecting a stove one must define its purpose—will the stove be used for heating, cooking or both? Stoves or heaters for large elements can be large and cumbersome. Stoves for smaller elements might just be used for cooking and making water, and are simple and lightweight. Stoves are a necessity in mountaineering for cooking and making water from snow and ice. When choosing a stove, factors that should be considered are weight, altitude and temperature where it will be used, fuel availability, and its reliability.

(1) There are many choices in stove design and in fuel types. White gas, kerosene, and butane are the common fuels used. All stoves require a means of pressurization to force the fuel to the burner. Stoves that burn white gas or kerosene have a hand pump to generate the pressurization and butane stoves have pressurized cartridges. All stoves need to vaporize the liquid fuel before it is burned. This can be accomplished by burning a small amount of fuel in the burner cup assembly, which will vaporize the fuel in the fuel line.

(2) Stoves should be tested and maintained prior to a mountaineering mission. They should be easy to clean and repair during an operation. The reliability of the stove has a huge impact on the success of the mission and the morale of personnel.

g. **Tents.** When selecting a tent, the mission must be defined to determine the number of people the tent will accommodate. The climate the tents will be used in is also of concern. A tent used for warmer temperatures will greatly differ from tents used in a colder, more harsh environment. Manufacturers of tents offer many designs of different sizes, weights, and materials.

(1) Mountaineering tents are made out of a breathable or weatherproof material. A single-wall tent allows for moisture inside the tent to escape through the tent's material. A double-wall tent has a second layer of material (referred to as a fly) that covers the tent. The fly protects against rain and snow and the space between the fly and tent helps moisture to escape from inside. Before using a new tent, the seams should be treated with seam sealer to prevent moisture from entering through the stitching.

(2) The frame of a tent is usually made of an aluminum or carbon fiber pole. The poles are connected with an elastic cord that allows them to extend, connect, and become long and rigid. When the tent poles are secured into the tent body, they create the shape of the tent.

(3) Tents are rated by a "relative strength factor," the speed of wind a tent can withstand before the frame deforms. Temperature and expected weather for the mission should be determined before choosing the tent.

h. **Skis.** Mountaineering skis are wide and short. They have a binding that pivots at the toe and allows for the heel to be free for uphill travel or locked for downhill. Synthetic skins with fibers on the bottom can be attached to the bottom of the ski and allow the ski to travel forward and prevent slipping backward. The skins aid in traveling uphill and slow down the rate of descents. Wax can be applied to the ski to aid in ascents instead of skins. Skis can decrease the time needed to reach an objective depending on the ability of the user. Skis can make crossing crevasses easier because of the load distribution, and they can become a makeshift stretcher for casualties. Ski techniques can be complicated and require thorough training for adequate proficiency.

i. **Snowshoes.** Snowshoes are the traditional aid to snow travel that attach to most footwear and have been updated into small, lightweight designs that are more efficient than older models. Snowshoes offer a large displacement area on top of soft snow preventing tiresome post-holing. Some snowshoes come equipped with a crampon like binding that helps in ascending steep snow and ice. Snowshoes are slower than skis, but are better suited for mixed terrain, especially if personnel are not experienced with the art of skiing. When carrying heavy packs, snowshoes can be easier to use than skis.

j. **Ski poles**. Ski poles were traditionally designed to assist in balance during skiing. They have become an important tool in mountaineering for aid in balance while hiking, snowshoeing, and carrying heavy packs. They can take some of the weight off of the lower body when carrying a heavy pack. Some ski poles are collapsible for ease of packing when not needed (Figure 3-30). The basket at the bottom prevents the pole from plunging deep into the snow and, on some models, can be detached so the pole becomes an avalanche or crevasse probe. Some ski poles come with a self-arrest grip, but should not be the only means of protection on technical terrain.

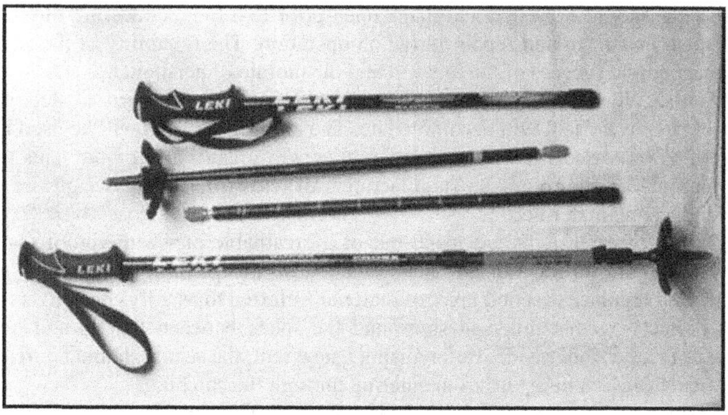

Figure 3-30. Collapsible ski poles.

k. **Sleds**. Sleds vary greatly in size, from the squad-size Ahkio, a component of the 10-man arctic tent system, to the one-person skow. Regardless of the size, sleds are an invaluable asset during mountainous operations when snow and ice is the primary surface on which to travel. Whichever sled is chosen, it must be attachable to the person or people that will be pulling it. Most sleds are constructed using fiberglass bottoms with or without exterior runners. Runners will aid the sleds ability to maintain a true track in the snow. The sled should also come with a cover of some sort—whether nylon or canvas, a cover is essential for keeping the components in the sled dry. Great care should be taken when packing the sled, especially when hauling fuel. Heavier items should be carried towards the rear of the sled and lighter items towards the front.

l. **Headlamps**. A headlamp is a small item that is not appreciated until it is needed. It is common to need a light source and the use of both hands during limited light conditions in mountaineering operations. A flashlight can provide light, but can be cumbersome when both hands are needed. Most headlamps attach to helmets by means of elastic bands.

(1) When choosing a headlamp, ensure it is waterproof and the battery apparatus is small. All components should be reliable in extreme weather conditions. When the light is being packed, care should be taken that the switch doesn't accidentally activate and use precious battery life.

(2) The battery source should compliment the resupply available. Most lights will accept alkaline, nickel-cadmium, or lithium batteries. Alkaline battery life diminishes quickly in cold temperatures, nickel-cadmium batteries last longer in cold but require a recharging unit, and lithium batteries have twice the voltage so modifications are required.

Section II. EQUIPMENT PACKING

Equipment brought on a mission is carried in the pack, worn on the body, or hauled in a sled (in winter). Obviously, the rucksack and sled (or Ahkio) can hold much more than a climber can carry. They would be used for major bivouac gear, food, water, first aid kits, climbing equipment, foul weather shells, stoves, fuel, ropes, and extra ammunition and demolition materials, if needed.

3-7. CHOICE OF EQUIPMENT

Mission requirements and unit SOP will influence the choice of gear carried but the following lists provide a sample of what should be considered during mission planning.

a. **Personal Gear.** Personal gear includes emergency survival kit containing signaling material, fire starting material, food procurement material, and water procurement material. Pocket items should include a knife, whistle, pressure bandage, notebook with pen or pencil, sunglasses, sunblock and lip protection, map, compass and or altimeter.

b. **Standard Gear.** Standard gear that can be individually worn or carried includes cushion sole socks; combat boots or mountain boots, if available; BDU and cap; LCE with canteens, magazine pouches, and first aid kit; individual weapon; a large rucksack containing waterproof coat and trousers, polypropylene top, sweater, or fleece top; helmet; poncho; and sleeping bag.

> **CAUTION**
> Cotton clothing, due to its poor insulating and moisture-wicking characteristics, is virtually useless in most mountain climates, the exception being hot, desert, or jungle mountain environments. Cotton clothing should be replaced with synthetic fabric clothing.

c. **Mountaineering Equipment and Specialized Gear.** This gear includes:

- Sling rope or climbing harness.
- Utility cord(s).
- Nonlocking carabiners.
- Locking carabiner(s).
- Rappelling gloves.
- Rappel/belay device.
- Ice ax.
- Crampons.
- Climbing rope, one per climbing team.
- Climbing rack, one per climbing team.

d. **Day Pack.** When the soldier plans to be away from the bivouac site for the day on a patrol or mountaineering mission, he carries a light day pack. This pack should contain the following items:

- Extra insulating layer: polypropylene, pile top, or sweater.
- Protective layer: waterproof jacket and pants, rain suit, or poncho.
- First aid kit.
- Flashlight or headlamp.
- Canteen.
- Cold weather hat or scarf.
- Rations for the time period away from the base camp.
- Survival kit.
- Sling rope or climbing harness.
- Carabiners.
- Gloves.
- Climbing rope, one per climbing team.
- Climbing rack, one per climbing team.

e. **Squad or Team Safety Pack.** When a squad-sized element leaves the bivouac site, squad safety gear should be carried in addition to individual day packs. This can either be loaded into one rucksack or cross-loaded among the squad members. In the event of an injury, casualty evacuation, or unplanned bivouac, these items may make the difference between success and failure of the mission.

- Sleeping bag.
- Sleeping mat.

- Squad stove.
- Fuel bottle.

f. **The Ten Essentials.** Regardless of what equipment is carried, the individual military mountaineer should always carry the "ten essentials" when moving through the mountains.

(1) ***Map.***

(2) ***Compass, Altimeter, and or GPS.***

(3) ***Sunglasses and Sunscreen.***

(a) In alpine or snow-covered sub-alpine terrain, sunglasses are a vital piece of equipment for preventing snow blindness. They should filter 95 to 100 percent of ultraviolet light. Side shields, which minimize the light entering from the side, should permit ventilation to help prevent lens fogging. At least one extra pair of sunglasses should be carried by each independent climbing team.

(b) Sunscreens should have an SPF factor of 15 or higher. For lip protection, a total UV blocking lip balm that resists sweating, washing, and licking is best. This lip protection should be carried in the chest pocket or around the neck to allow frequent reapplication.

(4) ***Extra Food.*** One day's worth extra of food should be carried in case of delay caused by bad weather, injury, or navigational error.

(5) ***Extra Clothing.*** The clothing used during the active part of a climb, and considered to be the basic climbing outfit, includes socks, boots, underwear, pants, blouse, sweater or fleece jacket, hat, gloves or mittens, and foul weather gear (waterproof, breathable outerwear or waterproof rain suit).

(a) Extra clothing includes additional layers needed to make it through the long, inactive hours of an unplanned bivouac. Keep in mind the season when selecting this gear.

- Extra underwear to switch out with sweat-soaked underwear.
- Extra hats or balaclavas.
- Extra pair of heavy socks.
- Extra pair of insulated mittens or gloves.
- In winter or severe mountain conditions, extra insulation for the upper body and the legs.

(b) To back up foul weather gear, bring a poncho or extra-large plastic trash bag. A reflective emergency space blanket can be used for hypothermia first aid and emergency shelter. Insulated foam pads prevent heat loss while sitting or lying on snow. Finally, a bivouac sack can help by protecting insulating layers from the weather, cutting the wind, and trapping essential body heat inside the sack.

(6) ***Headlamp and or Flashlight.*** Headlamps provide the climber a hands-free capability, which is important while climbing, working around the camp, and employing weapons systems. Miniature flashlights can be used, but commercially available headlamps are best. Red lens covers can be fabricated for tactical conditions. Spare batteries and spare bulbs should also be carried.

(7) ***First-aid Kit.*** Decentralized operations, the mountain environment—steep, slick terrain and loose rock combined with heavy packs, sharp tools, and fatigue—requires each climber to carry his own first-aid kit. Common mountaineering injuries that can be expected are punctures and abrasions with severe bleeding, a broken bone, serious sprain,

and blisters. Therefore, the kit should contain at least enough material to stabilize these conditions. Pressure dressings, gauze pads, elastic compression wrap, small adhesive bandages, butterfly bandages, moleskin, adhesive tape, scissors, cleanser, latex gloves and splint material (if above tree line) should all be part of the kit.

(8) ***Fire Starter.*** Fire starting material is key to igniting wet wood for emergency campfires. Candles, heat tabs, and canned heat all work. These can also be used for quick warming of water or soup in a canteen cup. In alpine zones above tree line with no available firewood, a stove works as an emergency heat source.

(9) ***Matches and Lighter.*** Lighters are handy for starting fires, but they should be backed up by matches stored in a waterproof container with a strip of sandpaper.

(10) ***Knife.*** A multipurpose pocket tool should be secured with cord to the belt, harness, or pack.

g. **Other Essential Gear.** Other essential gear may be carried depending on mission and environmental considerations.

(1) ***Water and Water Containers.*** These include wide-mouth water bottles for water collection; camel-back type water holders for hands-free hydration; and a small length of plastic tubing for water procurement at snow-melt seeps and rainwater puddles on bare rock.

(2) ***Ice Ax.*** The ice ax is essential for travel on snowfields and glaciers as well as snow-covered terrain in spring and early summer. It helps for movement on steep scree and on brush and heather covered slopes, as well as for stream crossings.

(3) ***Repair Kit.*** A repair kit should include:

- Stove tools and spare parts.
- Duct tape.
- Patches.
- Safety pins.
- Heavy-duty thread.
- Awl and or needles.
- Cord and or wire.
- Small pliers (if not carrying a multipurpose tool).
- Other repair items as needed.

(4) ***Insect Repellent.***

(5) ***Signaling Devices.***

(6) ***Snow Shovel.***

3-8. TIPS ON PACKING

When loading the internal frame pack the following points should be considered.

a. In most cases, speed and endurance are enhanced if the load is carried more by the hips (using the waist belt) and less by the shoulders and back. This is preferred for movement over trails or less difficult terrain. By packing the lighter, more compressible items (sleeping bag, clothing) in the bottom of the rucksack and the heavier gear (stove, food, water, rope, climbing hardware, extra ammunition) on top, nearer the shoulder blades, the load is held high and close to the back, thus placing the most weight on the hips.

b. In rougher terrain it pays to modify the pack plan. Heavy articles of gear are placed lower in the pack and close to the back, placing more weight on the shoulders and back. This lowers the climber's center of gravity and helps him to better keep his balance.

c. Equipment that may be needed during movement should be arranged for quick access using either external pockets or placing immediately underneath the top flap of the pack. As much as possible, this placement should be standardized across the team so that necessary items can be quickly reached without unnecessary unpacking of the pack in emergencies.

d. The pack and its contents should be soundly waterproofed. Clothing and sleeping bag are separately sealed and then placed in the larger wet weather bag that lines the rucksack. Zip-lock plastic bags can be used for small items, which are then organized into color-coded stuffsacks. A few extra-large plastic garbage bags should be carried for a variety of uses—spare waterproofing, emergency bivouac shelter, and water procurement, among others.

e. The ice ax, if not carried in hand, should be stowed on the outside of the pack with the spike up and the adze facing forward or to the outside, and be securely fastened. Mountaineering packs have ice ax loops and buckle fastening systems for this. If not, the ice ax is placed behind one of the side pockets, as stated above, and then tied in place.

f. Crampons should be secured to the outside rear of the pack with the points covered.

CHAPTER 4

ROPE MANAGEMENT AND KNOTS

The rope is a vital piece of equipment to the mountaineer. When climbing, rappelling, or building various installations, the mountaineer must know how to properly utilize and maintain this piece of equipment. If the rope is not managed or maintained properly, serious injury may occur. This chapter discusses common rope terminology, management techniques, care and maintenance procedures, and knots.

SECTION I. PREPARATION, CARE AND MAINTENANCE, INSPECTION, TERMINOLOGY

The service life of a rope depends on the frequency of use, applications (rappelling, climbing, rope installations), speed of descent, surface abrasion, terrain, climate, and quality of maintenance. Any rope may fail under extreme conditions (shock load, sharp edges, misuse).

4-1. PREPARATION

The mountaineer must select the proper rope for the task to be accomplished according to type, diameter, length, and tensile strength. It is important to prepare all ropes before departing on a mission. Avoid rope preparation in the field.

a. **Packaging.** New rope comes from the manufacturer in different configurations—boxed on a spool in various lengths, or coiled and bound in some manner. Precut ropes are usually packaged in a protective cover such as plastic or burlap. Do not remove the protective cover until the rope is ready for use.

b. **Securing the Ends of the Rope:** If still on a spool, the rope must be cut to the desired length. All ropes will fray at the ends unless they are bound or seared. Both static and dynamic rope ends are secured in the same manner. The ends must be heated to the melting point so as to attach the inner core strands to the outer sheath. By fusing the two together, the sheath cannot slide backward or forward. Ensure that this is only done to the ends of the rope. If the rope is exposed to extreme temperatures, the sheath could be weakened, along with the inner core, reducing overall tensile strength. The ends may also be dipped in enamel or lacquer for further protection.

4-2. CARE AND MAINTENANCE

The rope is a climber's lifeline. It must be cared for and used properly. These general guidelines should be used when handling ropes.

a. Do not step on or drag ropes on the ground unnecessarily. Small particles of dirt will be ground between the inner strands and will slowly cut them.

b. While in use, do not allow the rope to come into contact with sharp edges. Nylon rope is easily cut, particularly when under tension. If the rope must be used over a sharp edge, pad the edge for protection.

c. Always keep the rope as dry as possible. Should the rope become wet, hang it in large loops off the ground and allow it to dry. Never dry a rope with high heat or in direct sunlight.

d. Never leave a rope knotted or tightly stretched for longer than necessary. Over time it will reduce the strength and life of the rope.

e. Never allow one rope to continuously rub over or against another. Allowing rope-on-rope contact with nylon rope is extremely dangerous because the heat produced by the friction will cause the nylon to melt.

f. Inspect the rope before each use for frayed or cut spots, mildew or rot, or defects in construction (new rope).

g. The ends of the rope should be whipped or melted to prevent unraveling.

h. Do not splice ropes for use in mountaineering.

i. Do not mark ropes with paints or allow them to come in contact with oils or petroleum products. Some of these will weaken or deteriorate nylon.

j. Never use a mountaineering rope for any purpose except mountaineering.

k. Each rope should have a corresponding rope log (DA Form 5752-R, Rope History and Usage), which is also a safety record. It should annotate use, terrain, weather, application, number of falls, dates, and so on, and should be annotated each time the rope is used (Figure 4-1). DA Form 5752-R is authorized for local reproduction on 8 1/2- by 11-inch paper.

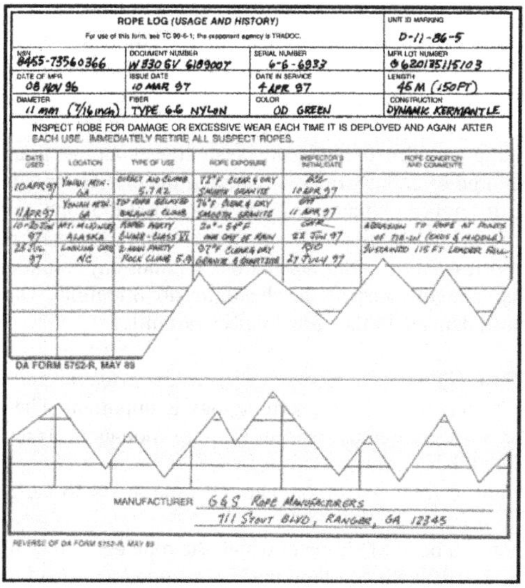

ROPE LOG (USAGE AND HISTORY)

For use of this form, see TC 90-6-1; the proponent agency is TRADOC.

NSN 8455-73560366	DOCUMENT NUMBER W530 6V 6189007	SERIAL NUMBER 6-6-6933	UNIT ID MARKING D-11-86-5 / MFR LOT NUMBER 0620175115103
DATE OF MFR 08 NOV 96	ISSUE DATE 10 MAR 97	DATE IN SERVICE 4 APR 97	LENGTH 45 M (150 FT)
DIAMETER 11 mm (7/16 inch)	FIBER TYPE 6.6 NYLON	COLOR OD GREEN	CONSTRUCTION DYNAMIC KERMANTLE

INSPECT ROBE FOR DAMAGE OR EXCESSIVE WEAR EACH TIME IT IS DEPLOYED AND AGAIN AFTER EACH USE. IMMEDIATELY RETIRE ALL SUSPECT ROPES.

DATE USED	LOCATION	TYPE OF USE	ROPE EXPOSURE	INSPECTOR'S INITIAL/DATE	ROPE CONDITION AND COMMENTS
10 APR 97	Yonah Mtn. GA	Direct aid climb 5.7 A2	72°F clear & dry smooth granite	[illegible] 10 APR 97	
11 APR 97	Yonah Mtn. GA	Top rope belayed balance climb	76°F clear & dry smooth granite	[illegible] 11 APR 97	
10-25 Jun 97	Mt. McKinley Alaska	Rope party climb - class VI	20°-54°F [illegible]	[illegible] 22 Jun 97	Abrasion to rope at points of tie-in (ends & middle)
25 Jul 97	Linville Gorge NC	2-man party rock climb 5.9	97°F clear & dry granite & quartzite	[illegible] 27 July 97	Sustained 115 ft leader fall

DA FORM 5752-R, MAY 89

MANUFACTURER G & S Rope Manufacturers
711 Stout Blvd, Ranger, GA 12345

REVERSE OF DA FORM 5752-R, MAY 89

Figure 4-1. Example of completed DA Form 5752-R.

l. Never subject the rope to high heat or flame. This will significantly weaken it.

m. All ropes should be washed periodically to remove dirt and grit, and rinsed thoroughly. Commercial rope washers are made from short pieces of modified pipe that connect to any faucet. Pinholes within the pipe force water to circulate around and scrub the rope as you slowly feed it through the washer. Another method is to machine wash, on a gentle cycle, in cold water with a nylon safe soap, never bleach or harsh cleansers. Ensure that only front loading washing machine are used to wash ropes.

n. Ultraviolet radiation (sunlight) tends to deteriorate nylon over long periods of time. This becomes important if rope installations are left in place over a number of months.

o. When not in use, ropes should be loosely coiled and hung on wooden pegs rather than nails or other metal objects. Storage areas should be relatively cool with low humidity levels to prevent mildew or rotting. Rope may also be loosely stacked and placed in a rope bag and stored on a shelf. Avoid storage in direct sunlight, as the ultraviolet radiation will deteriorate the nylon over long periods

4-3. INSPECTION

Ropes should be inspected before and after each use, especially when working around loose rock or sharp edges.

a. Although the core of the kernmantle rope cannot be seen, it is possible to damage the core without damaging the sheath. Check a kernmantle rope by carefully inspecting the sheath before and after use while the rope is being coiled. When coiling, be aware of how the rope feels as it runs through the hands. Immediately note and tie off any lumps or depressions felt.

b. Damage to the core of a kernmantle rope usually consists of filaments or yarn breakage that results in a slight retraction. If enough strands rupture, a localized reduction in the diameter of the rope results in a depression that can be felt or even seen.

c. Check any other suspected areas further by putting them under tension (the weight of one person standing on a Prusik tensioning system is about maximum). This procedure will emphasize the lump or depression by separating the broken strands and enlarging the dip. If a noticeable difference in diameter is obvious, retire the rope immediately.

d. Many dynamic kernmantle ropes are quite soft. They may retain an indention occasionally after an impact or under normal use without any trauma to the core. When damage is suspected, patiently inspect the sheath for abnormalities. Damage to the sheath does not always mean damage to the core. Inspect carefully.

4-4. TERMINOLOGY

When using ropes, understanding basic terminology is important. The terms explained in this section are the most commonly used in military mountaineering. (Figure 4-2, page 4-4, illustrates some of these terms.)

a. **Bight**. A bight of rope is a simple bend of rope in which the rope does not cross itself.

b. **Loop.** A loop is a bend of a rope in which the rope does cross itself.

c. **Half Hitch**. A half hitch is a loop that runs around an object in such a manner as to lock or secure itself.

d. **Turn.** A turn wraps around an object, providing 360-degree contact.

e. **Round Turn**. A round turn wraps around an object one and one-half times. A round turn is used to distribute the load over a small diameter anchor (3 inches or less). It may also be used around larger diameter anchors to reduce the tension on the knot, or provide added friction.

f. **Running End**. A running end is the loose or working end of the rope.

g. **Standing Part**. The standing part is the static, stationary, or nonworking end of the rope.

h. **Lay.** The lay is the direction of twist used in construction of the rope.

i. **Pigtail.** The pigtail (tail) is the portion of the running end of the rope between the safety knot and the end of the rope.

j. **Dress.** Dress is the proper arrangement of all the knot parts, removing unnecessary kinks, twists, and slack so that all rope parts of the knot make contact.

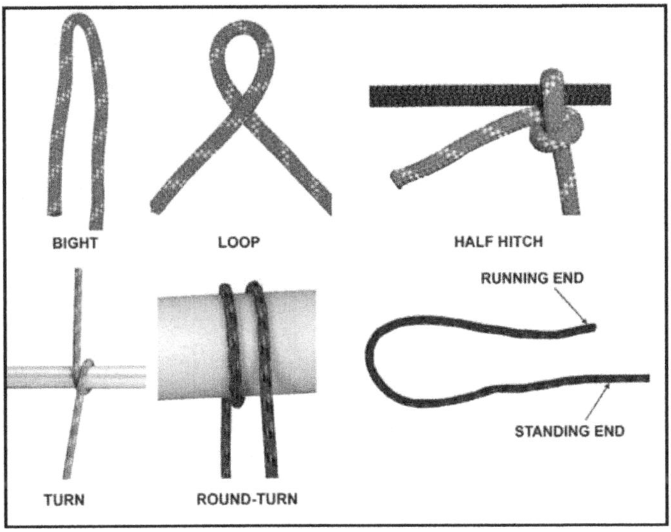

Figure 4-2. Examples of roping terminology.

Section II. COILING, CARRYING, THROWING

The ease and speed of rope deployment and recovery greatly depends upon technique and practice.

4-5. COILING AND CARRYING THE ROPE

Use the butterfly or mountain coil to coil and carry the rope. Each is easy to accomplish and results in a minimum amount of kinks, twists, and knots later during deployment.

a. **Mountain Coil.** To start a mountain coil, grasp the rope approximately 1 meter from the end with one hand. Run the other hand along the rope until both arms are outstretched. Grasping the rope firmly, bring the hands together forming a loop, which is laid in the hand closest to the end of the rope. This is repeated, forming uniform loops that run in a clockwise direction, until the rope is completely coiled. The rope may be given a 1/4 twist as each loop is formed to overcome any tendency for the rope to twist or form figure-eights.

(1) In finishing the mountain coil, form a bight approximately 30 centimeters long with the starting end of the rope and lay it along the top of the coil. Uncoil the last loop and, using this length of the rope, begin making wraps around the coil and the bight, wrapping toward the closed end of the bight and making the first wrap bind across itself so as to lock it into place. Make six to eight wraps to adequately secure the coil, and then route the end of the rope through the closed end of the bight. Pull the running end of the bight tight, securing the coil.

(2) The mountain coil may be carried either in the pack (by forming a figure eight), doubling it and placing it under the flap, or by placing it over the shoulder and under the opposite arm, slung across the chest. (Figure 4-3 shows how to coil a mountain coil.)

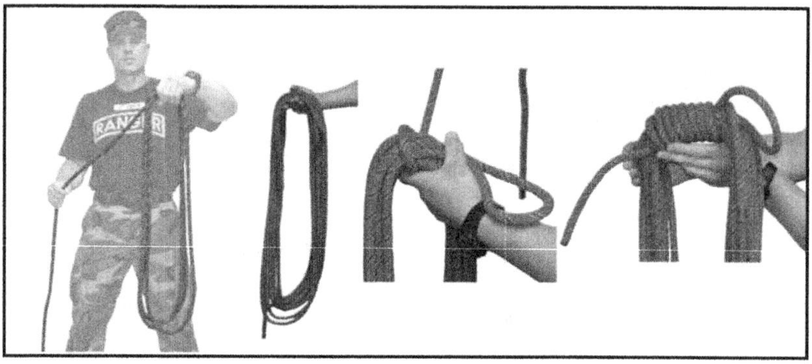

Figure 4-3. Mountain coil.

b. **Butterfly Coil.** The butterfly coil is the quickest and easiest technique for coiling (Figure 4-4).

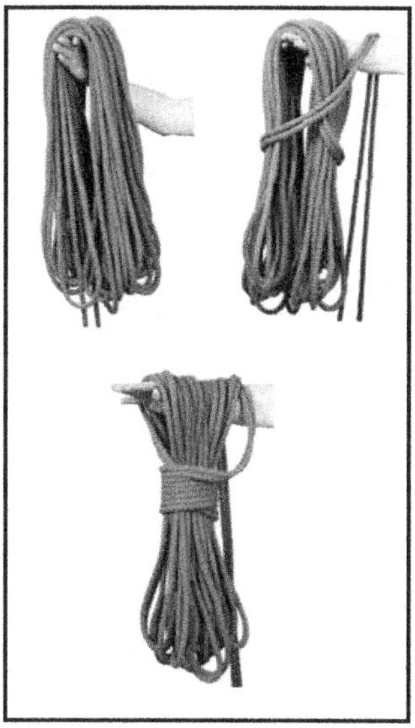

Figure 4-4. Butterfly coil.

(1) ***Coiling.*** To start the double butterfly, grasp both ends of the rope and begin back feeding. Find the center of the rope forming a bight. With the bight in the left hand, grasp both ropes and slide the right hand out until there is approximately one arms length of rope. Place the doubled rope over the head, draping it around the neck and on top of the shoulders. Ensure that it hangs no lower than the waist. With the rest of the doubled rope in front of you, make doubled bights placing them over the head in the same manner as the first bight. Coil alternating from side to side (left to right, right to left) while maintaining equal-length bights. Continue coiling until approximately two arm-lengths of rope remain. Remove the coils from the neck and shoulders carefully, and hold the center in one hand. Wrap the two ends around the coils a minimum of three doubled wraps, ensuring that the first wrap locks back on itself.

(2) ***Tie-off and Carrying.*** Take a doubled bight from the loose ends of rope and pass it through the apex of the coils. Pull the loose ends through the doubled bight and dress it down. Place an overhand knot in the loose ends, dressing it down to the apex of the bight securing coils. Ensure that the loose ends do not exceed the length of the coils. In this configuration the coiled rope is secure enough for hand carrying or carrying in a rucksack, or for storage. (Figure 4-5 shows a butterfly coil tie-off.)

Figure 4-5. Butterfly coil tie-off.

c. **Coiling Smaller Diameter Rope.** Ropes of smaller diameters may be coiled using the butterfly or mountain coil depending on the length of the rope. Pieces 25 feet and shorter (also known as cordage, sling rope, utility cord) may be coiled so that they can be hung from the harness. Bring the two ends of the rope together, ensuring no kinks are in the rope. Place the ends of the rope in the left hand with the two ends facing the body. Coil the doubled rope in a clockwise direction forming 6- to 8-inch coils (coils may be larger depending on the length of rope) until an approximate 12-inch bight is left. Wrap that bight around the coil, ensuring that the first wrap locks on itself. Make three or more wraps. Feed the bight up through the bights formed at the top of the coil. Dress it down tightly. Now the piece of rope may be hung from a carabiner on the harness.

e. **Uncoiling, Back-feeding, and Stacking.** When the rope is needed for use, it must be uncoiled and stacked on the ground properly to avoid kinks and snarls.

(1) Untie the tie-off and lay the coil on the ground. Back-feed the rope to minimize kinks and snarls. (This is also useful when the rope is to be moved a short distance and

coiling is not desired.) Take one end of the rope in the left hand and run the right hand along the rope until both arms are outstretched. Next, lay the end of the rope in the left hand on the ground. With the left hand, re-grasp the rope next to the right hand and continue laying the rope on the ground.

(2) The rope should be laid or stacked in a neat pile on the ground to prevent it from becoming tangled and knotted when throwing the rope, feeding it to a lead climber, and so on. This technique can also be started using the right hand.

4-6. THROWING THE ROPE

Before throwing the rope, it must be properly managed to prevent it from tangling during deployment. The rope should first be anchored to prevent complete loss of the rope over the edge when it is thrown. Several techniques can be used when throwing a rope. Personal preference and situational and environmental conditions should be taken into consideration when determining which technique is best.

a. Back feed and neatly stack the rope into coils beginning with the anchored end of the rope working toward the running end. Once stacked, make six to eight smaller coils in the left hand. Pick up the rest of the larger coils in the right hand. The arm should be generally straight when throwing. The rope may be thrown underhanded or overhanded depending on obstacles around the edge of the site. Make a few preliminary swings to ensure a smooth throw. Throw the large coils in the right hand first. Throw up and out. A slight twist of the wrist, so that the palm of the hand faces up as the rope is thrown, allows the coils to separate easily without tangling. A smooth follow through is essential. When a slight tug on the left hand is felt, toss the six to eight smaller coils out. This will prevent the ends of the rope from becoming entangled with the rest of the coils as they deploy. As soon as the rope leaves the hand, the thrower should sound off with a warning of "ROPE" to alert anyone below the site.

b. Another technique may also be used when throwing rope. Anchor, back feed, and stack the rope properly as described above. Take the end of the rope and make six to eight helmet-size coils in the right hand (more may be needed depending on the length of the rope). Assume a "quarterback" simulated stance. Aiming just above the horizon, vigorously throw the rope overhanded, up and out toward the horizon. The rope must be stacked properly to ensure smooth deployment.

c. When windy weather conditions prevail, adjustments must be made. In a strong cross wind, the rope should be thrown angled into the wind so that it will land on the desired target. The stronger the wind, the harder the rope must be thrown to compensate.

SECTION III. KNOTS

All knots used by a mountaineer are divided into four classes: Class I—joining knots, Class II—anchor knots, Class III—middle rope knots, and Class IV—special knots. The variety of knots, bends, bights, and hitches is almost endless. These classes of knots are intended only as a general guide since some of the knots discussed may be appropriate in more than one class. The skill of knot tying can perish if not used and practiced. With experience and practice, knot tying becomes instinctive and helps the mountaineer in many situations.

4-7. SQUARE KNOT

The square knot is used to tie the ends of two ropes of equal diameter (Figure 4-6). It is a joining knot.

a. **Tying the Knot.**

STEP 1. Holding one working end in each hand, place the working end in the right hand over the one in the left hand.

STEP 2. Pull it under and back over the top of the rope in the left hand.

STEP 3. Place the working end in the left hand over the one in the right hand and repeat STEP 2.

STEP 4. Dress the knot down and secure it with an overhand knot on each side of the square knot.

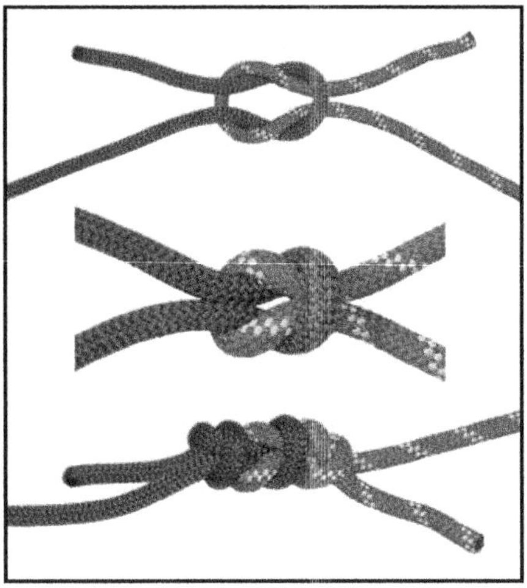

Figure 4-6. Square knot.

b. **Checkpoints**.

(1) There are two interlocking bights.

(2) The running end and standing part are on the same side of the bight formed by the other rope.

(3) The running ends are parallel to and on the same side of the standing ends with 4-inch minimum pig tails after the overhand safeties are tied.

4-8. FISHERMAN'S KNOT

The fisherman's knot is used to tie two ropes of the same or approximately the same diameter (Figure 4-7, page 4-10). It is a joining knot.

a. **Tying the Knot.**

STEP 1. Tie an overhand knot in one end of the rope.

STEP 2. Pass the working end of the other rope through the first overhand knot. Tie an overhand knot around the standing part of the first rope with the working end of the second rope.

STEP 3. Tightly dress down each overhand knot and tightly draw the knots together.

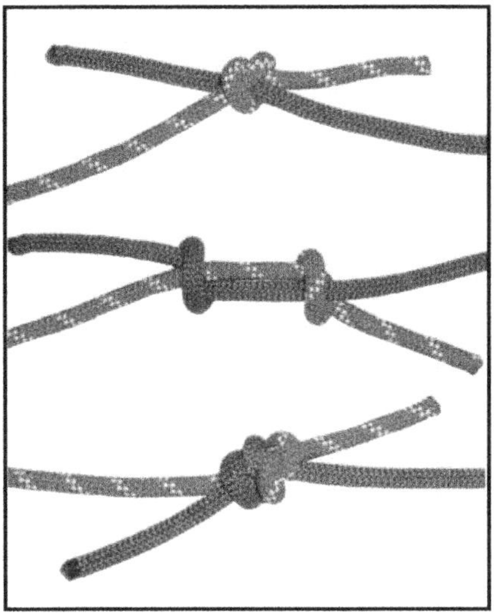

Figure 4-7. Fisherman's knot.

b. **Checkpoints.**

(1) The two separate overhand knots are tied tightly around the long, standing part of the opposing rope.

(2) The two overhand knots are drawn snug.

(3) Ends of rope exit knot opposite each other with 4-inch pigtails.

4-9. DOUBLE FISHERMAN'S KNOT

The double fisherman's knot (also called double English or grapevine) is used to tie two ropes of the same or approximately the same diameter (Figure 4-8). It is a joining knot.

a. **Tying the Knot.**

STEP 1. With the working end of one rope, tie two wraps around the standing part of another rope.

STEP 2. Insert the working end (STEP 1) back through the two wraps and draw it tight.

STEP 3. With the working end of the other rope, which contains the standing part (STEPS 1 and 2), tie two wraps around the standing part of the other rope (the working end in STEP 1). Insert the working end back through the two wraps and draw tight.

STEP 4. Pull on the opposing ends to bring the two knots together.

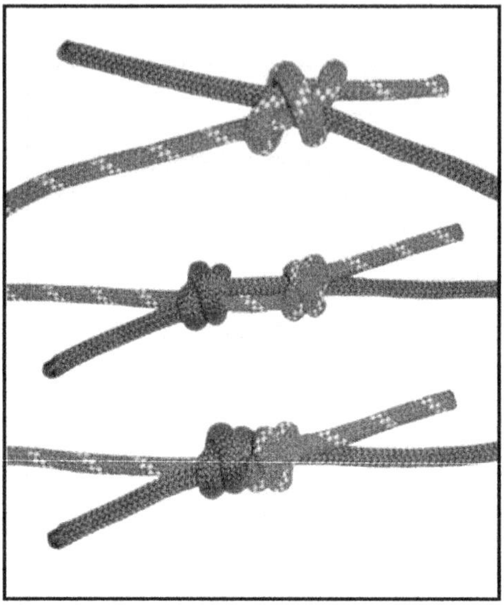

Figure 4-8. Double fisherman's knot.

b. **Checkpoints.**

(1) Two double overhand knots securing each other as the standing parts of the rope are pulled apart.

(2) Four rope parts on one side of the knot form two "x" patterns, four rope parts on the other side of the knot are parallel.

(3) Ends of rope exit knot opposite each other with 4-inch pigtails.

4-10. FIGURE-EIGHT BEND

The figure-eight bend is used to join the ends of two ropes of equal or unequal diameter within 5-mm difference (Figure 4-9, page 4-12).

a. **Tying the Knot.**

STEP 1. Grasp the top of a 2-foot bight.

STEP 2. With the other hand, grasp the running end (short end) and make a 360-degree turn around the standing end.

STEP 3. Place the running end through the loop just formed creating an in-line figure eight.

STEP 4. Route the running end of the other ripe back through the figure eight starting from the original rope's running end. Trace the original knot to the standing end.

STEP 5. Remove all unnecessary twists and crossovers. Dress the knot down.

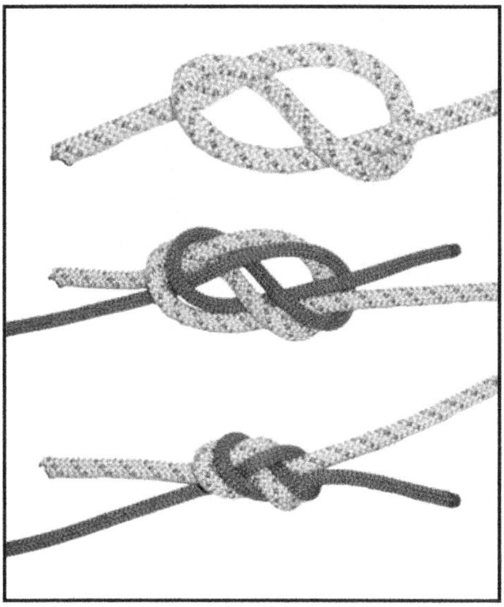

Figure 4-9. Figure-eight bend.

b. **Checkpoints.**

(1) There is a figure eight with two ropes running side by side.

(2) The running ends are on opposite sides of the knot.

(3) There is a minimum 4-inch pigtail.

4-11. WATER KNOT

The water knot is used to attach two webbing ends (Figure 4-10). It is also called a ring bend, overhand retrace, or tape knot. It is used in runners and harnesses and is a joining knot.

a. **Tying the Knot.**

STEP 1. Tie an overhand knot in one of the ends.

STEP 2. Feed the other end back through the knot, following the path of the first rope in reverse.

STEP 3. Draw tight and pull all of the slack out of the knot. The remaining tails must extend at least 4 inches beyond the knot in both directions.

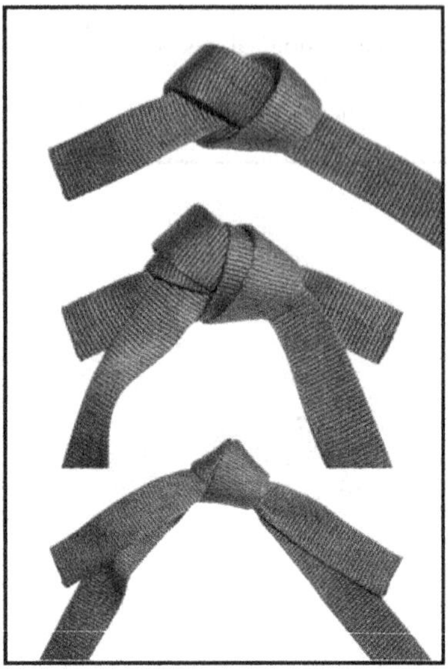

Figure 4-10. Water knot.

b. **Checkpoints.**

(1) There are two overhand knots, one retracing the other.

(2) There is no slack in the knot, and the working ends come out of the knot in opposite directions.

(3) There is a minimum 4-inch pigtail.

4-12. BOWLINE

The bowline is used to tie the end of a rope around an anchor. It may also be used to tie a single fixed loop in the end of a rope (Figure 4-11, page 4-14). It is an anchor knot.

a. **Tying the Knot.**

STEP 1. Bring the working end of the rope around the anchor, from right to left (as the climber faces the anchor).

STEP 2. Form an overhand loop in the standing part of the rope (on the climber's right) toward the anchor.

STEP 3. Reach through the loop and pull up a bight.

STEP 4. Place the working end of the rope (on the climber's left) through the bight, and bring it back onto itself. Now dress the knot down.

STEP 5. Form an overhand knot with the tail from the bight.

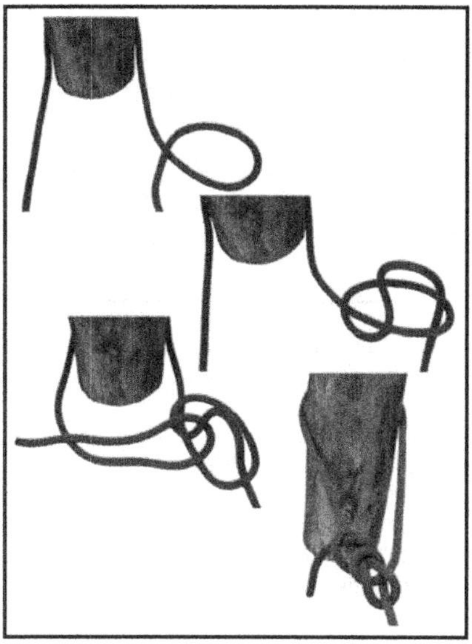

Figure 4-11. Bowline knot.

b. **Checkpoints.**

(1) The bight is locked into place by a loop.

(2) The short portion of the bight is on the inside and on the loop around the anchor (or inside the fixed loop).

(3) There is a minimum 4-inch pigtail after tying the overhand safety.

4-13. ROUND TURN AND TWO HALF HITCHES

This knot is used to tie the end of a rope to an anchor, and it must have constant tension (Figure 4-12). It is an anchor knot.

a. **Tying the Knot.**

STEP 1. Route the rope around the anchor from right to left and wrap down (must have two wraps in the rear of the anchor, and one in the front). Run the loop around the object to provide 360-degree contact, distributing the load over the anchor.

STEP 2. Bring the working end of the rope left to right and over the standing part, forming a half hitch (first half hitch).

STEP 3. Repeat STEP 2 (last half hitch has a 4 inch pigtail).

STEP 4. Dress the knot down.

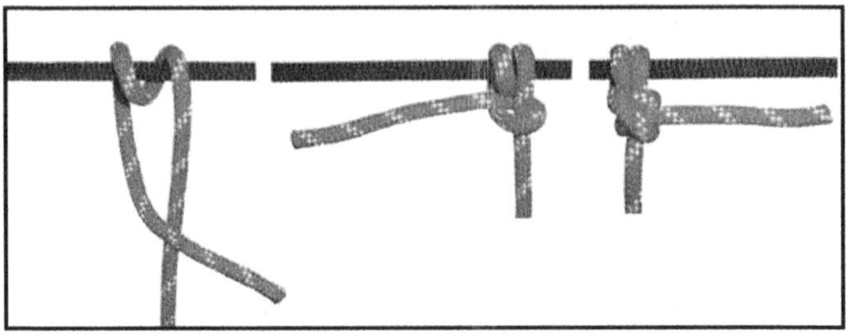

Figure 4-12. Round turn and two half hitches.

b. **Checkpoints.**

(1) A complete round turn should exist around the anchor with no crosses.

(2) Two half hitches should be held in place by a diagonal locking bar with no less than a 4-inch pigtail remaining.

4-14. FIGURE-EIGHT RETRACE (REROUTED FIGURE-EIGHT)

The figure-eight retrace knot produces the same result as a figure-eight loop. However, by tying the knot in a retrace, it can be used to fasten the rope to trees or to places where the loop cannot be used (Figure 4-13, page 4-16). It is also called a rerouted figure-eight and is an anchor knot.

a. **Tying the Knot.**

STEP 1. Use a length of rope long enough to go around the anchor, leaving enough rope to work with.

STEP 2. Tie a figure-eight knot in the standing part of the rope, leaving enough rope to go around the anchor. To tie a figure-eight knot form a loop in the rope, wrap the working end around the standing part, and route the working end through the loop. The finished knot is dressed loosely.

STEP 3. Take the working end around the anchor point.

STEP 4. With the working end, insert the rope back through the loop of the knot in reverse.

STEP 5. Keep the original figure eight as the outside rope and retrace the knot around the wrap and back to the long-standing part.

STEP 6. Remove all unnecessary twists and crossovers; dress the knot down.

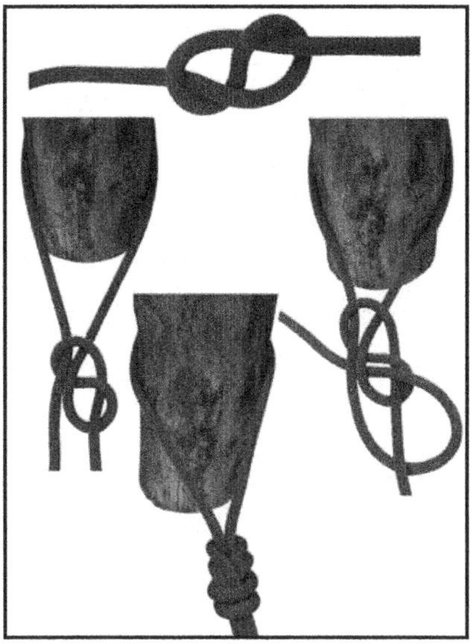

Figure 4-13. Figure-eight retrace.

b. **Checkpoints**

(1) A figure eight with a doubled rope running side by side, forming a fixed loop around a fixed object or harness.

(2) There is a minimum 4-inch pigtail.

4-15. CLOVE HITCH

The clove hitch is an anchor knot that can be used in the middle of the rope as well as at the end (Figure 4-14). The knot must have constant tension on it once tied to prevent slipping. It can be used as either an anchor or middle of the rope knot, depending on how it is tied.

a. **Tying the Knot.**

(1) ***Middle of the Rope.***

STEP 1. Hold rope in both hands, palms down with hands together. Slide the left hand to the left from 20 to 25 centimeters.

STEP 2. Form a loop away from and back toward the right.

STEP 3. Slide the right hand from 20 to 25 centimeters to the right. Form a loop inward and back to the left hand.

STEP 4. Place the left loop on top of the right loop. Place both loops over the anchor and pull both ends of the rope in opposite directions. The knot is tied.

(2) ***End of the Rope.***

Note: For instructional purposes, assume that the anchor is horizontal.

STEP 1. Place 76 centimeters of rope over the top of the anchor. Hold the standing end in the left hand. With the right hand, reach under the horizontal anchor, grasp the working end, and bring it inward.

STEP 2. Place the working end of the rope over the standing end (to form a loop). Hold the loop in the left hand. Place the working end over the anchor from 20 to 25 centimeters to the left of the loop.

STEP 3. With the right hand, reach down to the left hand side of the loop under the anchor. Grasp the working end of the rope. Bring the working end up and outward.

STEP 4. Dress down the knot.

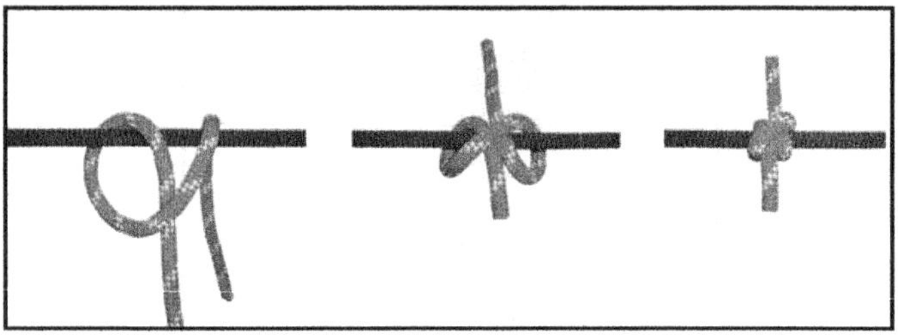

Figure 4-14. Clove hitch.

b. **Checkpoints.**

(1) The knot has two round turns around the anchor with a diagonal locking bar.

(2) The locking bar is facing 90 degrees from the direction of pull.

(3) The ends exit 180 degrees from each other.

(4) The knot has more than a 4-inch pigtail remaining.

4-16. WIREMAN'S KNOT

The wireman's knot forms a single, fixed loop in the middle of the rope (Figure 4-15, page 4-18). It is a middle rope knot.

a. **Tying the Knot.**

STEP 1. When tying this knot, face the anchor that the tie-off system will be tied to. Take up the slack from the anchor, and wrap two turns around the left hand (palm up) from left to right.

STEP 2. A loop of 30 centimeters is taken up in the second round turn to create the fixed loop of the knot.

STEP 3. Name the wraps from the palm to the fingertips: heel, palm, and fingertip.

STEP 4. Secure the palm wrap with the right thumb and forefinger, and place it over the heel wrap.

STEP 5. Secure the heel wrap and place it over the fingertip wrap.

STEP 6. Secure the fingertip wrap and place it over the palm wrap.

STEP 7. Secure the palm wrap and pull up to form a fixed loop.

STEP 8. Dress the knot down by pulling on the fixed loop and the two working ends.

STEP 9. Pull the working ends apart to finish the knot.

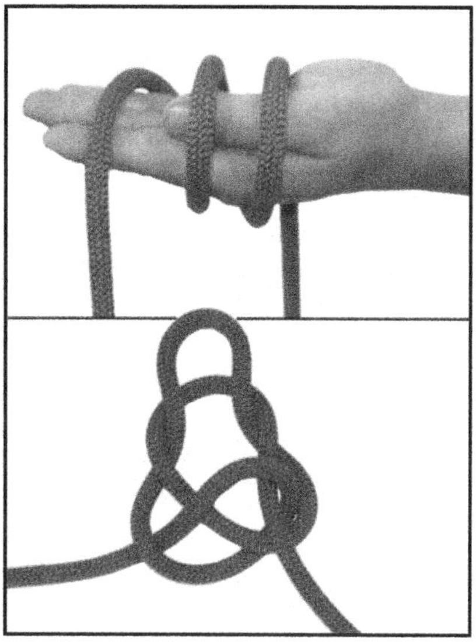

Figure 4-15. Wireman's knot.

b. **Checkpoints.**

(1) The completed knot should have four separate bights locking down on themselves with the fixed loop exiting from the top of the knot and laying toward the near side anchor point.

(2) Both ends should exit opposite each other without any bends.

4-17. DIRECTIONAL FIGURE-EIGHT

The directional figure-eight knot forms a single, fixed loop in the middle of the rope that lays back along the standing part of the rope (Figure 4-16). It is a middle rope knot.

a. **Tying the Knot.**

STEP 1. Face the far side anchor so that when the knot is tied, it lays inward.

STEP 2. Lay the rope from the far side anchor over the left palm. Make one wrap around the palm.

STEP 3. With the wrap thus formed, tie a figure-eight knot around the standing part that leads to the far side anchor.

STEP 4. When dressing the knot down, the tail and the bight must be together.

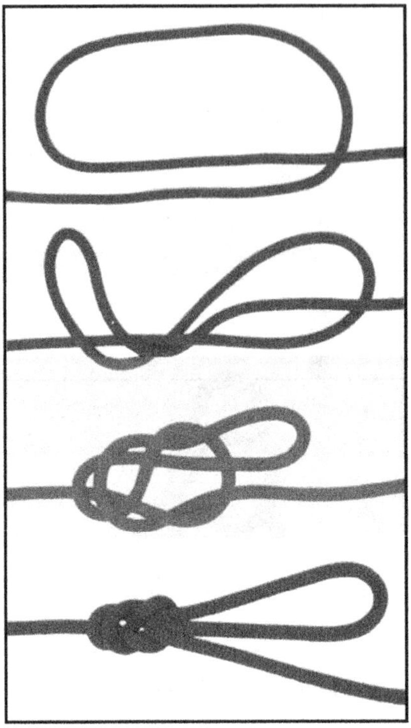

Figure 4-16. Directional figure-eight.

b. **Checkpoints.**

(1) The loop should be large enough to accept a carabiner but no larger than a helmet-size loop.

(2) The tail and bight must be together.

(3) The figure eight is tied tightly.

(4) The bight in the knot faces back toward the near side.

4-18. BOWLINE-ON-A-BIGHT (TWO-LOOP BOWLINE)

The bowline-on-a-bight is used to form two fixed loops in the middle of a rope (Figure 4-17, page 4-20). It is a middle rope knot.

a. **Tying the Knot.**

STEP 1. Form a bight in the rope about twice as long as the finished loops will be.

STEP 2. Tie an overhand knot on a bight.

STEP 3. Hold the overhand knot in the left hand so that the bight is running down and outward.

STEP 4. Grasp the bight with the right hand; fold it back over the overhand knot so that the overhand knot goes through the bight.

STEP 5. From the end (apex) of the bight, follow the bight back to where it forms the cross in the overhand knot. Grasp the two ropes that run down and outward and pull up, forming two loops.

STEP 6. Pull the two ropes out of the overhand knot and dress the knot down.

STEP 7. A final dress is required: grasp the ends of the two fixed loops and pull, spreading them apart to ensure the loops do not slip.

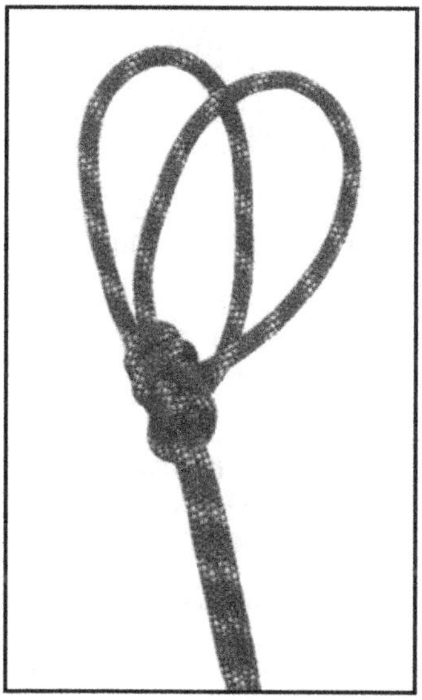

Figure 4-17. Bowline-on-a-bight.

b. **Checkpoints.**

(1) There are two fixed loops that will not slip.

(2) There are no twists in the knot.

(3) A double loop is held in place by a bight.

4-19. TWO-LOOP FIGURE-EIGHT

The two-loop figure-eight is used to form two fixed loops in the middle of a rope (Figure 4-18.) It is a middle rope knot.

a. **Tying the Knot.**

STEP 1. Using a doubled rope, form an 18-inch bight in the left hand with the running end facing to the left.

STEP 2. Grasp the bight with the right hand and make a 360-degree turn around the standing end in a counterclockwise direction.

STEP 3. With the working end, form another bight and place that bight through the loop just formed in the left hand.

STEP 4. Hold the bight with the left hand, and place the original bight (moving toward the left hand) over the knot.

STEP 5. Dress the knot down.

Figure 4-18. Two-loop figure-eight.

b. **Checkpoints.**

(1) There is a double figure-eight knot with two loops that share a common locking bar.

(2) The two loops must be adjustable by means of a common locking bar.

(3) The common locking bar is on the bottom of the double figure-eight knot.

4-20. FIGURE-EIGHT LOOP (FIGURE-EIGHT-ON-A-BIGHT)

The figure-eight loop, also called the figure-eight-on-a-bight, is used to form a fixed loop in a rope (Figure 4-19). It is a middle of the rope knot.

a. **Tying the Knot.**

STEP 1. Form a bight in the rope about as large as the diameter of the desired loop.

STEP 2. With the bight as the working end, form a loop in rope (standing part).

STEP 3. Wrap the working end around the standing part 360 degrees and feed the working end through the loop. Dress the knot tightly.

Figure 4-19. Figure-eight loop.

b. **Checkpoints.**

(1) The loop is the desired size.

(2) The ropes in the loop are parallel and do not cross over each other.

(3) The knot is tightly dressed.

4-21. PRUSIK KNOT

The Prusik knot is used to put a moveable rope on a fixed rope such as a Prusik ascent or a tightening system. This knot can be tied as a middle or end of the rope Prusik. It is a specialty knot.

a. **Tying the Knot.**

(1) ***Middle-of-the-Rope Prusik.*** The middle-of-the-rope Prusik knot can be tied with a short rope to a long rope as follows (Figure 4-20.):

STEP 1. Double the short rope, forming a bight, with the working ends even. Lay it over the long rope so that the closed end of the bight is 12 inches below the long rope and the remaining part of the rope (working ends) is the closest to the climber; spread the working end apart.

STEP 2. Reach down through the 12-inch bight. Pull up both of the working ends and lay them over the long rope. Repeat this process making sure that the working ends pass in the middle of the first two wraps. Now there are four wraps and a locking bar working across them on the long rope.

STEP 3. Dress the wraps and locking bar down to ensure they are tight and not twisted. Tying an overhand knot with both ropes will prevent the knot from slipping during periods of variable tension.

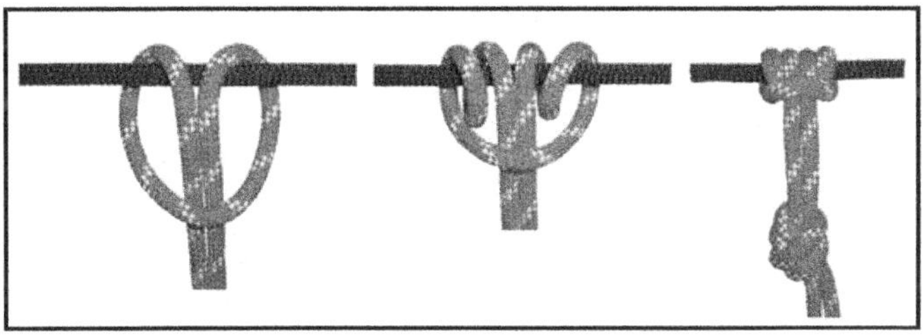

Figure 4-20. Middle-of-the-rope Prusik.

(2) ***End-of-the-Rope Prusik*** (Figure 4-21).

STEP 1. Using an arm's length of rope, and place it over the long rope.

STEP 2. Form a complete round turn in the rope.

STEP 3. Cross over the standing part of the short rope with the working end of the short rope.

STEP 4. Lay the working end under the long rope.

STEP 5. Form a complete round turn in the rope, working back toward the middle of the knot.

STEP 6. There are four wraps and a locking bar running across them on the long rope. Dress the wraps and locking bar down. Ensure they are tight, parallel, and not twisted.

STEP 7. Finish the knot with a bowline to ensure that the Prusik knot will not slip out during periods of varying tension.

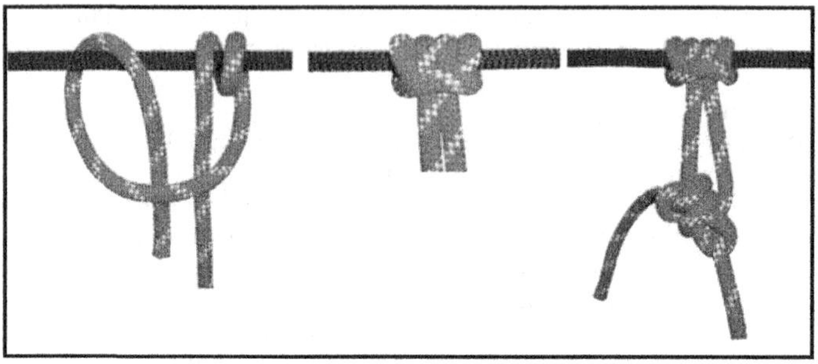

Figure 4-21. End-of-the-rope Prusik knot.

b. **Checkpoints.**

(1) Four wraps with a locking bar.

(2) The locking bar faces the climber.

(3) The knot is tight and dressed down with no ropes twisted or crossed.

(4) Other than a finger Prusik, the knot should contain an overhand or bowline to prevent slipping.

4-22. BACHMAN KNOT

The Bachman knot provides a means of using a makeshift mechanized ascender (Figure 4-22, page 4-24). It is a specialty knot.

a. **Tying the Knot.**

STEP 1. Find the middle of a utility rope and insert it into a carabiner.

STEP 2. Place the carabiner and utility rope next to a long climbing rope.

STEP 3. With the two ropes parallel from the carabiner, make two or more wraps around the climbing rope and through the inside portion of the carabiner.

Note: The rope can be tied into an etrier (stirrup) and used as a Prusik-friction principle ascender.

b. **Checkpoints.**

(1) The bight of the climbing rope is at the top of the carabiner.

(2) The two ropes run parallel without twisting or crossing.

(3) Two or more wraps are made around the long climbing rope and through the inside portion of the carabiner.

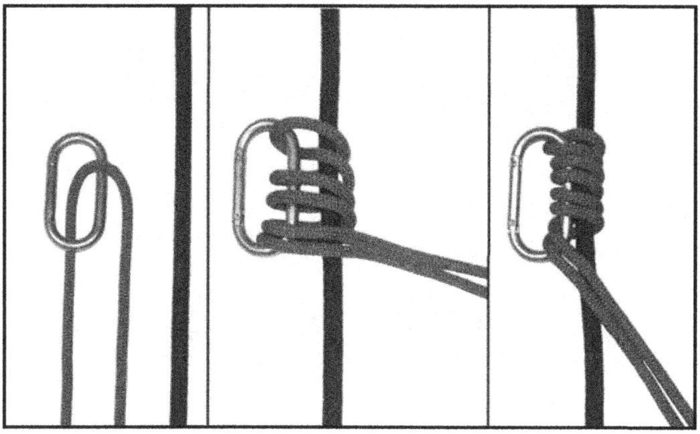

Figure 4-22. Bachman knot.

4-23. BOWLINE-ON-A-COIL

The bowline-on-a-coil is an expedient tie-in used by climbers when a climbing harness is not available (Figure 4-23). It is a specialty knot.

a. **Tying the Knot.**

STEP 1. With the running end, place 3 feet of rope over your right shoulder. The running end is to the back of the body.

STEP 2. Starting at the bottom of your rib cage, wrap the standing part of the rope around your body and down in a clockwise direction four to eight times.

STEP 3. With the standing portion of the rope in your left hand, make a clockwise loop toward the body. The standing portion is on the bottom.

STEP 4. Ensuring the loop does not come uncrossed, bring it up and under the coils between the rope and your body.

STEP 5. Using the standing part, bring a bight up through the loop. Grasp the running end of the rope with the right hand. Pass it through the bight from right to left and back on itself.

STEP 6. Holding the bight loosely, dress the knot down by pulling on the standing end.

STEP 7. Safety the bowline with an overhand around the top, single coil. Then, tie an overhand around all coils, leaving a minimum 4-inch pigtail.

b. **Checkpoints.**

(1) A minimum of four wraps, not crossed, with a bight held in place by a loop.

(2) The loop must be underneath all wraps.

(3) A minimum 4-inch pigtail after the second overhand safety is tied.

(4) Must be centered on the mid-line of the body.

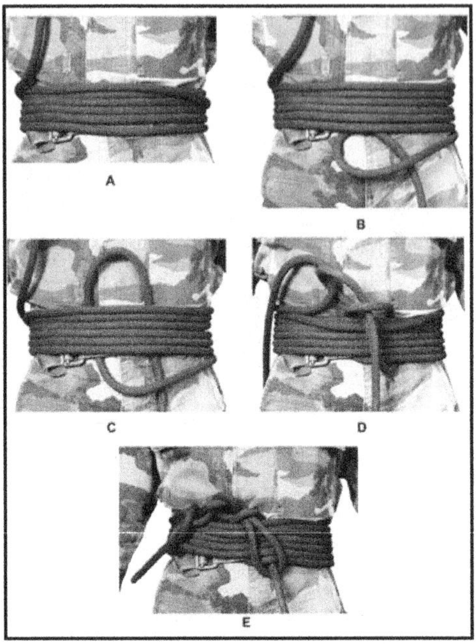

Figure 4-23. Bowline-on-a-coil.

4-24. THREE-LOOP BOWLINE

The three-loop bowline is used to form three fixed loops in the middle of a rope (Figure 4-24, page 4-26). It is used in a self-equalizing anchor system. It is a specialty knot.

a. **Tying the Knot.**

STEP 1. Form an approximate 24-inch bight.

STEP 2. With the right thumb facing toward the body, form a doubled loop in the standing part by turning the wrist clockwise. Lay the loops to the right.

STEP 3. With the right hand, reach down through the loops and pull up a doubled bight from the standing part of the rope.

STEP 4. Place the running end (bight) of the rope (on the left) through the doubled bight from left to right and bring it back on itself. Hold the running end loosely and dress the knot down by pulling on the standing parts.

STEP 5. Safety it off with a doubled overhand knot.

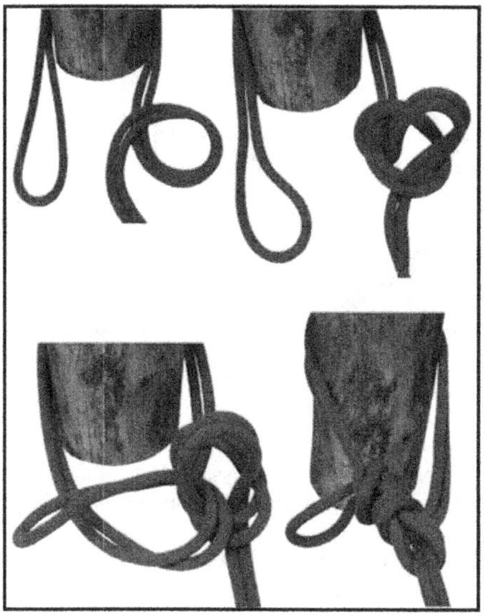

Figure 4-24. Three-loop bowline.

b. **Checkpoints.**

(1) There are two bights held in place by two loops.

(2) The bights form locking bars around the standing parts.

(3) The running end (bight) must be on the inside of the fixed loops.

(4) There is a minimum 4-inch pigtail after the double overhand safety knot is tied.

4-25. FIGURE-EIGHT SLIP KNOT

The figure eight slip knot forms an adjustable bight in a rope (Figure 4-25). It is a specialty knot.

a. **Tying the Knot.**

STEP 1. Form a 12-inch bight in the end of the rope.

STEP 2. Hold the center of the bight in the right hand. Hold the two parallel ropes from the bight in the left hand about 12 inches up the rope.

STEP 3. With the center of the bight in the right hand, twist two complete turns clockwise.

STEP 4. Reach through the bight and grasp the long, standing end of the rope. Pull another bight (from the long standing end) back through the original bight.

STEP 5. Pull down on the short working end of the rope and dress the knot down.

STEP 6. If the knot is to be used in a transport tightening system, take the working end of the rope and form a half hitch around the loop of the figure eight knot.

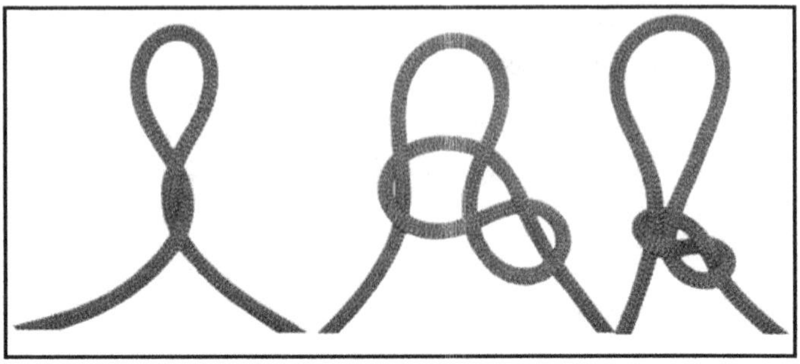

Figure 4-25. Figure-eight slip knot.

b. **Checkpoints.**

(1) The knot is in the shape of a figure eight.

(2) Both ropes of the bight pass through the same loop of the figure eight.

(3) The sliding portion of the rope is the long working end of the rope.

4-26. TRANSPORT KNOT (OVERHAND SLIP KNOT/MULE KNOT)

The transport knot is used to secure the transport tightening system (Figure 4-26, page 4-28). It is simply an overhand slip knot.

a. **Tying the Knot.**

STEP 1. Pass the running end of the rope around the anchor point passing it back under the standing portion (leading to the far side anchor) forming a loop.

STEP 2. Form a bight with the running end of the rope. Pass over the standing portion and down through the loop and dress it down toward the anchor point.

STEP 3. Secure the knot by tying a half hitch around the standing portion with the bight.

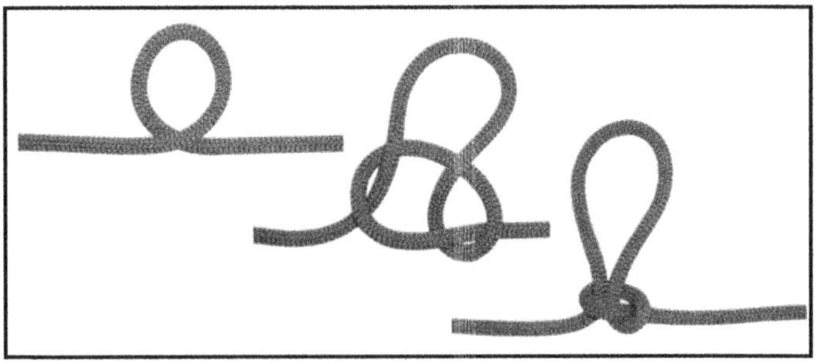

Figure 4-26. Transport knot.

b. **Check Points.**

(1) There is a single overhand slip knot.

(2) The knot is secured using a half hitch on a bight.
(3) The bight is a minimum of 12 inches long.

4-27. KLEIMHIEST KNOT

The Kleimhiest knot provides a moveable, easily adjustable, high-tension knot capable of holding extremely heavy loads while being pulled tight (Figure 4-27). It is a special-purpose knot.

a. **Tying the Knot.**

STEP 1. Using a utility rope or webbing offset the ends by 12 inches. With the ends offset, find the center of the rope and form a bight. Lay the bight over a horizontal rope.

STEP 2. Wrap the tails of the utility rope around the horizontal rope back toward the direction of pull. Wrap at least four complete turns.

STEP 3. With the remaining tails of the utility rope, pass them through the bight (see STEP 1).

STEP 4. Join the two ends of the tail with a joining knot.

STEP 5. Dress the knot down tightly so that all wraps are touching.

Note: Spectra should not be used for the Kleimhiest knot. It has a low melting point and tends to slip .

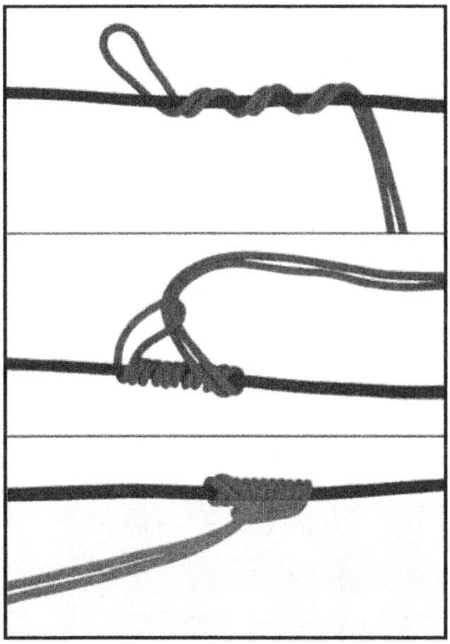

Figure 4-27. Kleimhiest knot.

b. **Checkpoints.**

(1) The bight is opposite the direction of pull.

(2) All wraps are tight and touching.

(3) The ends of the utility rope are properly secured with a joining knot.

4-28. FROST KNOT

The frost knot is used when working with webbing (Figure 4-28, page 4-30). It is used to create the top loop of an etrier. It is a special-purpose knot.

a. **Tying the Knot.**

STEP 1. Lap one end (a bight) of webbing over the other about 10 to 12 inches.

STEP 2. Tie an overhand knot with the newly formed triple-strand webbing; dress tightly.

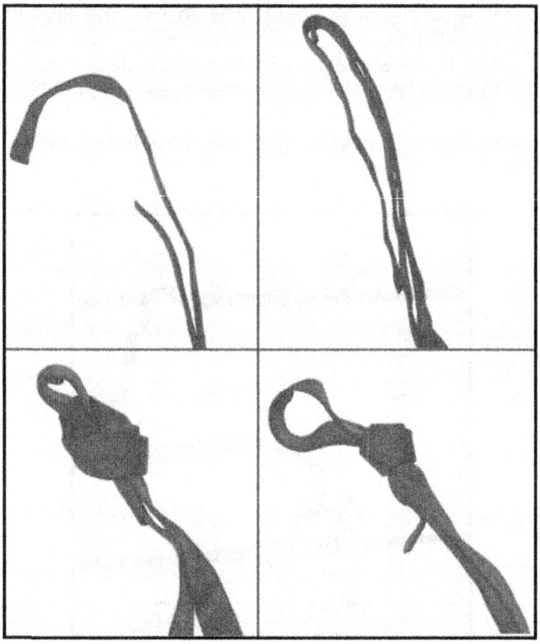

Figure 4-28. Frost knot.

b. **Checkpoints.**

(1) The tails of the webbing run in opposite directions.

(2) Three strands of webbing are formed into a tight overhand knot.

(3) There is a bight and tail exiting the top of the overhand knot.

4-29. GIRTH HITCH

The girth hitch is used to attach a runner to an anchor or piece of equipment (Figure 4-29). It is a special-purpose knot.

a. **Tying the Knot.**

STEP 1: Form a bight.
STEP 2: Bring the runner back through the bight.
STEP 3: Cinch the knot tightly.

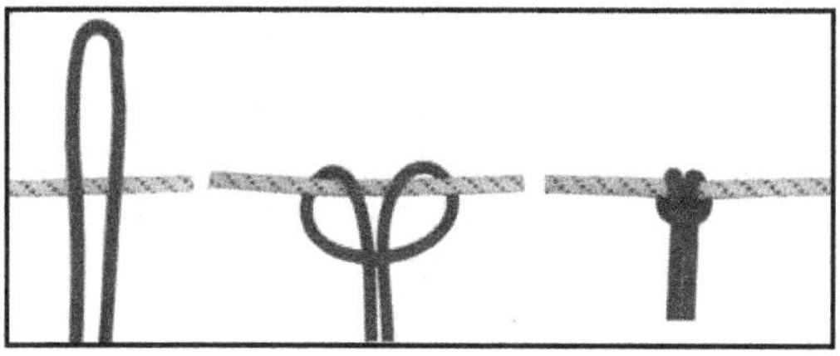

Figure 4-29. Girth hitch.

b. **Checkpoint.**

(1) Two wraps exist with a locking bar running across the wraps.
(2) The knot is dressed tightly.

4-30. MUNTER HITCH

The munter hitch, when used in conjunction with a pear-shaped locking carabiner, is used to form a mechanical belay (Figure 4-30).

a. **Tying the Knot.**

STEP 1. Hold the rope in both hands, palms down about 12 inches apart.
STEP 2. With the right hand, form a loop away from the body toward the left hand. Hold the loop with the left hand.
STEP 3. With the right hand, place the rope that comes from the bottom of the loop over the top of the loop.
STEP 4. Place the bight that has just been formed around the rope into the pear shaped carabiner. Lock the locking mechanism.

b. **Check Points.**

(1) A bight passes through the carabiner, with the closed end around the standing or running part of the rope.

(2) The carabiner is locked.

Figure 4-30. Munter hitch.

4-31. RAPPEL SEAT

The rappel seat is an improvised seat rappel harness made of rope (Figure 4-31, page 4-32). It usually requires a sling rope 14 feet or longer.

a. **Tying the Knot.**

STEP 1. Find the middle of the sling rope and make a bight.

STEP 2. Decide which hand will be used as the brake hand and place the bight on the opposite hip.

STEP 3. Reach around behind and grab a single strand of rope. Bring it around the waist to the front and tie two overhands on the other strand of rope, thus creating a loop around the waist.

STEP 4. Pass the two ends between the legs, ensuring they do not cross.

STEP 5. Pass the two ends up under the loop around the waist, bisecting the pocket flaps on the trousers. Pull up on the ropes, tightening the seat.

STEP 6. From rear to front, pass the two ends through the leg loops creating a half hitch on both hips.

STEP 7. Bring the longer of the two ends across the front to the nonbrake hand hip and secure the two ends with a square knot safetied with overhand knots. Tuck any excess rope in the pocket below the square knot.

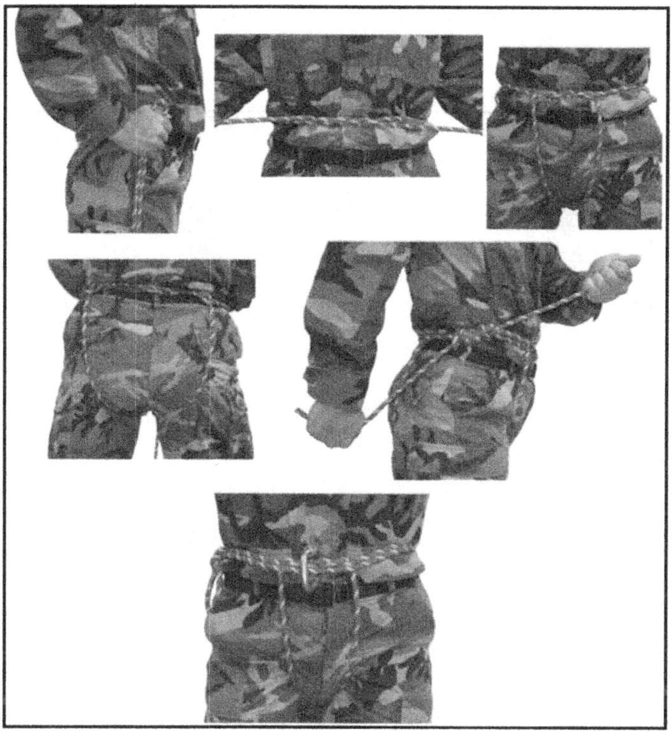

Figure 4-31. Rappel seat.

b. **Check Points.**

(1) There are two overhand knots in the front.

(2) The ropes are not crossed between the legs.

(3) A half hitch is formed on each hip.

(4) Seat is secured with a square knot with overhand safeties on the non-brake hand side.

(5) There is a minimum 4-inch pigtail after the overhand safeties are tied.

4-32. GUARDE KNOT

The guarde knot (ratchet knot, alpine clutch) is a special purpose knot primarily used for hauling systems or rescue (Figure 4-32). The knot works in only one direction and cannot be reversed while under load.

a. **Tying the Knot.**

STEP 1. Place a bight of rope into the two anchored carabiners (works best with two like carabiners, preferably ovals).

STEP 2. Take a loop of rope from the non-load side and place it down into the opposite cararabiner so that the rope comes out between the two carabiners.

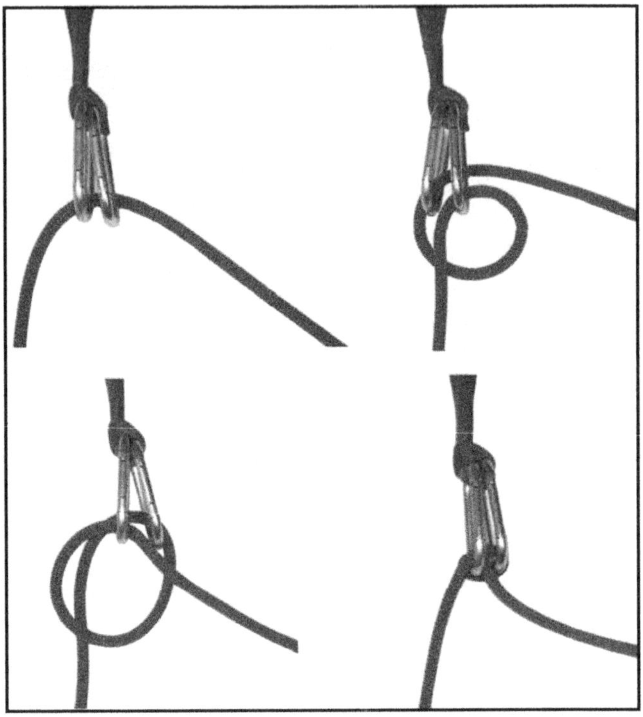

Figure 4-32. Guarde knot.

b. **Check Points.**

(1) When properly dressed, rope can only be pulled in one direction.

(2) The knot will not fail when placed under load.

CHAPTER 5

ANCHORS

This chapter discusses different types of anchors and their application in rope systems and climbing. Proper selection and placement of anchors is a critical skill that requires a great deal of practice. Failure of any system is most likely to occur at the anchor point. If the anchor is not strong enough to support the intended load, it will fail. Failure is usually the result of poor terrain features selected for the anchor point, or the equipment used in rigging the anchor was placed improperly or in insufficient amounts.

When selecting or constructing anchors, always try to make sure the anchor is "bombproof." A bombproof anchor is stronger than any possible load that could be placed on it. An anchor that has more strength than the climbing rope is considered bombproof.

Section I. NATURAL ANCHORS

Natural anchors should be considered for use first. They are usually strong and often simple to construct with minimal use of equipment. Trees, boulders, and other terrain irregularities are already in place and simply require a method of attaching the rope. However, natural anchors should be carefully studied and evaluated for stability and strength before use. Sometimes the climbing rope is tied directly to the anchor, but under most circumstances a sling is attached to the anchor and then the climbing rope is attached to the sling with a carabiner(s). (See paragraph 5-7 for slinging techniques.)

5-1. TREES

Trees are probably the most widely used of all natural anchors depending on the terrain and geographical region (Figure 5-1). However, trees must be carefully checked for suitability.

a. In rocky terrain, trees usually have a shallow root system. This can be checked by pushing or tugging on the tree to see how well it is rooted. Anchoring as low as possible to prevent excess leverage on the tree may be necessary.

b. Use padding on soft, sap producing trees to keep sap off ropes and slings.

Figure 5-1. Trees used as anchors.

5-2. BOULDERS

Boulders and rock nubbins make ideal anchors (Figure 5-2). The rock can be firmly tapped with a piton hammer to ensure it is solid. Sedimentary and other loose rock formations are not stable. Talus and scree fields are an indicator that the rock in the area is not solid. All areas around the rock formation that could cut the rope or sling should be padded.

Figure 5-2. Boulders used as anchors.

5-3. CHOCKSTONES

A chockstone is a rock that is wedged in a crack because the crack narrows downward (Figure 5-3). Chockstones should be checked for strength, security, and crumbling and should always be tested before use. All chockstones must be solid and strong enough to support the load. They must have maximum surface contact and be well tapered with the surrounding rock to remain in position.

a. Chockstones are often directional—they are secure when pulled in one direction but may pop out if pulled in another direction.

b. A creative climber can often make his own chockstone by wedging a rock into position, tying a rope to it, and clipping on a carabiner.

c. Slings should not be wedged between the chockstone and the rock wall since a fall could cut the webbing runner.

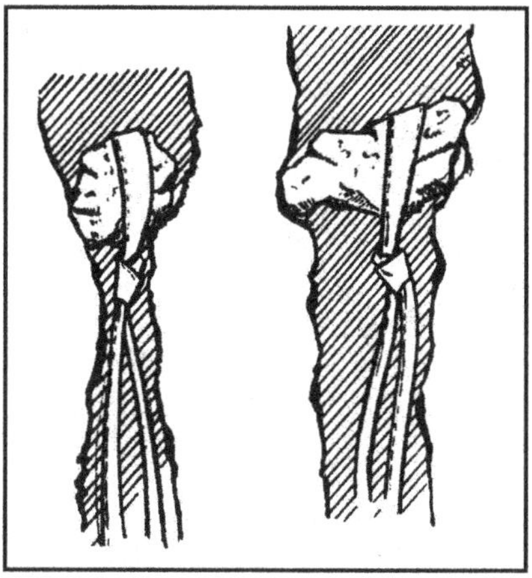

Figure 5-3. Chockstones.

5-4. ROCK PROJECTIONS

Rock projections (sometimes called nubbins) often provide suitable protection (Figure 5-4). These include blocks, flakes, horns, and spikes. If rock projections are used, their firmness is important. They should be checked for cracks or weathering that may impair their firmness. If any of these signs exist, the projection should be avoided.

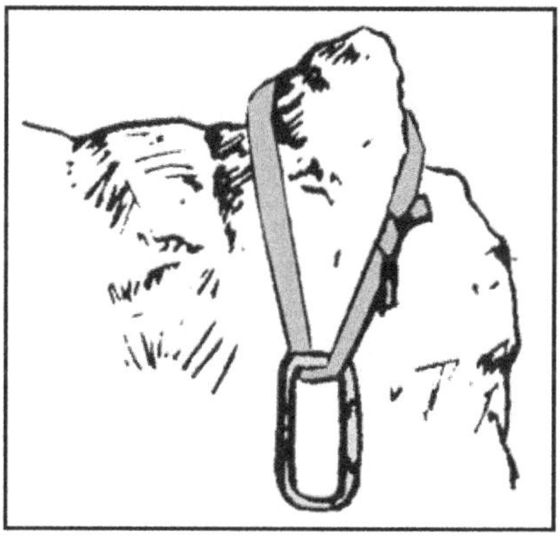

Figure 5-4. Rock projections.

5-5. TUNNELS AND ARCHES

Tunnels and arches are holes formed in solid rock and provide one of the more secure anchor points because they can be pulled in any direction. A sling is threaded through the opening hole and secured with a joining knot or girth hitch. The load-bearing hole must be strong and free of sharp edges (pad if necessary).

5-6. BUSHES AND SHRUBS

If no other suitable anchor is available, the roots of bushes can be used by routing a rope around the bases of several bushes (Figure 5-5). As with trees, the anchoring rope is placed as low as possible to reduce leverage on the anchor. All vegetation should be healthy and well rooted to the ground.

Figure 5-5. Bushes and shrubs.

5-7. SLINGING TECHNIQUES

Three methods are used to attach a sling to a natural anchor—drape, wrap, and girth. Whichever method is used, the knot is set off to the side where it will not interfere with normal carabiner movement. The carabiner gate should face away from the ground and open away from the anchor for easy insertion of the rope. When a locking carabiner cannot be used, two carabiners are used with gates opposed. Correctly opposed gates should open on opposite sides and form an "X" when opened (Figure 5-6).

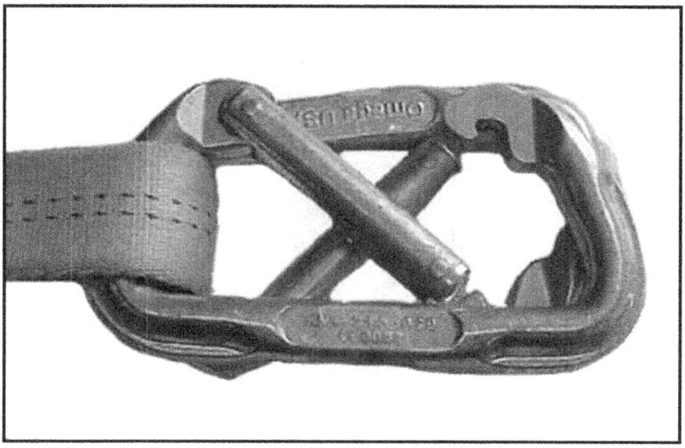

Figure 5-6. Correctly opposed carabiners.

a. **Drape.** Drape the sling over the anchor (Figure 5-7). Untying the sling and routing it around the anchor and then retying is still considered a drape.

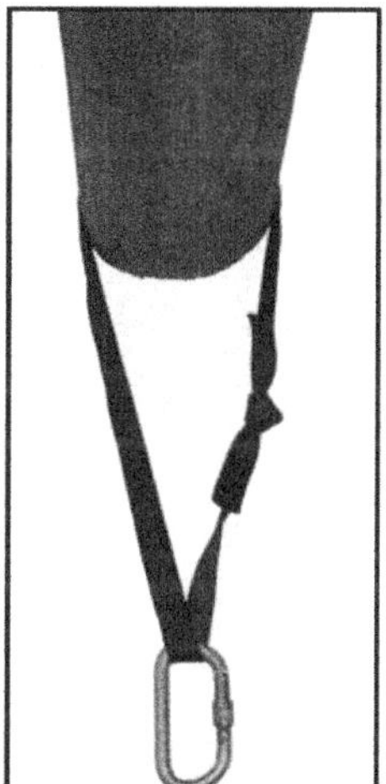

Figure 5-7. Drape.

b. **Wrap.** Wrap the sling around the anchor and connect the two ends together with a carabiner(s) or knot (Figure 5-8).

Figure 5-8. Wrap.

c. **Girth.** Tie the sling around the anchor with a girth hitch (Figure 5-9). Although a girth hitch reduces the strength of the sling, it allows the sling to remain in position and not slide on the anchor.

Figure 5-9. Girth.

Section II. ANCHORING WITH THE ROPE

The climbing or installation rope can be tied directly to the anchor using several different techniques. This requires less equipment, but also sacrifices some rope length to tie the anchor. The rope can be tied to the anchor using an appropriate anchor knot such as a bowline or a rerouted figure eight. Round turns can be used to help keep the rope in position on the anchor. A tensionless anchor can be used in high-load installations where tension on the attachment point and knot is undesirable.

5-8. ROPE ANCHOR

When tying the climbing or installation rope around an anchor, the knot should be placed approximately the same distance away from the anchor as the diameter of the anchor (Figure 5-10). The knot shouldn't be placed up against the anchor because this can stress and distort the knot under tension.

Figure 5-10. Rope tied to anchor with anchor knot.

5-9. TENSIONLESS ANCHOR

The tensionless anchor is used to anchor the rope on high-load installations such as bridging and traversing (Figure 5-11, page 5-8). The wraps of the rope around the anchor absorb the tension of the installation and keep the tension off the knot and carabiner. The anchor is usually tied with a minimum of four wraps, more if necessary, to absorb the tension. A smooth anchor may require several wraps, whereas a rough barked tree might only require a few. The rope is wrapped from top to bottom. A fixed loop is placed into the end of the rope and attached loosely back onto the rope with a carabiner.

Figure 5-11. Tensionless anchor.

Section III. ARTIFICIAL ANCHORS

Using artificial anchors becomes necessary when natural anchors are unavailable. The art of choosing and placing good anchors requires a great deal of practice and experience. Artificial anchors are available in many different types such as pitons, chocks, hexcentrics, and SLCDs. Anchor strength varies greatly; the type used depends on the terrain, equipment, and the load to be placed on it.

5-10. DEADMAN

A "deadman" anchor is any solid object buried in the ground and used as an anchor.

a. An object that has a large surface area and some length to it works best. (A hefty timber, such as a railroad tie, would be ideal.) Large boulders can be used, as well as a bundle of smaller tree limbs or poles. As with natural anchors, ensure timbers and tree limbs are not dead or rotting and that boulders are solid. Equipment, such as skis, ice axes, snowshoes, and ruck sacks, can also be used if necessary.

b. In extremely hard, rocky terrain (where digging a trench would be impractical, if not impossible) a variation of the deadman anchor can be constructed by building above the ground. The sling is attached to the anchor, which is set into the ground as deeply as possible. Boulders are then stacked on top of it until the anchor is strong enough for the load. Though normally not as strong as when buried, this method can work well for light-load installations as in anchoring a hand line for a stream crossing.

Note: Artificial anchors, such as pitons and bolts, are not widely accepted for use in all areas because of the scars they leave on the rock and the environment. Often they are left in place and become unnatural, unsightly fixtures in the natural environment. For training planning, local laws and courtesies should be taken into consideration for each area of operation.

5-11. PITONS

Pitons have been in use for over 100 years. Although still available, pitons are not used as often as other types of artificial anchors due primarily to their impact on the environment. Most climbers prefer to use chocks, SLCDs and other artificial anchors rather than pitons because they do not scar the rock and are easier to remove. Eye protection should always be worn when driving a piton into rock.

Note: The proper use and placement of pitons, as with any artificial anchor, should be studied, practiced, and tested while both feet are firmly on the ground and there is no danger of a fall.

a. **Advantages.** Some advantages in using pitons are:

- Depending on type and placement, pitons can support multiple directions of pull.
- Pitons are less complex than other types of artificial anchors.
- Pitons work well in thin cracks where other types of artificial anchors do not.

b. **Disadvantages.** Some disadvantages in using pitons are:

- During military operations, the distinct sound created when hammering pitons is a tactical disadvantage.
- Due to the expansion force of emplacing a piton, the rock could spread apart or break causing an unsafe condition.
- Pitons are more difficult to remove than other types of artificial anchors.
- Pitons leave noticeable scars on the rock.
- Pitons are easily dropped if not tied off when being used.

c. **Piton Placement.** The proper positioning or placement of pitons is critical. (Figure 5-12, page 5-10, shows examples of piton placement.) Usually a properly sized piton for a rock crack will fit one half to two thirds into the crack before being driven with the piton hammer. This helps ensure the depth of the crack is adequate for the size piton selected. As pitons are driven into the rock the pitch or sound that is made will change with each hammer blow, becoming higher pitched as the piton is driven in.

(1) Test the rock for soundness by tapping with the hammer. Driving pitons in soft or rotten rock is not recommended. When this type of rock must be used, clear the loose rock, dirt, and debris from the crack before driving the piton completely in.

(2) While it is being driven, attach the piton to a sling with a carabiner (an old carabiner should be used, if available) so that if the piton is knocked out of the crack, it will not be lost. The greater the resistance overcome while driving the piton, the firmer the anchor will be. The holding power depends on the climber placing the piton in a sound crack, and on the type of rock. The piton should not spread the rock, thereby loosening the emplacement.

Note: Pitons that have rings as attachment points might not display much change in sound as they are driven in as long as the ring moves freely.

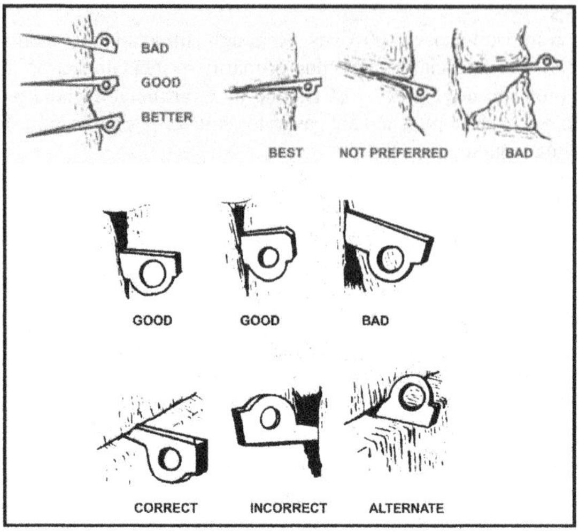

Figure 5-12. Examples of piton placements.

(3) Military mountaineers should practice emplacing pitons using either hand. Sometimes a piton cannot be driven completely into a crack, because the piton is too long. Therefore, it should be tied off using a hero-loop (an endless piece of webbing) (Figure 5-13). Attach this loop to the piton using a girth hitch at the point where the piton enters the rock so that the girth hitch is snug against the rock. Clip a carabiner into the loop.

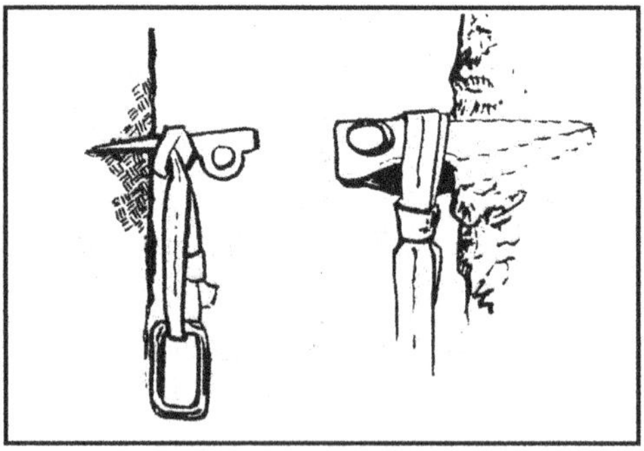

Figure 5-13. Hero-loop.

d. **Testing.** To test pitons pull up about 1 meter of slack in the climbing rope or use a sling. Insert this rope into a carabiner attached to the piton, then grasp the rope at least 1/2 meter from the carabiner. Jerk vigorously upward, downward, to each side, and then outward while observing the piton for movement. Repeat these actions as many times as necessary. Tap the piton to determine if the pitch has changed. If the pitch has changed greatly, drive the piton in as far as possible. If the sound regains its original pitch, the emplacement is probably safe. If the piton shows any sign of moving or if, upon driving it, there is any question of its soundness, drive it into another place. Try to be in a secure position before testing. This procedure is intended for use in testing an omni-directional anchor (one that withstands a pull in any direction). When a directional anchor (pull in one direction) is used, as in most free and direct-aid climbing situations, and when using chocks, concentrate the test in the direction that force will be applied to the anchor.

e. **Removing Pitons.** Attach a carabiner and sling to the piton before removal to eliminate the chance of dropping and losing it. Tap the piton firmly along the axis of the crack in which it is located. Alternate tapping from both sides while applying steady pressure. Pulling out on the attached carabiner eventually removes the piton (Figure 5-14).

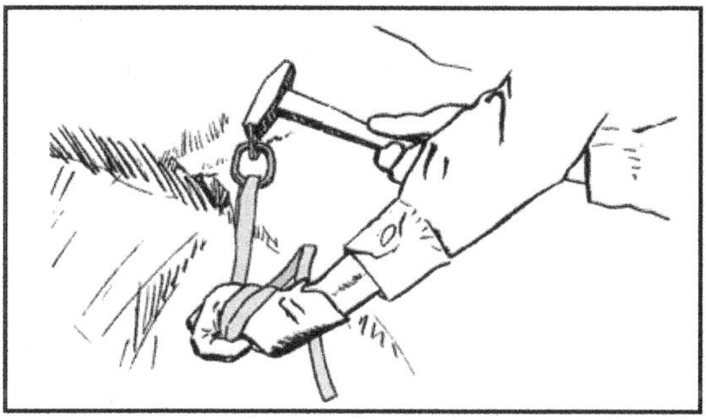

Figure 5-14. Piton removal.

f. **Reusing Pitons.** Soft iron pitons that have been used, removed, and straightened may be reused, but they must be checked for strength. In training areas, pitons already in place should not be trusted since weather loosens them in time. Also, they may have been driven poorly the first time. Before use, test them as described above and drive them again until certain of their soundness.

5-12. CHOCKS

Chock craft has been in use for many decades. A natural chockstone, having fallen and wedged in a crack, provides an excellent anchor point. Sometimes these chockstones are in unstable positions, but can be made into excellent anchors with little adjustment. Chock craft is an art that requires time and technique to master—simple in theory, but complex in practice. Imagination and resourcefulness are key principles to chock craft. The skilled climber must understand the application of mechanical advantage, vectors, and other forces that affect the belay chain in a fall.

a. **Advantages.** The advantages of using chocks are:

- Tactically quiet installation and recovery.
- Usually easy to retrieve and, unless severely damaged, are reusable.
- Light to carry.
- Easy to insert and remove.
- Minimal rock scarring as opposed to pitons.
- Sometimes can be placed where pitons cannot (expanding rock flakes where pitons would further weaken the rock).

b. **Disadvantages.** The disadvantages of using chocks are:

- May not fit in thin cracks, which may accept pitons.
- Often provide only one direction of pull.
- Practice and experience necessary to become proficient in proper placement.

c. **Placement.** The principles of placing chocks are to find a crack with a constriction at some point, place a chock of appropriate size above and behind the constriction, and set the chock by jerking down on the chock loop (Figure 5-15). Maximum surface contact with a tight fit is critical. Chocks are usually good for a single direction of pull.

(1) Avoid cracks that have crumbly (soft) or deteriorating rock, if possible. Some cracks may have loose rock, grass, and dirt, which should be removed before placing the chock. Look for a constriction point in the crack, then select a chock to fit it.

(2) When selecting a chock, choose one that has as much surface area as possible in contact with the rock. A chock resting on one small crystal or point of rock is likely to be unsafe. A chock that sticks partly out of the crack is avoided. Avoid poor protection. Ensure that the chock has a wire or runner long enough; extra ropes, cord, or webbing may be needed to extend the length of the runner.

(3) End weighting of the placement helps to keep the protection in position. A carabiner often provides enough weight

(4) Parallel-sided cracks without constrictions are a problem. Chocks designed to be used in this situation rely on camming principles to remain emplaced. Weighting the emplacement with extra hardware is often necessary to keep the chocks from dropping out.

(a) Emplace the wedge-shaped chock above and behind the constriction; seat it with a sharp downward tug.

(b) Place a camming chock with its narrow side into the crack, then rotate it to the attitude it will assume under load; seat it with a sharp downward tug.

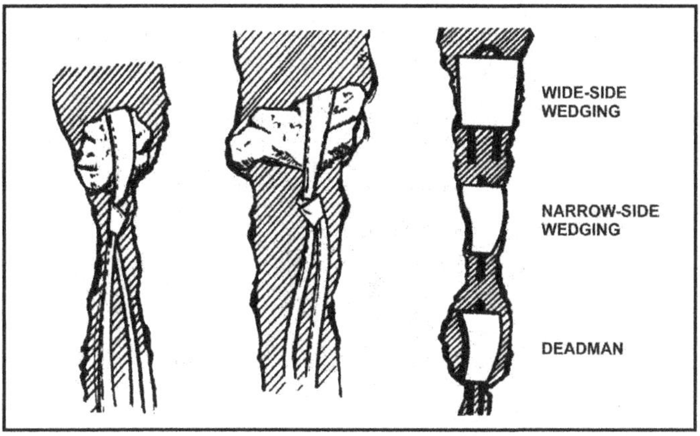

Figure 5-15. Chock placements.

d. **Testing.** After seating a chock, test it to ensure it remains in place. A chock that falls out when the climber moves past it is unsafe and offers no protection. To test it, firmly pull the chock in every anticipated direction of pull. Some chock placements fail in one or more directions; therefore, use pairs of chocks in opposition.

5-13. SPRING-LOADED CAMMING DEVICE

The SLCD offers quick and easy placement of artificial protection. It is well suited in awkward positions and difficult placements, since it can be emplaced with one hand. It can usually be placed quickly and retrieved easily (Figure 5-16, page 5-14).

a. To emplace an SLCD hold the device in either hand like a syringe, pull the retractor bar back, place the device into a crack, and release the retractor bar. The SLCD holds well in parallel-sided hand- and fist-sized cracks. Smaller variations are available for finger-sized cracks.

b. Careful study of the crack should be made before selecting the device for emplacement. It should be placed so that it is aligned in the direction of force applied to it. It should not be placed any deeper than is needed for secure placement, since it may be impossible to reach the extractor bar for removal. An SLCD should be extended with a runner and placed so that the direction of pull is parallel to the shaft; otherwise, it may rotate and pull out. The versions that have a semi-rigid wire cable shaft allow for greater flexibility and usage, without the danger of the shaft snapping off in a fall.

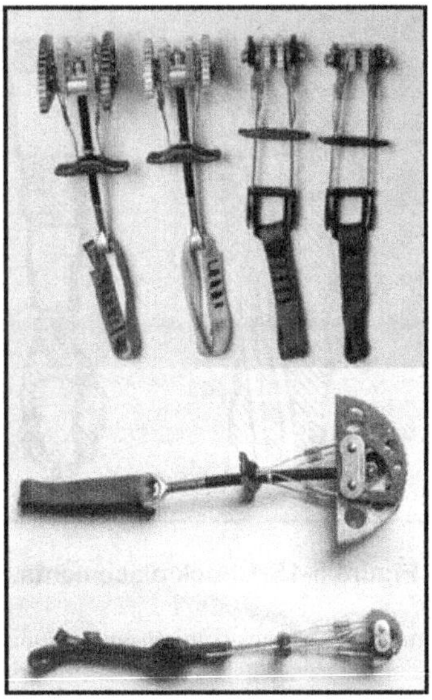

Figure 5-16. SLCD placements.

5-14. BOLTS

Bolts are often used in fixed-rope installations and in aid climbing where cracks are not available.

a. Bolts provide one of the most secure means of establishing protection. The rock should be inspected for evidence of crumbling, flaking, or cracking, and should be tested with a hammer. Emplacing a bolt with a hammer and a hand drill is a time-consuming and difficult process that requires drilling a hole in the rock deeper than the length of the bolt. This normally takes more than 20 minutes for one hole. Electric or even gas-powered drills can be used to greatly shorten drilling time. However, their size and weight can make them difficult to carry on the climbing route.

b. A hanger (carrier) and nut are placed on the bolt, and the bolt is inserted and then driven into the hole. A climber should never hammer on a bolt to test or "improve" it, since this permanently weakens it. Bolts should be used with carriers, carabiners, and runners.

c. When using bolts, the climber uses a piton hammer and hand drill with a masonry bit for drilling holes. Some versions are available in which the sleeve is hammered and turned into the rock (self-drilling), which bores the hole. Split bolts and expanding sleeves are common bolts used to secure hangers and carriers (Figure 5-17). Surgical tubing is useful in blowing dust out of the holes. Nail type bolts are emplaced by driving the nail with a hammer to expand the sleeve against the wall of the drilled hole. Safety glasses should always be worn when emplacing bolts.

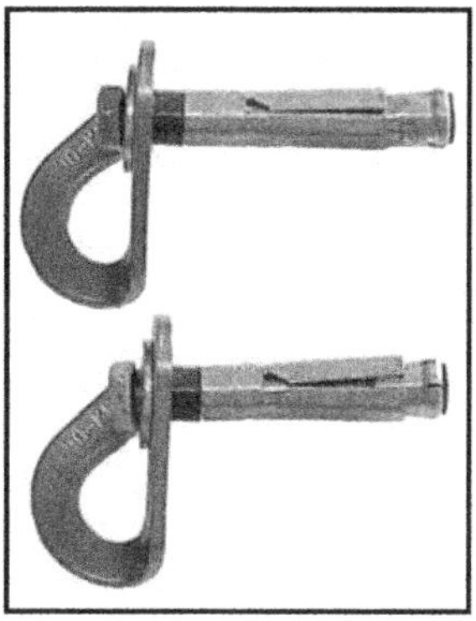

Figure 5-17. Bolt with expanding sleeve.

5-15. EQUALIZING ANCHORS

Equalizing anchors are made up of more than one anchor point joined together so that the intended load is shared equally. This not only provides greater anchor strength, but also adds redundancy or backup because of the multiple points.

a. **Self-equalizing Anchor.** A self-equalizing anchor will maintain an equal load on each individual point as the direction of pull changes (Figure 5-18). This is sometimes used in rappelling when the route must change left or right in the middle of the rappel. A self-equalizing anchor should only be used when necessary because if any one of the individual points fail, the anchor will extend and shock-load the remaining points or even cause complete anchor failure.

Figure 5-18. Self-equalizing anchors.

b. **Pre-equalized Anchor.** A pre-equalized anchor distributes the load equally to each individual point (Figure 5-19). It is aimed in the direction of the load. A pre-equalized anchor prevents extension and shock-loading of the anchor if an individual point fails. An anchor is pre-equalized by tying an overhand or figure-eight knot in the webbing or sling.

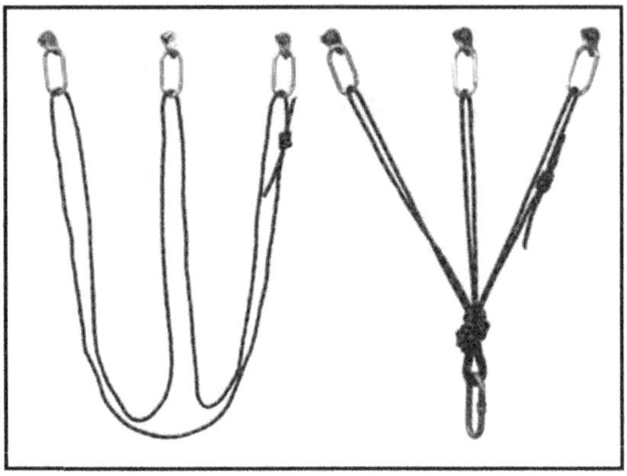

Figure 5-19. Pre-equalized anchor.

Note: When using webbing or slings, the angles of the webbing or slings directly affect the load placed on an anchor. An angle greater than 90 degrees can result in anchor failure (Figure 5-20).

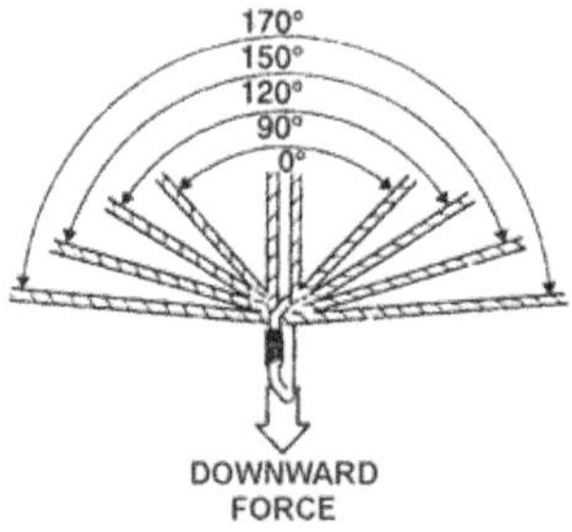

TENSION IN MULTIPLE ANCHOR RIGGING	
ANGLE	RESULTING LEG TENSION*
170°	1150%
150°	200%
120°	100%
90°	70%
0°	50%
*On each leg relative to downward force.	

Figure 5-20. Effects of angles on an anchor.

CHAPTER 6

CLIMBING

A steep rock face is a terrain feature that can be avoided most of the time through prior planning and good route selection. Rock climbing can be time consuming, especially for a larger unit with a heavy combat load. It can leave the climbing party totally exposed to weather, terrain hazards, and the enemy for the length of the climb.

Sometimes steep rock cannot be avoided. Climbing relatively short sections of steep rock (one or two pitches) may prove quicker and safer than using alternate routes. A steep rock route would normally be considered an unlikely avenue of approach and, therefore, might be weakly defended or not defended at all.

All personnel in a unit preparing for deployment to mountainous terrain should be trained in the basics of climbing. Forward observers, reconnaissance personnel, and security teams are a few examples of small units who may require rock climbing skills to gain their vantage points in mountainous terrain. Select personnel demonstrating the highest degree of skill and experience should be trained in roped climbing techniques. These personnel will have the job of picking and "fixing" the route for the rest of the unit.

Rock climbing has evolved into a specialized "sport" with a wide range of varying techniques and styles. This chapter focuses on the basics most applicable to military operations.

Section I. CLIMBING FUNDAMENTALS

A variety of refined techniques are used to climb different types of rock formations. The foundation for all of these styles is the art of climbing. Climbing technique stresses climbing with the weight centered over the feet, using the hands primarily for balance. It can be thought of as a combination of the balanced movement required to walk a tightrope and the technique used to ascend a ladder. No mountaineering equipment is required; however, the climbing technique is also used in roped climbing.

6-1. ROUTE SELECTION

The experienced climber has learned to climb with the "eyes." Even before getting on the rock, the climber studies all possible routes, or "lines," to the top looking for cracks, ledges, nubbins, and other irregularities in the rock that will be used for footholds and handholds, taking note of any larger ledges or benches for resting places. When picking the line, he mentally climbs the route, rehearsing the step-by-step sequence of movements that will be required to do the climb, ensuring himself that the route has an adequate number of holds and the difficulty of the climb will be well within the limit of his ability.

6-2. TERRAIN SELECTION FOR TRAINING

Route selection for military climbing involves picking the easiest and quickest possible line for all personnel to follow. However, climbing skill and experience can only be developed by increasing the length and difficulty of routes as training progresses. In the training environment, beginning lessons in climbing should be performed CLOSE to the ground on lower-angled rock with plenty of holds for the hands and feet. Personnel not climbing can act as "spotters" for those climbing. In later lessons, a "top-rope" belay can be used for safety, allowing the individual to increase the length and difficulty of the climb under the protection of the climbing rope.

6-3. PREPARATION

In preparation for climbing, the boot soles should be dry and clean. A small stick can be used to clean out dirt and small rocks that might be caught between the lugs of the boot sole. If the soles are wet or damp, dry them off by stomping and rubbing the soles on clean, dry rock. All jewelry should be removed from the fingers. Watches and bracelets can interfere with hand placements and may become damaged if worn while climbing. Helmets should be worn to protect the head from injury if an object, such as a rock or climbing gear, falls from climbers above. Most climbing helmets are not designed to provide protection from impact to the head if the wearer falls, but will provide a minimal amount of protection if a climber comes in contact with the rock during climbing.

CAUTION

Rings can become caught on rock facial features and or lodged into cracks, which could cause injuries during a slip or fall.

6-4. SPOTTING

Spotting is a technique used to add a level of safety to climbing without a rope. A second man stands below and just outside of the climbers fall path and helps (spots) the climber to land safely if he should fall. Spotting is only applicable if the climber is not going above the spotters head on the rock. Beyond that height a roped climbing should be conducted. If an individual climbs beyond the effective range of the spotter(s), he has climbed TOO HIGH for his own safety. The duties of the spotter are to help prevent the falling climber from impacting the head and or spine, help the climber land feet first, and reduce the impact of a fall.

CAUTION

The spotter should not catch the climber against the rock because additional injuries could result. If the spotter pushes the falling climber into the rock, deep abrasions of the skin or knee may occur. Ankle joints could be twisted by the fall if the climber's foot remained high on the rock. The spotter might be required to fully support the weight of the climber causing injury to the spotter.

6-5. CLIMBING TECHNIQUE

Climbing involves linking together a series of movements based on foot and hand placement, weight shift, and movement. When this series of movements is combined correctly, a smooth climbing technique results. This technique reduces excess force on the limbs, helping to minimize fatigue. The basic principle is based on the five body parts described here.

a. **Five Body Parts.** The five body parts used for climbing are the right hand, left hand, right foot, left foot, and body (trunk). The basic principle to achieve smooth climbing is to move only one of the five body parts at a time. The trunk is not moved in conjunction with a foot or in conjunction with a hand, a hand is not moved in conjunction with a foot, and so on. Following this simple technique forces both legs to do all the lifting simultaneously.

b. **Stance or Body Position.** Body position is probably the single most important element to good technique. A relaxed, comfortable stance is essential. (Figure 6-1 shows a correct climbing stance, and Figure 6-2, page 6-4, shows an incorrect stance.) The body should be in a near vertical or erect stance with the weight centered over the feet. Leaning in towards the rock will cause the feet to push outward, away from the rock, resulting in a loss of friction between the boot sole and rock surface. The legs are straight and the heels are kept low to reduce fatigue. Bent legs and tense muscles tire quickly. If strained for too long, tense muscles may vibrate uncontrollably. This vibration, known as "Elvis-ing" or "sewing-machine leg" can be cured by straightening the leg, lowering the heel, or moving on to a more restful position. The hands are used to maintain balance. Keeping the hands between waist and shoulder level will reduce arm fatigue.

Figure 6-1. Correct climbing stance—balanced over both feet.

Figure 6-2. Incorrect stance—stretched out.

(1) Whenever possible, three points of contact are maintained with the rock. Proper positioning of the hips and shoulders is critical. When using two footholds and one handhold, the hips and shoulders should be centered over both feet. In most cases, as the climbing progresses, the body is resting on one foot with two handholds for balance. The hips and shoulders must be centered over the support foot to maintain balance, allowing the "free" foot to maneuver.

(2) The angle or steepness of the rock also determines how far away from the rock the hips and shoulders should be. On low-angle slopes, the hips are moved out away from the rock to keep the body in balance with the weight over the feet. The shoulders can be moved closer to the rock to reach handholds. On steep rock, the hips are pushed closer to the rock. The shoulders are moved away from the rock by arching the back. The body is still in balance over the feet and the eyes can see where the hands need to go. Sometimes, when footholds are small, the hips are moved back to increase friction between the foot and the rock. This is normally done on quick, intermediate holds. It should be avoided in the rest position as it places more weight on the arms and hands. When weight must be placed on handholds, the arms should be kept straight to reduce fatigue. Again, flexed muscles tire quickly.

c. **Climbing Sequence.** The steps defined below provide a complete sequence of events to move the entire body on the rock. These are the basic steps to follow for a smooth climbing technique. Performing these steps in this exact order will not always be necessary because the nature of the route will dictate the availability of hand and foot placements. The basic steps are weight, shift, and movement (movement being either the foot, hand, or body). (A typical climbing sequence is shown in Figure 6-3, pages 6-6 through 6-8.)

STEP ONE: Shift the weight from both feet to one foot. This will allow lifting of one foot with no effect on the stance.

STEP TWO:	Lift the unweighted foot and place it in a new location, within one to two feet of the starting position, with no effect on body position or balance (higher placement will result in a potentially higher lift for the legs to make, creating more stress, and is called a high step) The trunk does not move during foot movement.
STEP THREE:	Shift the weight onto both feet. (Repeat steps 1 through 3 for remaining foot.)
STEP FOUR:	Lift the body into a new stance with both legs.
STEP FIVE:	Move one hand to a new position between waist and head height. During this movement, the trunk should be completely balanced in position and the removed hand should have no effect on stability.
STEP SIX:	Move the remaining hand as in Step 5.

Now the entire body is in a new position and ready to start the process again. Following these steps will prevent lifting with the hands and arms, which are used to maintain stance and balance. If both legs are bent, leg extension can be performed as soon as one foot has been moved. Hand movements can be delayed until numerous foot movements have been made, which not only creates shorter lifts with the legs, but may allow a better choice for the next hand movements because the reach will have increased.

Figure 6-3. Typical climbing sequence.

Figure 6-3. Typical climbing sequence (continued).

Figure 6-3. Typical climbing sequence (continued).

(1) Many climbers will move more than one body part at a time, usually resulting in lifting the body with one leg or one leg and both arms. This type of lifting is inefficient, requiring one leg to perform the work of two or using the arms to lift the body. Proper climbing technique is lifting the body with the legs, not the arms, because the legs are much stronger.

(2) When the angle of the rock increases, these movements become more critical. Holding or pulling the body into the rock with the arms and hands may be necessary as the angle increases (this is still not lifting with the arms). Many climbing routes have angles greater than ninety degrees (overhanging) and the arms are used to support partial body weight. The same technique applies even at those angles.

(3) The climber should avoid moving on the knees and elbows. Other than being uncomfortable, even painful, to rest on, these bony portions of the limbs offer little friction and "feel" on the rock.

6-6. SAFETY PRECAUTIONS

The following safety precautions should be observed when rock climbing.

a. While ascending a seldom or never traveled route, you may encounter precariously perched rocks. If the rock will endanger your second, it may be possible to remove it from the route and trundle it, tossing it down. This is extremely dangerous to climbers below and should not be attempted unless you are absolutely sure no men are below. If you are not sure that the flight path is clear, do not do it. Never dislodge loose rocks carelessly. Should a rock become loose accidentally, immediately shout the warning "ROCK" to alert climbers below. Upon hearing the warning, personnel should seek immediate cover behind any rock bulges or overhangs available, or flatten themselves against the rock to minimize exposure.

b. Should a climber fall, he should do his utmost to maintain control and not panic. If on a low-angle climb, he may be able to arrest his own fall by staying in contact with the rock, grasping for any possible hold available. He should shout the warning "FALLING" to alert personnel below.

CAUTION

Grasping at the rock in a fall can result in serious injuries to the upper body. If conducting a roped climb, let the rope provide protection.

c. When climbing close to the ground and without a rope, a spotter can be used for safety. The duties of the spotter are to ensure the falling climber does not impact the head or spine, and to reduce the impact of a fall.

d. Avoid climbing directly above or below other climbers (with the exception of spotters). When personnel must climb at the same time, following the same line, a fixed rope should be installed.

e. Avoid climbing with gloves on because of the decreased "feel" for the rock. The use of gloves in the training environment is especially discouraged, while their use in the mountains is often mandatory when it is cold. A thin polypropylene or wool glove is best for rock climbing, although heavier cotton or leather work gloves are often used for belaying.

f. Be extremely careful when climbing on wet or moss-covered rock; friction on holds is greatly reduced.

g. Avoid grasping small vegetation for handholds; the root systems can be shallow and will usually not support much weight.

6-7. MARGIN OF SAFETY

Besides observing the standard safety precautions, the climber can avoid catastrophe by climbing with a wide margin of safety. The margin of safety is a protective buffer the climber places between himself and potential climbing hazards. Both subjective (personnel-related) and objective (environmental) hazards must be considered when applying the margin of safety. The leader must apply the margin of safety taking into account the strengths and weaknesses of the entire team or unit.

a. When climbing, the climber increases his margin of safety by selecting routes that are well within the limit of his ability. When leading a group of climbers, he selects a route well within the ability of the weakest member.

b. When the rock is wet, or when climbing in other adverse weather conditions, the climber's ability is reduced and routes are selected accordingly. When the climbing becomes difficult or exposed, the climber knows to use the protection of the climbing rope and belays. A lead climber increases his margin of safety by placing protection along the route to limit the length of a potential fall.

Section II. USE OF HOLDS

The climber should check each hold before use. This may simply be a quick, visual inspection if he knows the rock to be solid. When in doubt, he should grab and tug on the hold to test it for soundness BEFORE depending on it. Sometimes, a hold that appears weak can actually be solid as long as minimal force is applied to it, or the force is applied in a direction that strengthens it. A loose nubbin might not be strong enough to support the climber's weight, but it may serve as an adequate handhold. Be especially careful when climbing on weathered, sedimentary-type rock.

6-8. CLIMBING WITH THE FEET

"Climb with the feet and use the hands for balance" is extremely important to remember. In the early learning stages of climbing, most individuals will rely heavily on the arms, forgetting to use the feet properly. It is true that solid handholds and a firm grip are needed in some combination techniques; however, even the most strenuous techniques require good footwork and a quick return to a balanced position over one or both feet. Failure to climb any route, easy or difficult, is usually the result of poor footwork.

a. The beginning climber will have a natural tendency to look up for handholds. Try to keep the hands low and train your eyes to look down for footholds. Even the smallest irregularity in the rock can support the climber once the foot is positioned properly and weight is committed to it.

b. The foot remains on the rock as a result of friction. Maximum friction is obtained from a correct stance over a properly positioned foot. The following describes a few ways the foot can be positioned on the rock to maximize friction.

(1) ***Maximum Sole Contact.*** The principle of using full sole contact, as in mountain walking, also applies in climbing. Maximum friction is obtained by placing as much of the boot sole on the rock as possible. Also, the leg muscles can relax the most when the entire foot is placed on the rock. (Figure 6-4 shows examples of maximum and minimum sole contact.)

(a) Smooth, low-angled rock (slab) and rock containing large "bucket" holds and ledges are typical formations where the entire boot sole should be used.

(b) On some large holds, like bucket holds that extend deep into the rock, the entire foot cannot be used. The climber may not be able to achieve a balanced position if the foot is stuck too far underneath a bulge in the rock. In this case, placing only part of the foot on the hold may allow the climber to achieve a balanced stance. The key is to use as much of the boot sole as possible. Remember to keep the heels low to reduce strain on the lower leg muscles.

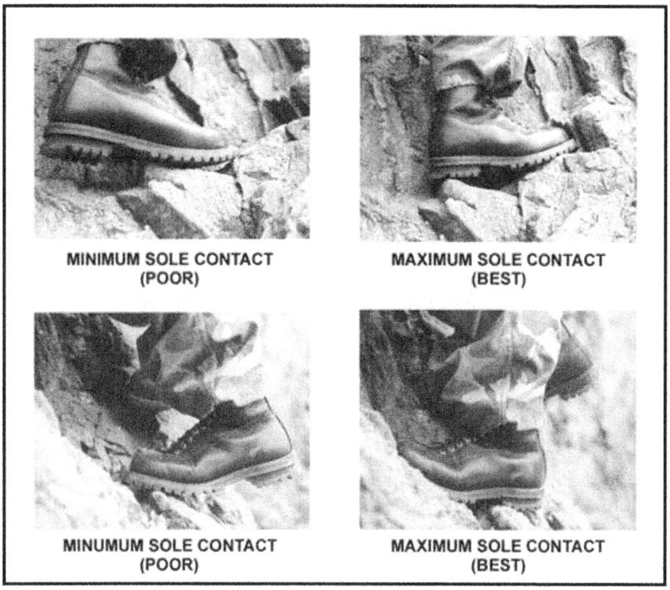

Figure 6-4. Examples of maximum and minimum sole contact.

(2) ***Edging.*** The edging technique is used where horizontal crack systems and other irregularities in the rock form small, well-defined ledges. The edge of the boot sole is placed on the ledge for the foothold. Usually, the inside edge of the boot or the edge area around the toes is used. Whenever possible, turn the foot sideways and use the entire inside edge of the boot. Again, more sole contact equals more friction and the legs can rest more when the heel is on the rock. (Figure 6-5, page 6-12, shows examples of the edging technique.)

(a) On smaller holds, edging with the front of the boot, or toe, may be used. Use of the toe is most tiring because the heel is off the rock and the toes support the climber's weight. Remember to keep the heel low to reduce fatigue. Curling and stiffening the toes in the boot increases support on the hold. A stronger position is usually obtained on small ledges by turning the foot at about a 45-degree angle, using the strength of the big toe and the ball of the foot.

(b) Effective edging on small ledges requires stiff-soled footwear. The stiffer the sole, the better the edging capability. Typical mountain boots worn by the US military have a relatively flexible lugged sole and, therefore, edging ability on smaller holds will be somewhat limited.

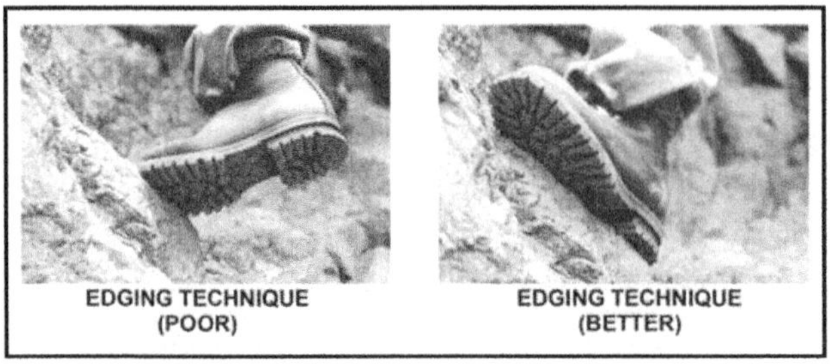

Figure 6-5. Examples of edging technique.

(3) ***Smearing.*** When footholds are too small to use a good edging technique, the ball of the foot can be "smeared" over the hold. The smearing technique requires the boot to adhere to the rock by deformation of the sole and by friction. Rock climbing shoes are specifically designed to maximize friction for smearing; some athletic shoes also work well. The Army mountain boot, with its softer sole, usually works better for smearing than for edging. Rounded, down-sloping ledges and low-angled slab rock often require good smearing technique. (Figure 6-6 shows examples of the smearing technique.)

(a) Effective smearing requires maximum friction between the foot and the rock. Cover as much of the hold as possible with the ball of the foot. Keeping the heel low will not only reduce muscle strain, but will increase the amount of surface contact between the foot and the rock.

(b) Sometimes flexing the ankles and knees slightly will place the climber's weight more directly over the ball of the foot and increase friction; however, this is more tiring and should only be used for quick, intermediate holds. The leg should be kept straight whenever possible.

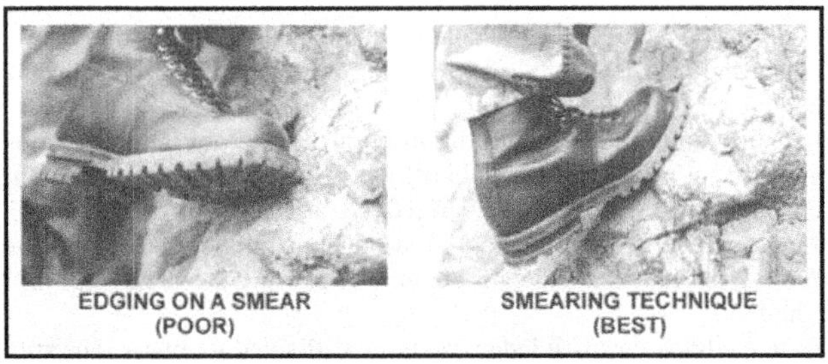

Figure 6-6. Examples of the smearing technique.

(4) ***Jamming.*** The jamming technique works on the same principal as chock placement. The foot is set into a crack in such a way that it "jams" into place, resisting a downward pull. The jamming technique is a specialized skill used to climb vertical or

near vertical cracks when no other holds are available on the rock face. The technique is not limited to just wedging the feet; fingers, hands, arms, even the entire leg or body are all used in the jamming technique, depending on the size of the crack. Jam holds are described in this text to broaden the range of climbing skills. Jamming holds can be used in a crack while other hand/foot holds are used on the face of the rock. Many cracks will have facial features, such as edges, pockets, and so on, inside and within reach. Always look or feel for easier to use features. (Figure 6-7 shows examples of jamming.)

(a) The foot can be jammed in a crack in different ways. It can be inserted above a constriction and set into the narrow portion, or it can be placed in the crack and turned, like a camming device, until it locks in place tight enough to support the climber's weight. Aside from these two basic ideas, the possibilities are endless. The toes, ball of the foot, or the entire foot can be used. Try to use as much of the foot as possible for maximum surface contact. Some positions are more tiring, and even more painful on the foot, than others. Practice jamming the foot in various ways to see what offers the most secure, restful position.

(b) Some foot jams may be difficult to remove once weight has been committed to them, especially if a stiffer sole boot is used. The foot is less likely to get stuck when it is twisted or "cammed" into position. When removing the boot from a crack, reverse the way it was placed to prevent further constriction.

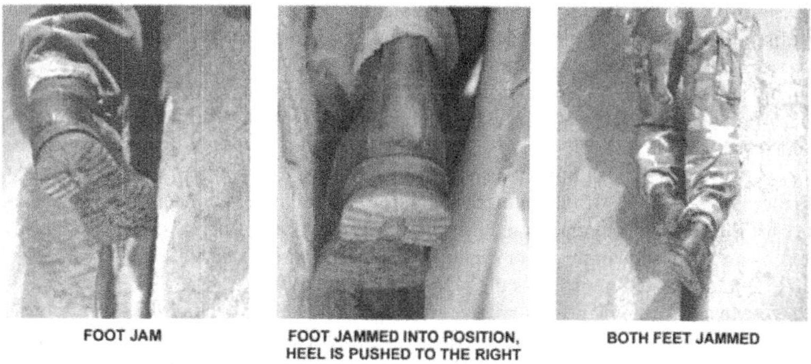

Figure 6-7. Examples of jamming.

6-9. USING THE HANDS

The hands can be placed on the rock in many ways. Exactly how and where to position the hands and arms depends on what holds are available, and what configuration will best support the current stance as well as the movement to the next stance. Selecting handholds between waist and shoulder level helps in different ways. Circulation in the arms and hands is best when the arms are kept low. Secondly, the climber has less tendency to "hang" on his arms when the handholds are at shoulder level and below. Both of these contribute to a relaxed stance and reduce fatigue in the hands and arms.

a. As the individual climbs, he continually repositions his hands and arms to keep the body in balance, with the weight centered over the feet. On lower-angled rock, he may simply need to place the hands up against the rock and extend the arm to maintain balance; just like using an ice ax as a third point of contact in mountain walking. Sometimes, he will be able to push directly down on a large hold with the palm of the hand. More often though, he will need to "grip" the rock in some fashion and then push or pull against the hold to maintain balance.

b. As stated earlier, the beginner will undoubtedly place too much weight on the hands and arms. If we think of ourselves climbing a ladder, our body weight is on our legs. Our hands grip, and our arms pull on each rung only enough to maintain our balance and footing on the ladder. Ideally, this is the amount of grip and pull that should be used in climbing. Of course, as the size and availability of holds decreases, and the steepness of the rock approaches the vertical, the grip must be stronger and more weight might be placed on the arms and handholds for brief moments. The key is to move quickly from the smaller, intermediate holds to the larger holds where the weight can be placed back on the feet allowing the hands and arms to relax. The following describes some of the basic handholds and how the hand can be positioned to maximize grip on smaller holds.

(1) ***Push Holds.*** Push holds rely on the friction created when the hand is pushed against the rock. Most often a climber will use a push hold by applying "downward pressure" on a ledge or nubbin. This is fine, and works well; however, the climber should not limit his use of push holds to the application of down pressure. Pushing sideways, and on occasion, even upward on less obvious holds can prove quite secure. Push holds often work best when used in combination with other holds. Pushing in opposite directions and "push-pull" combinations are excellent techniques. (Figure 6-8 shows examples of push holds.)

(a) An effective push hold does not necessarily require the use of the entire hand. On smaller holds, the side of the palm, the fingers, or the thumb may be all that is needed to support the stance. Some holds may not feel secure when the hand is initially placed on them. The hold may improve or weaken during the movement. The key is to try and select a hold that will improve as the climber moves past it.

(b) Most push holds do not require much grip; however, friction might be increased by taking advantage of any rough surfaces or irregularities in the rock. Sometimes the strength of the hold can be increased by squeezing, or "pinching," the rock between the thumb and fingers (see paragraph on pinch holds).

Figure 6-8. Examples of push holds.

(2) ***Pull Holds.*** Pull holds, also called "cling holds," which are grasped and pulled upon, are probably the most widely used holds in climbing. Grip plays more of a role in a pull hold, and, therefore, it normally feels more secure to the climber than a push hold. Because of this increased feeling of security, pull holds are often overworked. These are the holds the climber has a tendency to hang from. Most pull holds do not require great strength, just good technique. Avoid the "death grip" syndrome by climbing with the feet. (Figure 6-9, page 6-16, shows examples of pull holds.)

(a) Like push holds, pressure on a pull hold can be applied straight down, sideways, or upward. Again, these are the holds the climber tends to stretch and reach for, creating an unbalanced stance. Remember to try and keep the hands between waist and shoulder level, making use of intermediate holds instead of reaching for those above the head.

(b) Pulling sideways on vertical cracks can be very secure. There is less tendency to hang from "side-clings" and the hands naturally remain lower. The thumb can often push against one side of the crack, in opposition to the pull by the fingers, creating a stronger hold. Both hands can also be placed in the same crack, with the hands pulling in opposite directions. The number of possible combinations is limited only by the imagination and experience of the climber.

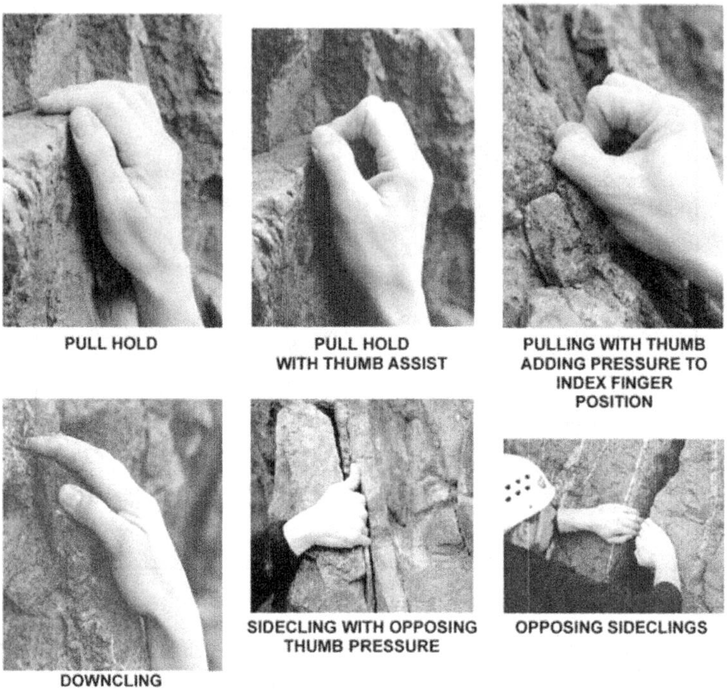

Figure 6-9. Examples of pull holds.

(c) Friction and strength of a pull hold can be increased by the way the hand grips the rock. Normally, the grip is stronger when the fingers are closed together; however, sometimes more friction is obtained by spreading the fingers apart and placing them between irregularities on the rock surface. On small holds, grip can often be improved by bending the fingers upward, forcing the palm of the hand to push against the rock. This helps to hold the finger tips in place and reduces muscle strain in the hand. Keeping the forearm up against the rock also allows the arm and hand muscles to relax more.

(d) Another technique that helps to strengthen a cling hold for a downward pull is to press the thumb against the side of the index finger, or place it on top of the index finger and press down. This hand configuration, known as a "ring grip," works well on smaller holds.

(3) ***Pinch Holds.*** Sometimes a small nubbin or protrusion in the rock can be "squeezed" between the thumb and fingers. This technique is called a pinch hold. Friction is applied by increasing the grip on the rock. Pinch holds are often overlooked by the

novice climber because they feel insecure at first and cannot be relied upon to support much body weight. If the climber has his weight over his feet properly, the pinch hold will work well in providing balance. The pinch hold can also be used as a gripping technique for push holds and pull holds. (Figure 6-10 shows examples of pinch holds.)

Figure 6-10. Examples of pinch holds.

(4) ***Jam Holds.*** Like foot jams, the fingers and hands can be wedged or cammed into a crack so they resist a downward or outward pull. Jamming with the fingers and hands can be painful and may cause minor cuts and abrasions to tender skin. Cotton tape can be used to protect the fingertips, knuckles, and the back of the hand; however, prolonged jamming technique requiring hand taping should be avoided. Tape also adds friction to the hand in jammed position. (Figure 6-11, page 6-18, shows examples of jam holds.)

(a) The hand can be placed in a crack a number of ways. Sometimes an open hand can be inserted and wedged into a narrower portion of the crack. Other times a clenched fist will provide the necessary grip. Friction can be created by applying cross pressure between the fingers and the back of the hand. Another technique for vertical cracks is to place the hand in the crack with the thumb pointed either up or down. The hand is then clenched as much as possible. When the arm is straightened, it will twist the hand and tend to cam it into place. This combination of clenching and camming usually produces the most friction, and the most secure hand jam in vertical cracks.

(b) In smaller cracks, only the fingers will fit. Use as many fingers as the crack will allow. The fingers can sometimes be stacked in some configuration to increase friction. The thumb is usually kept outside the crack in finger jams and pressed against the rock to increase friction or create cross pressure. In vertical cracks it is best to insert the fingers with the thumb pointing down to make use of the natural camming action of the fingers that occurs when the arm is twisted towards a normal position.

(c) Jamming technique for large cracks, or "off widths," requiring the use of arm, leg, and body jams, is another technique. To jam or cam an arm, leg, or body into an off width, the principle is the same as for fingers, hands, or feet—you are making the jammed appendage "fatter" by folding or twisting it inside the crack. For off widths, you may place your entire arm inside the crack with the arm folded and the palm pointing outward. The leg can be used, from the calf to the thigh, and flexed to fit the crack. Routes requiring this type of climbing should be avoided as the equipment normally used for protection might not be large enough to protect larger cracks and openings. However, sometimes a narrower section may be deeper in the crack allowing the use of "normal" size protection.

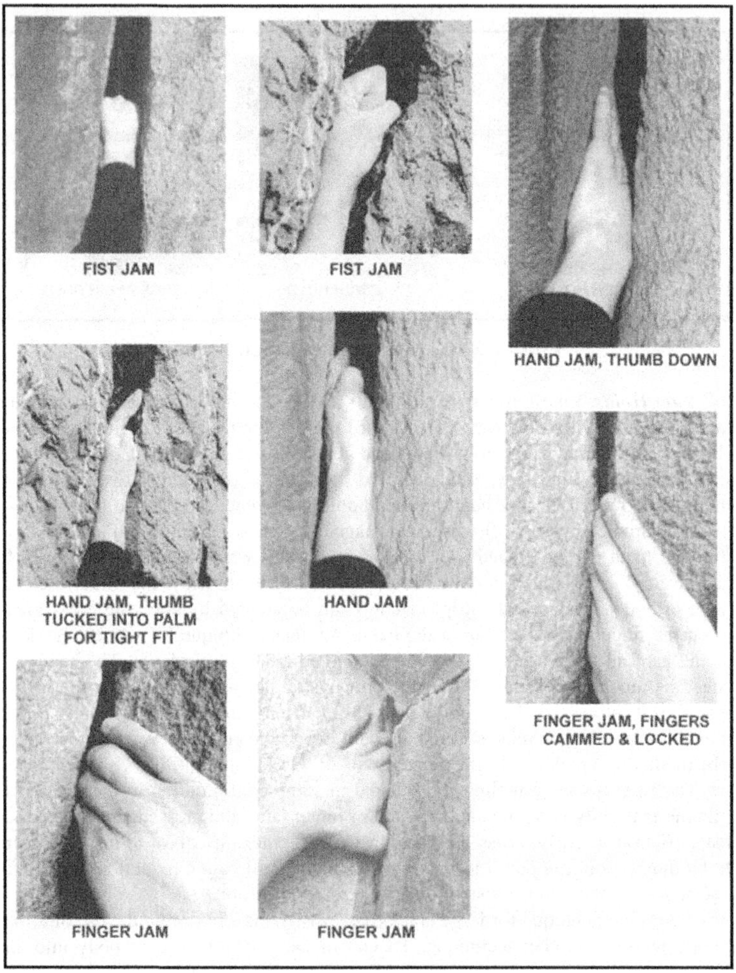

Figure 6-11. Examples of jam holds.

6-10. COMBINATION TECHNIQUES

The positions and holds previously discussed are the basics and the ones most common to climbing. From these fundamentals, numerous combination techniques are possible. As the climber gains experience, he will learn more ways to position the hands, feet, and body in relation to the holds available; however, he should always strive to climb with his weight on his feet from a balanced stance.

a. Sometimes, even on an easy route, the climber may come upon a section of the rock that defies the basic principles of climbing. Short of turning back, the only alternative is to figure out some combination technique that will work. Many of these type problems require the hands and feet to work in opposition to one another. Most will place more weight on the hands and arms than is desirable, and some will put the climber in an "out of balance" position. To make the move, the climber may have to "break the rules" momentarily. This is not a problem and is done quite frequently by experienced climbers. The key to using these type of combination techniques is to plan and execute them deliberately, without lunging or groping for holds, yet quickly, before the hands, arms, or other body parts tire. Still, most of these maneuvers require good technique more than great strength, though a certain degree of hand and arm strength certainly helps.

b. Combination possibilities are endless. The following is a brief description of some of the more common techniques.

(1) ***Change Step.*** The change step, or hop step, can be used when the climber needs to change position of the feet. It is commonly used when traversing to avoid crossing the feet, which might put the climber in an awkward position. To prevent an off balance situation, two solid handholds should be used. The climber simply places his weight on his handholds while he repositions the feet. He often does this with a quick "hop," replacing the lead foot with the trail foot on the same hold. Keeping the forearms against the rock during the maneuver takes some of the strain off the hands, while at the same time strengthening the grip on the holds.

(2) ***Mantling.*** Mantling is a technique that can be used when the distance between the holds increases and there are no immediate places to move the hands or feet. It does require a ledge (mantle) or projection in the rock that the climber can press straight down upon. (Figure 6-12, page 6-20, shows the mantling sequence.)

(a) When the ledge is above head height, mantling begins with pull holds, usually "hooking" both hands over the ledge. The climber pulls himself up until his head is above the hands, where the pull holds become push holds. He elevates himself until the arms are straight and he can lock the elbows to relax the muscles. Rotating the hands inward during the transition to push holds helps to place the palms more securely on the ledge. Once the arms are locked, a foot can be raised and placed on the ledge. The climber may have to remove one hand to make room for the foot. Mantling can be fairly strenuous; however, most individuals should be able to support their weight, momentarily, on one arm if they keep it straight and locked. With the foot on the ledge, weight can be taken off the arms and the climber can grasp the holds that were previously out of reach. Once balanced over the foot, he can stand up on the ledge and plan his next move.

(b) Pure mantling uses arm strength to raise the body; however, the climber can often smear the balls of the feet against the rock and "walk" the feet up during the maneuver to take some of the weight off the arms. Sometimes edges will be available for short steps in the process.

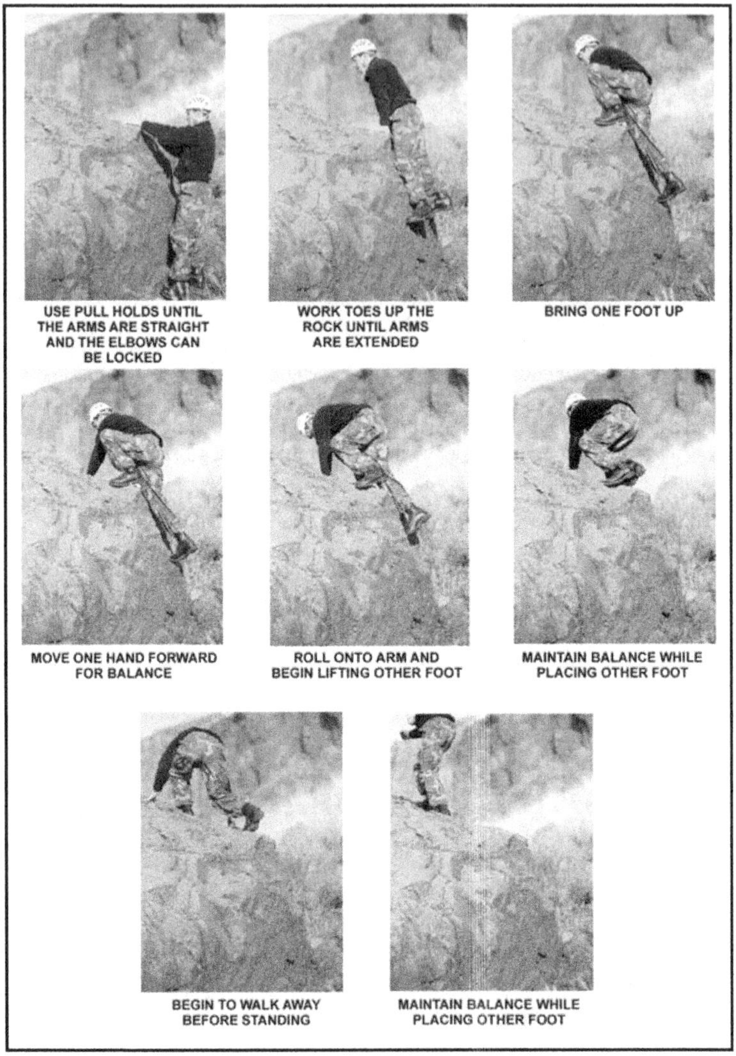

Figure 6-12. Mantling sequence.

(3) ***Undercling.*** An "undercling" is a classic example of handholds and footholds working in opposition (Figure 6-13). It is commonly used in places where the rock

projects outward, forming a bulge or small overhang. Underclings can be used in the tops of buckets, also. The hands are placed "palms-up" underneath the bulge, applying an upward pull. Increasing this upward pull creates a counterforce, or body tension, which applies more weight and friction to the footholds. The arms and legs should be kept as straight as possible to reduce fatigue. The climber can often lean back slightly in the undercling position, enabling him to see above the overhang better and search for the next hold.

Figure 6-13. Undercling.

(4) ***Lieback.*** The "lieback" is another good example of the hands working in opposition to the feet. The technique is often used in a vertical or diagonal crack separating two rock faces that come together at, more or less, a right angle (commonly referred to as a dihedral). The crack edge closest to the body is used for handholds while the feet are pressed against the other edge. The climber bends at the waist, putting the body into an L-shaped position. Leaning away from the crack on two pull holds, body tension creates friction between the feet and the hands. The feet must be kept relatively high to maintain weight, creating maximum friction between the sole and the rock surface. Either full sole contact or the smearing technique can be used, whichever seems to produce the most friction.

(a) The climber ascends a dihedral by alternately shuffling the hands and feet upward. The lieback technique can be extremely tiring, especially when the dihedral is near vertical. If the hands and arms tire out before completing the sequence, the climber will likely fall. The arms should be kept straight throughout the entire maneuver so the climber's weight is pulling against bones and ligaments, rather than muscle. The legs should be straightened whenever possible.

(b) Placing protection in a lieback is especially tiring. Look for edges or pockets for the feet in the crack or on the face for a better position to place protection from, or for a rest position. Often, a lieback can be avoided with closer examination of the available face features. The lieback can be used alternately with the jamming technique, or vice versa, for variation or to get past a section of a crack with difficult or nonexistent jam possibilities. The lieback can sometimes be used as a face maneuver (Figure 6-14).

Figure 6-14. Lieback on a face.

(5) ***Stemming.*** When the feet work in opposition from a relatively wide stance, the maneuver is known as stemming. The stemming technique can sometimes be used on faces, as well as in a dihedral in the absence of solid handholds for the lieback (Figure 6-15).

Figure 6-15. Stemming on a face.

(a) The classic example of stemming is when used in combination with two opposing push holds in wide, parallel cracks, known as chimneys. Chimneys are cracks in which the walls are at least 1 foot apart and just big enough to squeeze the body into. Friction is created by pushing outward with the hands and feet on each side of the crack. The climber ascends the chimney by alternately moving the hands and feet up the crack (Figure 6-16). Applying pressure with the back and bottom is usually necessary in wider chimneys. Usually, full sole contact of the shoes will provide the most friction, although smearing may work best in some instances. Chimneys that do not allow a full stemming position can be negotiated using the arms, legs, or body as an integral contact point. This technique will often feel more secure since there is more body to rock contact.

Figure 6-16. Chimney sequence.

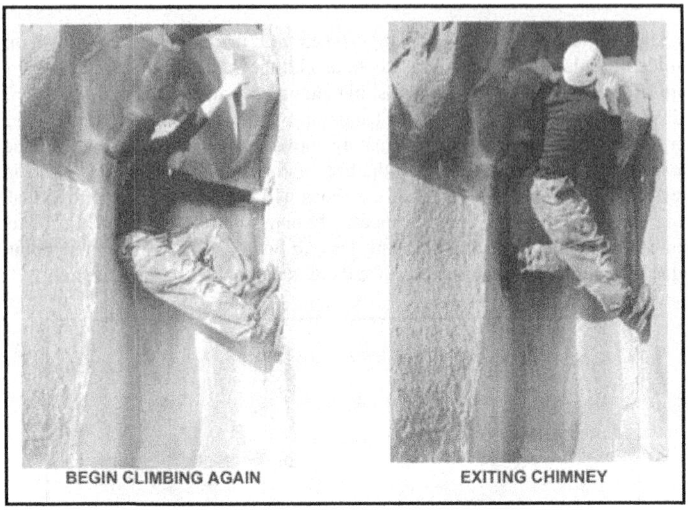

Figure 6-16. Chimney sequence (continued).

(b) The climber can sometimes rest by placing both feet on the same side of the crack, forcing the body against the opposing wall. The feet must be kept relatively high up under the body so the force is directed sideways against the walls of the crack. The arms should be straightened with the elbows locked whenever possible to reduce muscle strain. The climber must ensure that the crack does not widen beyond the climbable width before committing to the maneuver. Remember to look for face features inside chimneys for more security in the climb.

(c) Routes requiring this type of climbing should be avoided as the equipment normally used for protection might not be large enough to protect chimneys. However, face features, or a much narrower crack in one or both corners, may sometimes be found deeper in the chimney allowing the use of normal size protection.

(6) ***Slab Technique.*** A slab is a relatively smooth, low-angled rock formation that requires a slightly modified climbing technique (Figure 6-17). Since slab rock normally contains few, if any holds, the technique requires maximum friction and perfect balance over the feet.

(a) On lower-angled slab, the climber can often stand erect and climb using full sole contact and other mountain walking techniques. On steeper slab, the climber will need to apply good smearing technique. Often, maximum friction cannot be attained on steeper slab from an erect stance. The climber will have to flex the ankles and knees so his weight is placed more directly over the balls of the feet. He may then have to bend at the waist to place the hands on the rock, while keeping the hips over his feet.

(b) The climber must pay attention to any changes in slope angle and adjust his body accordingly. Even the slightest change in the position of the hips over the feet can mean the difference between a good grip or a quick slip. The climber should also take advantage of any rough surfaces, or other irregularities in the rock he can place his hands or feet on, to increase friction.

Figure 6-17. Slab technique.

(7) ***Down Climbing.*** Descending steep rock is normally performed using a roped method; however, the climber may at some point be required to down climb a route. Even if climbing ropes and related equipment are on hand, down climbing easier terrain is often quicker than taking the time to rig a rappel point. Also, a climber might find himself confronted with difficulties part way up a route that exceed his climbing ability, or the abilities of others to follow. Whatever the case may be, down climbing is a skill well worth practicing.

CAUTIONS

1. Down climbing can inadvertently lead into an unforeseen dangerous position on a descent. When in doubt, use a roped descent.
2. Down climbing is accomplished at a difficulty level well below the ability of the climber. When in doubt, use a roped descent.

(a) On easier terrain, the climber can face outward, away from the rock, enabling him to see the route better and descend quickly. As the steepness and difficulty increase, he can often turn sideways, still having a good view of the descent route, but being better able to use the hands and feet on the holds available. On the steepest terrain, the climber will have to face the rock and down climb using good climbing techniques.

(b) Down climbing is usually more difficult than ascending a given route. Some holds will be less visible when down climbing, and slips are more likely to occur. The climber must often lean well away from the rock to look for holds and plan his movements. More weight is placed on the arms and handholds at times to accomplish this, as well as to help lower the climber to the next foothold. Hands should be moved to holds as low as waist level to give the climber more range of movement with each step. If the handholds are too high, he may have trouble reaching the next foothold. The climber must be careful not to overextend himself, forcing a release of his handholds before reaching the next foothold.

CAUTION

Do not drop from good handholds to a standing position. A bad landing could lead to injured ankles or a fall beyond the planned landing area.

(c) Descending slab formations can be especially tricky. The generally lower angle of slab rock may give the climber a false sense of security, and a tendency to move too quickly. Down climbing must be slow and deliberate, as in ascending, to maintain perfect balance and weight distribution over the feet. On lower-angle slab the climber may be able to stand more or less erect, facing outward or sideways, and descend using good flat foot technique. The climber should avoid the tendency to move faster, which can lead to uncontrollable speed.

(d) On steeper slab, the climber will normally face the rock and down climb, using the same smearing technique as for ascending. An alternate method for descending slab is to face away from the rock in a "crab" position (Figure 6-18). Weight is still concentrated over the feet, but may be shifted partly onto the hands to increase overall friction. The climber is able to maintain full sole contact with the rock and see the entire descent route. Allowing the buttocks to "drag behind" on the rock will decrease the actual weight on the footholds, reducing friction, and leading to the likelihood of a slip. Facing the rock, and down-climbing with good smearing technique, is usually best on steeper slab.

Figure 6-18. Descending slab in the crab position.

Section III. ROPED CLIMBING

When the angle, length, and difficulty of the proposed climbing route surpasses the ability of the climbers' safety margin (possibly on class 4 and usually on class 5 terrain), ropes must be used to proceed. Roped climbing is only safe if accomplished correctly. Reading this manual does not constitute skill with ropes—much training and practice is necessary. Many aspects of roped climbing take time to understand and learn. Ropes are normally not used in training until the basic principles of climbing are covered.

Note: A rope is completely useless for climbing unless the climber knows how to use it safely.

6-11. TYING-IN TO THE CLIMBING ROPE

Over the years, climbers have developed many different knots and procedures for tying-in to the climbing rope. Some of the older methods of tying directly into the rope require minimal equipment and are relatively easy to inspect; however, they offer little support to the climber, may induce further injuries, and may even lead to strangulation in a severe fall. A severe fall, where the climber might fall 20 feet or more and be left dangling on the end of the rope, is highly unlikely in most instances, especially for most personnel involved in military climbing. Tying directly into the rope is perfectly safe for many roped party climbs used in training on lower-angled rock. All climbers should know how to properly tie the rope around the waist in case a climbing harness is unavailable.

6-12. PRESEWN HARNESSES

Although improvised harnesses are made from readily available materials and take little space in the pack or pocket, presewn harnesses provide other aspects that should be considered. No assembly is required, which reduces preparation time for roped movement. All presewn harnesses provide a range of adjustability. These harnesses have a fixed buckle that, when used correctly, will not fail before the nylon materials connected to it. However, specialized equipment, such as a presewn harness, reduce the flexibility of gear. Presewn harness are bulky, also.

a. **Seat Harness.** Many presewn seat harnesses are available with many different qualities separating them, including cost.

(1) The most notable difference will be the amount and placement of padding. The more padding the higher the price and the more comfort. Gear loops sewn into the waist belt on the sides and in the back are a common feature and are usually strong enough to hold quite a few carabiners and or protection. The gear loops will vary in number from one model/manufacturer to another.

(2) Although most presewn seat harnesses have a permanently attached belay loop connecting the waist belt and the leg loops, the climbing rope should be run around the waist belt and leg loop connector. The presewn belay loop adds another link to the chain of possible failure points and only gives one point of security whereas running the rope through the waist belt and leg loop connector provides two points of contact.

(3) If more than two men will be on the rope, connect the middle position(s) to the rope with a carabiner routed the same as stated in the previous paragraph.

(4) Many manufactured seat harnesses will have a presewn loop of webbing on the rear. Although this loop is much stronger than the gear loops, it is not for a belay anchor. It is a quick attachment point to haul an additional rope.

b. **Chest Harness.** The chest harness will provide an additional connecting point for the rope, usually in the form of a carabiner loop to attach a carabiner and rope to. This type of additional connection will provide a comfortable hanging position on the rope, but otherwise provides no additional protection from injury during a fall (if the seat harness is fitted correctly).

(1) A chest harness will help the climber remain upright on the rope during rappelling or ascending a fixed rope, especially while wearing a heavy pack. (If rappelling or ascending long or multiple pitches, let the pack hang on a drop cord below the feet and attached to the harness tie-in point.)

(2) The presewn chest harnesses available commercially will invariably offer more comfort or performance features, such as padding, gear loops, or ease of adjustment, than an improvised chest harness.

c. **Full-Body Harness.** Full-body harnesses incorporate a chest and seat harness into one assembly. This is the safest harness to use as it relocates the tie-in point higher, at the chest, reducing the chance of an inverted position when hanging on the rope. This is especially helpful when moving on ropes with heavy packs. A full-body harness only affects the body position when hanging on the rope and will not prevent head injury in a fall.

CAUTION

This type of harness does not prevent the climber from falling head first. Body position during a fall is affected only by the forces that generated the fall, and this type of harness promotes an upright position only when hanging on the rope from the attachment point.

6-13. IMPROVISED HARNESSES

Without the use of a manufactured harness, many methods are still available for attaching oneself to a rope. Harnesses can be improvised using rope or webbing and knots.

a. **Swami Belt.** The swami belt is a simple, belt-only harness created by wrapping rope or webbing around the waistline and securing the ends. One-inch webbing will provide more comfort. Although an effective swami belt can be assembled with a minimum of one wrap, at least two wraps are recommended for comfort, usually with approximately ten feet of material. The ends are secured with an appropriate knot.

b. **Bowline-on-a-Coil.** Traditionally, the standard method for attaching oneself to the climbing rope was with a bowline-on-a-coil around the waist. The extra wraps distribute the force of a fall over a larger area of the torso than a single bowline would, and help prevent the rope from riding up over the rib cage and under the armpits. The knot must be tied snugly around the narrow part of the waist, just above the bony portions of the hips (pelvis). Avoid crossing the wraps by keeping them spread over the waist area. "Sucking in the gut" a bit when making the wraps will ensure a snug fit.

(1) The bowline-on-a-coil can be used to tie-in to the end of the rope (Figure 6-19). The end man should have a minimum of four wraps around the waist before completing the knot.

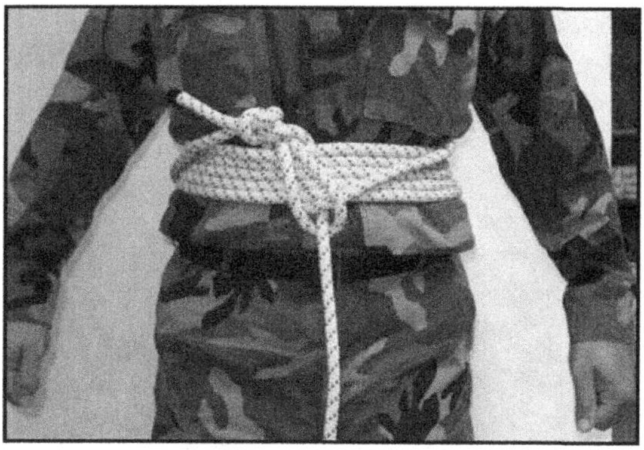

Figure 6-19. Tying-in with a bowline-on-a-coil.

(2) The bowline-on-a-coil is a safe and effective method for attaching to the rope when the terrain is low-angled, WITHOUT THE POSSIBILITY OF A SEVERE FALL. When the terrain becomes steeper, a fall will generate more force on the climber and this will be felt through the coils of this type of attachment. A hard fall will cause the coils to ride up against the ribs. In a severe fall, any tie-in around the waist only could place a "shock load" on the climber's lower back. Even in a relatively short fall, if the climber ends up suspended in mid-air and unable to regain footing on the rock, the rope around the waist can easily cut off circulation and breathing in a relatively short time.

(3) The climbing harness distributes the force of a fall over the entire pelvic region, like a parachute harness. Every climber should know how to tie some sort of improvised climbing harness from sling material. A safe, and comfortable, seat/chest combination harness can be tied from one-inch tubular nylon.

c. **Improvised Seat Harness.** A seat harness can be tied from a length of webbing approximately 25 feet long (Figure 6-20, page 6-30).

(1) Locate the center of the rope. Off to one side, tie two fixed loops approximately 6 inches apart (overhand loops). Adjust the size of the loops so they fit snugly around the thigh. The loops are tied into the sling "off center" so the remaining ends are different lengths. The short end should be approximately 4 feet long (4 to 5 feet for larger individuals).

(2) Slip the leg loops over the feet and up to the crotch, with the knots to the front. Make one complete wrap around the waist with the short end, wrapping to the outside, and hold it in place on the hip. Keep the webbing flat and free of twists when wrapping.

(3) Make two to three wraps around the waist with the long end in the opposite direction (wrapping to the outside), binding down on the short end to hold it in place. Grasping both ends, adjust the waist wraps to a snug fit. Connect the ends with the

appropriate knot between the front and one side so you will be able to see what you are doing.

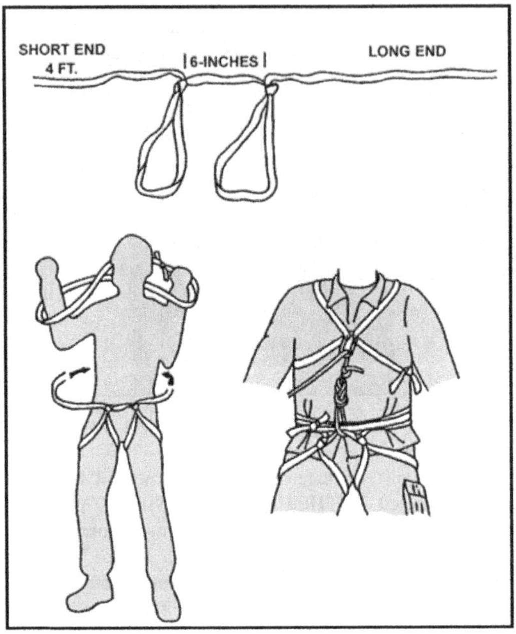

Figure 6-20. Improvised seat and chest harness.

d. **Improvised Chest Harness.** The chest harness can be tied from rope or webbing, but remember that with webbing, wider is better and will be more comfortable when you load this harness. Remember as you tie this harness that the remaining ends will need to be secured so choose the best length. Approximately 6 to 10 feet usually works.

(1) Tie the ends of the webbing together with the appropriate knot, making a sling 3 to 4 feet long.

(2) Put a single twist into the sling, forming two loops.

(3) Place an arm through each loop formed by the twist, just as you would put on a jacket, and drape the sling over the shoulders. The twist, or cross, in the sling should be in the middle of the back.

(4) Join the two loops at the chest with a carabiner. The water knot should be set off to either side for easy inspection (if a pack is to be worn, the knot will be uncomfortable if it gets between the body and the pack). The chest harness should fit just loose enough to allow necessary clothing and not to restrict breathing or circulation. Adjust the size of the sling if necessary.

e. **Improvised Full-Body Harness.** Full-body harnesses incorporate a chest and seat harness into one assembly.

(1) The full-body harness is the safest harness because it relocates the tie-in point higher, at the chest, reducing the chance of an inverted hanging position on the rope. This is especially helpful when moving on ropes with heavy packs. A full-body harness affects the body position only when hanging on the rope.

CAUTION

A full-body harness does not prevent falling head first; body position in a fall is caused by the forces that caused the fall.

(2) Although running the rope through the carabiner of the chest harness does, in effect, create a type of full-body harness, it is not a true full-body harness until the chest harness and the seat harness are connected as one piece. A true full-body harness can be improvised by connecting the chest harness to the seat harness, but not by just tying the rope into both—the two harnesses must be "fixed" as one harness. Fix them together with a short loop of webbing or rope so that the climbing rope can be connected directly to the chest harness and your weight is supported by the seat harness through the connecting material.

f. **Attaching the Rope to the Improvised Harness.** The attachment of the climbing rope to the harness is a CRITICAL LINK. The strength of the rope means nothing if it is attached poorly, or incorrectly, and comes off the harness in a fall. The climber ties the end of the climbing rope to the seat harness with an appropriate knot. If using a chest harness, the standing part of the rope is then clipped into the chest harness carabiner. The seat harness absorbs the main force of the fall, and the chest harness helps keep the body upright.

CAUTION

The knot must be tied around all the waist wraps and the 6-inch length of webbing between the leg loops.

(1) A middleman must create a fixed loop to tie in to. A rethreaded figure-eight loop tied on a doubled rope or the three loop bowline can be used. If using the three loop bowline, ensure the end, or third loop formed in the knot, is secured around the tie-in loops with an overhand knot. The standing part of the rope going to the lead climber is clipped into the chest harness carabiner.

Note: The climbing rope is not clipped into the chest harness when belaying.

(2) The choice of whether to tie-in with a bowline-on-a-coil or into a climbing harness depends entirely on the climber's judgment, and possibly the equipment

available. A good rule of thumb is: "Wear a climbing harness when the potential for severe falls exists and for all travel over snow-covered glaciers because of the crevasse fall hazard."

(3) Under certain conditions many climbers prefer to attach the rope to the seat harness with a locking carabiner, rather than tying the rope to it. This is a common practice for moderate snow and ice climbing, and especially for glacier travel where wet and frozen knots become difficult to untie.

CAUTION

Because the carabiner gate may be broken or opened by protruding rocks during a fall, tie the rope directly to the harness for maximum safety.

Section IV. BELAY TECHNIQUES

Tying-in to the climbing rope and moving as a member of a rope team increases the climber's margin of safety on difficult, exposed terrain. In some instances, such as when traveling over snow-covered glaciers, rope team members can often move at the same time, relying on the security of a tight rope and "team arrest" techniques to halt a fall by any one member. On steep terrain, however, simultaneous movement only helps to ensure that if one climber falls, he will jerk the other rope team members off the slope. For the climbing rope to be of any value on steep rock climbs, the rope team must incorporate "belays" into the movement.

Belaying is a method of managing the rope in such a way that, if one person falls, the fall can be halted or "arrested" by another rope team member (belayer). One person climbs at a time, while being belayed from above or below by another. The belayer manipulates the rope so that friction, or a "brake," can be applied to halt a fall. Belay techniques are also used to control the descent of personnel and equipment on fixed rope installations, and for additional safety on rappels and stream crossings.

Belaying is a skill that requires practice to develop proficiency. Setting up a belay may at first appear confusing to the beginner, but with practice, the procedure should become "second nature." If confronted with a peculiar problem during the setup of a belay, try to use common sense and apply the basic principles stressed throughout this text. Remember the following key points:

- Select the best possible terrain features for the position and use terrain to your advantage.
- Use a well braced, sitting position whenever possible.
- Aim and anchor the belay for all possible load directions.
- Follow the "minimum" rule for belay anchors—2 for a downward pull, 1 for an upward pull.
- Ensure anchor attachments are aligned, independent, and snug.
- Stack the rope properly.
- Choose a belay technique appropriate for the climbing.

- Use a guide carabiner for rope control in all body belays.
- Ensure anchor attachments, guide carabiner (if applicable), and rope running to the climber are all on the guidehand side.
- The brake hand remains on the rope when belaying.

CAUTION
Never remove the brake hand from the rope while belaying. If the brake hand is removed, there is no belay.

- Ensure you are satisfied with your position before giving the command "BELAY ON."
- The belay remains in place until the climber gives the command "OFF BELAY."

CAUTION
The belay remains in place the from the time the belayer commands *"BELAY ON"* until the climber commands *"OFF BELAY."*

6-14. PROCEDURE FOR MANAGING THE ROPE

A number of different belay techniques are used in modern climbing, ranging from the basic "body belays" to the various "mechanical belays," which incorporate some type of friction device.

a. Whether the rope is wrapped around the body, or run through a friction device, the rope management procedure is basically the same. The belayer must be able to perform three basic functions: manipulate the rope to give the climber slack during movement, take up rope to remove excess slack, and apply the brake to halt a fall.

b. The belayer must be able to perform all three functions while maintaining "total control" of the rope at all times. Total control means the brake hand is NEVER removed from the rope. When giving slack, the rope simply slides through the grasp of the brake hand, at times being fed to the climber with the other "feeling" or guide hand. Taking up rope, however, requires a certain technique to ensure the brake hand remains on the rope at all times. The following procedure describes how to take up excess rope and apply the brake in a basic body belay.

(1) Grasping the rope with both hands, place it behind the back and around the hips. The hand on the section of rope between the belayer and the climber would be the guide hand. The other hand is the brake hand.

(2) Take in rope with the brake hand until the arm is fully extended. The guide hand can also help to pull in the rope (Figure 6-21, step 1, page 6-34).

(3) Holding the rope in the brake hand, slide the guide hand out, extending the arm so the guide hand is father away from the body than the brake hand (Figure 6-21, step 2).

(4) Grasp both parts of the rope, to the front of the brake hand, with the guide hand (Figure 6-21, step 3).

(5) Slide the brake hand back towards the body (Figure 6-21, step 4).

(6) Repeat step 5 of Figure 6-21. The brake can be applied at any moment during the procedure. It is applied by wrapping the rope around the front of the hips while increasing grip with the brake hand (Figure 6-21, step 6).

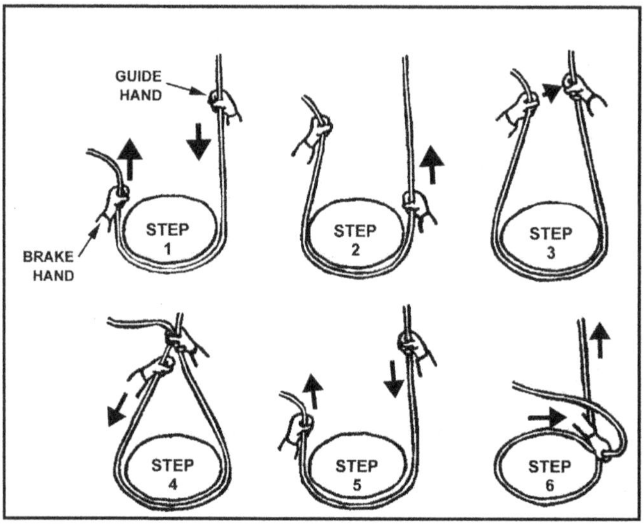

Figure 6-21. Managing the rope.

6-15. CHOOSING A BELAY TECHNIQUE

The climber may choose from a variety of belay techniques. A method that works well in one situation may not be the best choice in another. The choice between body belays and mechanical belays depends largely on equipment available, what the climber feels most comfortable with, and the amount of load, or fall force, the belay may have to absorb. The following describes a few of the more widely used techniques, and the ones most applicable to military mountaineering.

a. **Body Belay.** The basic body belay is the most widely used technique on moderate terrain. It uses friction between the rope and the clothed body as the rope is pressured across the clothing. It is the simplest belay, requiring no special equipment, and should be the first technique learned by all climbers. A body belay gives the belayer the greatest "feel" for the climber, letting him know when to give slack or take up rope. Rope management in a body belay is quick and easy, especially for beginners, and is effective in snow and ice climbing when ropes often become wet, stiff, and frozen. The body belay, in its various forms, will hold low to moderate impact falls well. It has been known to arrest some severe falls, although probably not without inflicting great pain on the belayer.

CAUTION

The belayer must ensure he is wearing adequate clothing to protect his body from rope burns when using a body belay. Heavy duty cotton or leather work gloves can also be worn to protect the hands.

(1) ***Sitting Body Belay.*** The sitting body belay is the preferred position and is usually the most secure (Figure 6-22). The belayer sits facing the direction where the force of a fall will likely come from, using terrain to his advantage, and attempts to brace both feet against the rock to support his position. It is best to sit in a slight depression, placing the buttocks lower than the feet, and straightening the legs for maximum support. When perfectly aligned, the rope running to the climber will pass between the belayer's feet, and both legs will equally absorb the force of a fall. Sometimes, the belayer may not be able to sit facing the direction he would like, or both feet cannot be braced well. The leg on the "guide hand" side should then point towards the load, bracing the foot on the rock when possible. The belayer can also "straddle" a large tree or rock nubbin for support, as long as the object is solid enough to sustain the possible load.

Figure 6-22. Sitting body belay.

(2) ***Standing Body Belay.*** The standing body belay is used on smaller ledges where there is no room for the belayer to sit (Figure 6-23). What appears at first to be a fairly unstable position can actually be quite secure when belay anchors are placed at or above shoulder height to support the stance when the force will be downward.

Figure 6-23. Standing body belay.

(a) For a body belay to work effectively, the belayer must ensure that the rope runs around the hips properly, and remains there under load when applying the brake. The rope should run around the narrow portion of the pelvic girdle, just below the bony high points of the hips. If the rope runs too high, the force of a fall could injure the belayer's midsection and lower rib cage. If the rope runs too low, the load may pull the rope below the buttocks, dumping the belayer out of position. It is also possible for a strong upward or downward pull to strip the rope away from the belayer, rendering the belay useless.

(b) To prevent any of these possibilities from happening, the belay rope is clipped into a carabiner attached to the guide hand side of the seat harness (or bowline-on-a-coil). This "guide carabiner" helps keep the rope in place around the hips and prevents loss of control in upward or downward loads (Figure 6-24).

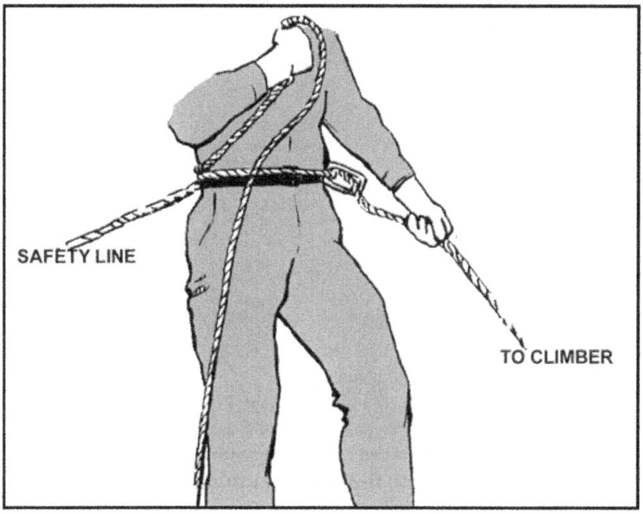

Figure 6-24. Guide carabiner for rope control in a body belay.

b. **Mechanical Belay.** A mechanical belay **must** be used whenever there is potential for the lead climber to take a severe fall. The holding power of a belay device is vastly superior to any body belay under high loads. However, rope management in a mechanical belay is more difficult to master and requires more practice. For the most part, the basic body belay should be totally adequate on a typical military route, as routes used during military operations should be the easiest to negotiate.

(1) ***Munter Hitch.*** The Munter hitch is an excellent mechanical belay technique and requires only a rope and a carabiner (Figure 6-25, page 6-38). The Munter is actually a two-way friction hitch. The Munter hitch will flip back and forth through the carabiner as the belayer switches from giving slack to taking up rope. The carabiner must be large enough, and of the proper design, to allow this function. The locking pear-shaped carabiner, or pearabiner, is designed for the Munter hitch.

(a) The Munter hitch works exceptionally well as a lowering belay off the anchor. As a climbing belay, the carabiner should be attached to the front of the belayer's seat harness. The hitch is tied by forming a loop and a bight in the rope, attaching both to the carabiner. It's fairly easy to place the bight on the carabiner backwards, which forms an obvious, useless hitch. Put some tension on the Munter to ensure it is formed correctly, as depicted in the following illustrations.

(b) The Munter hitch will automatically "lock-up" under load as the brake hand grips the rope. The brake is increased by pulling the slack rope away from the body, towards the load. The belayer must be aware that flipping the hitch DOES NOT change the function of the hands. The hand on the rope running to the climber, or load, is always the guide hand.

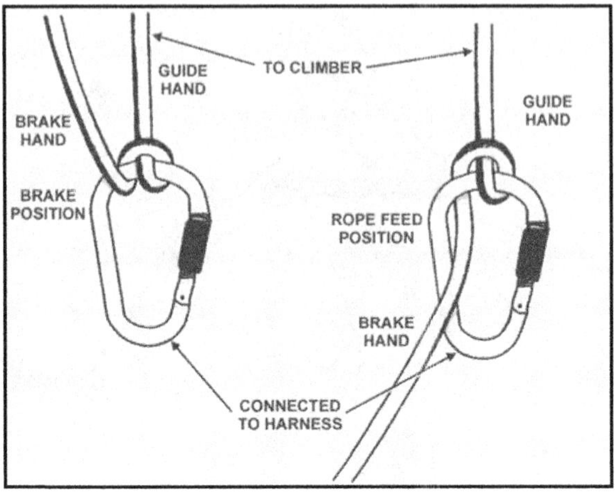

Figure 6-25. Munter hitch.

(2) ***Figure-Eight Device.*** The figure-eight device is a versatile piece of equipment and, though developed as a rappel device, has become widely accepted as an effective mechanical belay device (Figure 6-26). The advantage of any mechanical belay is friction required to halt a fall is applied on the rope through the device, rather than around the belayer's body. The device itself provides rope control for upward and downward pulls and excellent friction for halting severe falls. The main principle behind the figure-eight device in belay mode is the friction developing on the rope as it reaches and exceeds the 90-degree angle between the rope entering the device and leaving the device. As a belay device, the figure-eight works well for both belayed climbing and for lowering personnel and equipment on fixed-rope installations.

(a) As a climbing belay, a bight placed into the climbing rope is run through the "small eye" of the device and attached to a locking carabiner at the front of the belayer's seat harness. A short, small diameter safety rope is used to connect the "large eye" of the figure eight to the locking carabiner for control of the device. The guide hand is placed on the rope running to the climber. Rope management is performed as in a body belay. The brake is applied by pulling the slack rope in the brake hand towards the body, locking the rope between the device and the carabiner.

(b) As a lowering belay, the device is normally attached directly to the anchor with the rope routed as in rappelling.

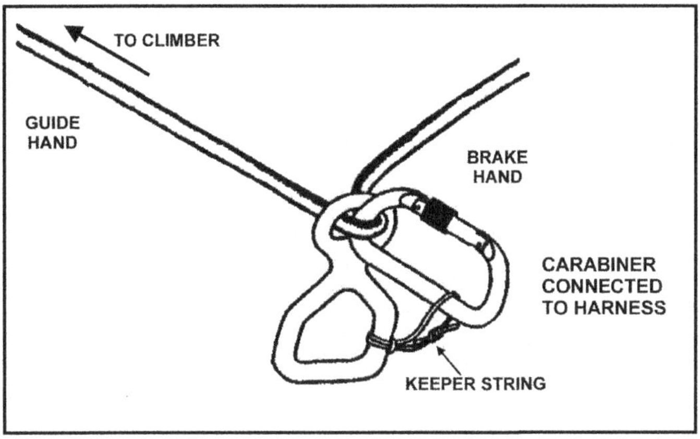

Figure 6-26. Figure-eight device.

Note: Some figure-eight descenders should not be used as belay devices due to their construction and design. Always refer to manufacturer's specifications and directions before use.

(3) ***Mechanical Camming Device.*** The mechanical camming device has an internal camming action that begins locking the rope in place as friction is increased. Unlike the other devices, the mechanical camming device can stop a falling climber without any input from the belayer. A few other devices perform similarly to this, but have no moving parts. Some limitations to these type devices are minimum and maximum rope diameters.

(4) ***Other Mechanical Belay Devices.*** There are many other commercially available mechanical belay devices. Most of these work with the same rope movement direction and the same braking principle. The air traffic controller (ATC), slotted plate, and other tube devices are made in many different shapes. These all work on the same principle as the figure-eight device—friction increases on the rope as it reaches and exceeds the 90-degree angle between the rope entering the device and leaving the device.

6-16. ESTABLISHING A BELAY

A belay can be established using either a direct or indirect connection. Each type has advantages and disadvantages. The choice will depend on the intended use of the belay.

a. **Direct Belay.** The direct belay removes any possible forces from the belayer and places this force completely on the anchor. Used often for rescue installations or to bring a second climber up to a new belay position in conjunction with the Munter hitch, the belay can be placed above the belayer's stance, creating a comfortable position and ease of applying the brake. Also, if the second falls or weights the rope, the belayer is not locked into a position. Direct belays provide no shock-absorbing properties from the belayer's attachment to the system as does the indirect belay; therefore, the belayer is apt to pay closer attention to the belaying process.

b. **Indirect Belay.** An indirect belay, the most commonly used, uses a belay device attached to the belayer's harness. This type of belay provides dynamic shock or weight absorption by the belayer if the climber falls or weights the rope, which reduces the direct force on the anchor and prevents a severe shock load to the anchor.

6-17. SETTING UP A BELAY

In rock climbing, climbers must sometimes make do with marginal protection placements along a route, but belay positions must be made as "bombproof" as possible. Additionally, the belayer must set up the belay in relation to where the fall force will come from and pay strict attention to proper rope management for the belay to be effective. All belay positions are established with the anchor connection to the front of the harness. If the belay is correctly established, the belayer will feel little or no force if the climber falls or has to rest on the rope. Regardless of the actual belay technique used, five basic steps are required to set up a sound belay.

a. **Select Position and Stance.** Once the climbing line is picked, the belayer selects his position. It's best if the position is off to the side of the actual line, putting the belayer out of the direct path of a potential fall or any rocks kicked loose by the climber. The position should allow the belayer to maintain a comfortable, relaxed stance, as he could be in the position for a fairly long time. Large ledges that allow a well braced, sitting stance are preferred. Look for belay positions close to bombproof natural anchors. The position must at least allow for solid artificial placements.

b. **Aim the Belay.** With the belay position selected, the belay must now be "aimed." The belayer determines where the rope leading to the climber will run and the direction the force of a fall will likely come from. When a lead climber begins placing protection, the fall force on the belayer will be in some upward direction, and in line with the first protection placement. If this placement fails under load, the force on the belay could be straight down again. The belayer must aim his belay for all possible load directions, adjusting his position or stance when necessary. The belay can be aimed through an anchor placement to immediately establish an upward pull; however, the belayer must always be prepared for the more severe downward fall force in the event intermediate protection placements fail.

c. **Anchor the Belay.** For a climbing belay to be considered bombproof, the belayer must be attached to a solid anchor capable of withstanding the highest possible fall force. A solid natural anchor would be ideal, but more often the belayer will have to place pitons or chocks. A single artificial placement should never be considered adequate for anchoring a belay (except at ground level). Multiple anchor points capable of supporting both upward and downward pulls should be placed. The rule of thumb is to place two anchors for a downward pull and one anchor for an upward pull as a MINIMUM. The following key points also apply to anchoring belays.

(1) Each anchor must be placed in line with the direction of pull it is intended to support.

(2) Each anchor attachment must be rigged "independently" so a failure of one will not shock load remaining placements or cause the belayer to be pulled out of position.

(3) The attachment between the anchor and the belayer must be snug to support the stance. Both belayer's stance and belay anchors should absorb the force of a fall.

(4) It is best for the anchors to be placed relatively close to the belayer with short attachments. If the climber has to be tied-off in an emergency, say after a severe fall, the belayer can attach a Prusik sling to the climbing rope, reach back, and connect the sling to one of the anchors. The load can be placed on the Prusik and the belayer can come out of the system to render help.

(5) The belayer can use either a portion of the climbing rope or slings of the appropriate length to connect himself to the anchors. It's best to use the climbing rope whenever possible, saving the slings for the climb. The rope is attached using either figure eight loops or clove hitches. Clove hitches have the advantage of being easily adjusted. If the belayer has to change his stance at some point, he can reach back with the guide hand and adjust the length of the attachment through the clove hitch as needed.

(6) The anchor attachments should also help prevent the force of a fall from "rotating" the belayer out of position. To accomplish this, the climbing rope must pass around the "guide-hand side" of the body to the anchors. Sling attachments are connected to the belayer's seat harness (or bowline-on-a-coil) on the guide-hand side.

(7) Arrangement of rope and sling attachments may vary according to the number and location of placements. Follow the guidelines set forth and remember the key points for belay anchors; "in line", "independent", and "snug". Figure 6-27 shows an example of a common arrangement, attaching the rope to the two "downward" anchors and a sling to the "upward" anchor. Note how the rope is connected from one of the anchors back to the belayer. This is not mandatory, but often helps "line-up" the second attachment.

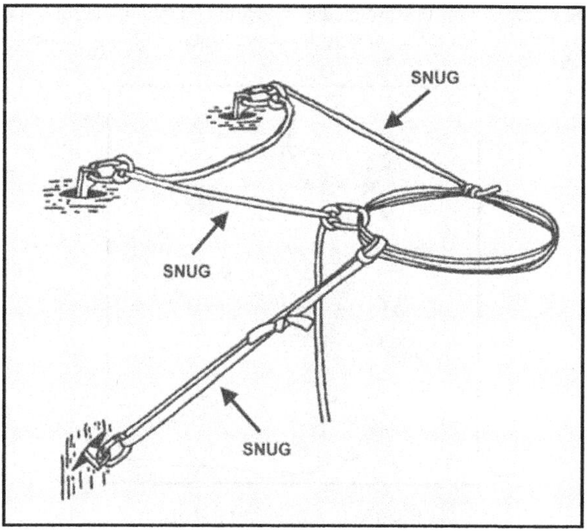

Figure 6-27. Anchoring a belay.

d. **Stack the Rope.** Once the belayer is anchored into position, he must stack the rope to ensure it is free of twists and tangles that might hinder rope management in the belay. The rope should be stacked on the ground, or on the ledge, where it will not get caught in cracks or nubbins as it is fed out to the climber.

(1) On small ledges, the rope can be stacked on top of the anchor attachments if there is no other place to lay it, but make sure to stack it carefully so it won't tangle with the anchored portion of the rope or other slings. The belayer must also ensure that the rope will not get tangled around his legs or other body parts as it "feeds" out.

(2) The rope should never be allowed to hang down over the ledge. If it gets caught in the rock below the position, the belayer may have to tie-off the climber and come out of the belay to free the rope; a time-consuming and unnecessary task. The final point to remember is the rope must be stacked "from the belayer's end" so the rope running to the climber comes off the "top" of the stacked pile.

e. **Attach the Belay.** The final step of the procedure is to attach the belay. With the rope properly stacked, the belayer takes the rope coming off the top of the pile, removes any slack between himself and the climber, and applies the actual belay technique. If using a body belay, ensure the rope is clipped into the guide carabiner.

(1) The belayer should make one quick, final inspection of his belay. If the belay is set up correctly, the anchor attachments, guide carabiner if applicable, and the rope running to the climber will all be on the "guide hand" side, which is normally closest to the rock (Figure 6-28). If the climber takes a fall, the force, if any, should not have any negative effect on the belayer's involvement in the system. The brake hand is out away from the slope where it won't be jammed between the body and the rock. The guide hand can be placed on the rock to help support the stance when applying the brake.

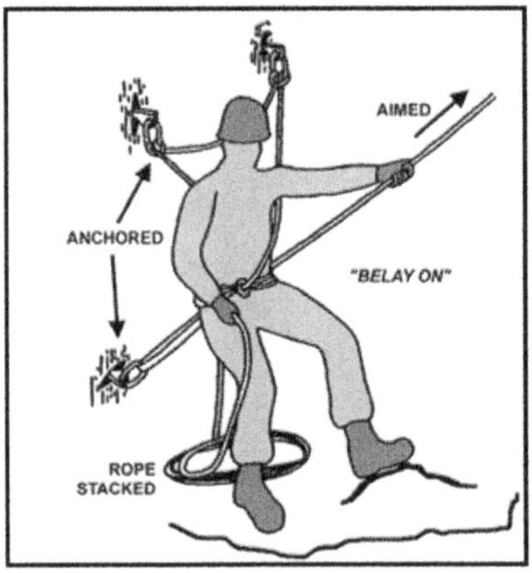

Figure 6-28. Belay setup.

(2) When the belayer is satisfied with his position, he gives the signal, "BELAY ON!". When belaying the "second", the same procedure is used to set up the belay. Unless the belay is aimed for an upward pull, the fall force is of course downward and the

belayer is usually facing away from the rock, the exception being a hanging belay on a vertical face. If the rope runs straight down to the climber and the anchors are directly behind the position, the belayer may choose to brake with the hand he feels most comfortable with. Anchor attachments, guide carabiner, and rope running to the climber through the guide hand must still be aligned on the same side to prevent the belayer from being rotated out of position, unless the belayer is using an improvised harness and the anchor attachment is at the rear.

6-18. TOP-ROPE BELAY

A "top-rope" is a belay setup used in training to protect a climber while climbing on longer, exposed routes. A solid, bombproof anchor is required at the top of the pitch. The belayer is positioned either on the ground with the rope running through the top anchor and back to the climber, or at the top at the anchor. The belayer takes in rope as the climber proceeds up the rock. If this is accomplished with the belayer at the bottom, the instructor can watch the belayer while he coaches the climber through the movements.

CAUTION

Do not use a body belay for top-rope climbing. The rope will burn the belayer if the climber has to be lowered.

Section V. CLIMBING COMMANDS

Communication is often difficult during a climb. As the distance between climber and belayer increases, it becomes harder to distinguish one word from another and the shortest sentence may be heard as nothing more than jumbled syllables. A series of standard voice commands were developed over the years to signal the essential rope management functions in a belayed climb. Each command is concise and sounds a bit different from another to reduce the risk of a misunderstanding between climber and belayer. They must be pronounced clearly and loudly so they can be heard and understood in the worst conditions.

6-19. VERBAL COMMANDS

Table 6-1, page 6-44, lists standard rope commands and their meanings in sequence as they would normally be used on a typical climb. (Note how the critical "BELAY" commands are reversed so they sound different and will not be confused.)

BELAYER	CLIMBER	MEANING/ACTION TAKEN
"BELAY ON"		The belay is on; you may climb when ready; the rope will be managed as needed.
	"CLIMBING" (as a courtesy)	I am ready to climb.
"CLIMB" (as a courtesy)		Proceed, and again, the rope will be managed as necessary.
"ROCK"	"ROCK"	**PROTECT YOURSELF FROM FALLING OBJECTS.** Signal will be echoed by all climbers in the area. If multipitch climbing, ensure climbers below hear.
	"TAKE ROPE"	Take in excess rope between us without pulling me off the route. Belayer takes in rope.
	"SLACK"	Release all braking/tension on the rope so I can have slack without pulling the rope. Belayer removes brake/tension.
	"TENSION"	Take all the slack, apply brake, and hold me. My weight will be on the rope. Belayer removes slack and applies brake.
	"FALLING"	I am falling. Belayer applies brake to arrest the fall.
"TWEN-TY-FIVE"		You have approximately 25 feet of rope left. Start looking for the next belay position. Climber selects a belay position.
"FIF-TEEN"		You have approximately 15 feet of rope left. Start looking for the next belay position. Climber selects a belay position within the next few feet.
"FIVE"	Set up the belay.	You have 5 feet of rope left. Set up the belay position. You have no more rope. Climber sets up the belay.
Removes the belay, remains anchored. Prepares to climb.	"OFF BELAY"	I have finished climbing and I am anchored. You may remove the belay. Belayer removes the belay and, remaining anchored, prepares to climb.

Table 6-1. Rope commands.

6-20. ROPE TUG COMMANDS

Sometimes the loudest scream cannot be heard when the climber and belayer are far apart. This is especially true in windy conditions, or when the climber is around a corner, above an overhang, or at the back of a ledge. It may be necessary to use a series of "tugs" on the rope in place of the standard voice commands. To avoid any possible confusion with interpretation of multiple rope tug commands, use only one.

a. While a lead climb is in progress, the most important command is "BELAY ON." This command is given only by the climber when the climber is anchored and is prepared for the second to begin climbing. With the issue of this command, the second knows the climber is anchored and the second prepares to climb.

b. For a rope tug command, the leader issues three distinct tugs on the rope AFTER anchoring and putting the second on belay. This is the signal for "BELAY ON" and signals the second to climb when ready. The new belayer keeps slack out of the rope.

Section VI. ROPED CLIMBING METHODS

In military mountaineering, the primary mission of a roped climbing team is to "fix" a route with some type of rope installation to assist movement of less trained personnel in the unit. This duty falls upon the most experienced climbers in the unit, usually working in two- or three-man groups or teams called assault climbing teams. Even if the climbing is for another purpose, roped climbing should be performed whenever the terrain becomes difficult and exposed.

6-21. TOP-ROPED CLIMBING

Top-roped climbing is used for training purposes only. This method of climbing is not used for movement due to the necessity of pre-placing anchors at the top of a climb. If you can easily access the top of a climb, you can easily avoid the climb itself.

a. For training, top-roped climbing is valuable because it allows climbers to attempt climbs above their skill level and or to hone present skills without the risk of a fall. Top-roped climbing may be used to increase the stamina of a climber training to climb longer routes as well as for a climber practicing protection placements.

b. The belayer is positioned either at the base of a climb with the rope running through the top anchor and back to the climber or at the top at the anchor. The belayer takes in rope as the climber moves up the rock, giving the climber the same protection as a belay from above. If this is accomplished with the belayer at the bottom, the instructor is able to keep an eye on the belayer while he coaches the climber through the movements.

6-22. LEAD CLIMBING

A lead climb consists of a belayer, a leader or climber, rope(s), and webbing or hardware used to establish anchors or protect the climb. As he climbs the route, the leader emplaces "intermediate" anchors, and the climbing rope is connected to these anchors with a carabiner. These "intermediate" anchors protect the climber against a fall-thus the term "protecting the climb."

Note: Intermediate anchors are commonly referred to as "protection," "pro," "pieces," "pieces of pro," "pro placements," and so on. For standardization within this publication, these specific anchors will be referred to as "protection;" anchors established for other purposes, such as rappel points, belays, or other rope installations, will be referred to as "anchors."

CAUTION

During all lead climbing, each climber in the team is either anchored or being belayed.

a. Lead climbing with two climbers is the preferred combination for movement on technically difficult terrain. Two climbers are at least twice as fast as three climbers, and are efficient for installing a "fixed rope," probably the most widely used rope installation in the mountains. A group of three climbers are typically used on moderate snow, ice, and snow-covered glaciers where the rope team can often move at the same time, stopping occasionally to set up belays on particularly difficult sections. A group or team of three climbers is sometimes used in rock climbing because of an odd number of personnel, a shortage of ropes (such as six climbers and only two ropes), or to protect and assist an individual who has little or no experience in climbing and belaying. Whichever technique is chosen, a standard roped climbing procedure is used for maximum speed and safety.

b. When the difficulty of the climbing is within the "leading ability" of both climbers, valuable time can be saved by "swinging leads." This is normally the most efficient method for climbing multipitch routes. The second finishes cleaning the first pitch and continues climbing, taking on the role of lead climber. Unless he requires equipment from the other rack or desires a break, he can climb past the belay and immediately begin leading. The belayer simply adjusts his position, re-aiming the belay once the new leader begins placing protection. Swinging leads, or "leap frogging," should be planned before starting the climb so the leader knows to anchor the upper belay for both upward and downward pulls during the setup.

c. The procedures for conducting a lead climb with a group of two are relatively simple. The most experienced individual is the "lead" climber or leader, and is responsible for selecting the route. The leader must ensure the route is well within his ability and the ability of the second. The lead climber carries most of the climbing equipment in order to place protection along the route and set up the next belay. The leader must also ensure that the second has the necessary equipment, such as a piton hammer, nut tool, etc., to remove any protection that the leader may place.

(1) The leader is responsible for emplacing protection frequently enough and in such a manner that, in the event that either the leader or the second should fall, the fall will be neither long enough nor hard enough to result in injury. The leader must also ensure that the rope is routed in a way that will allow it to run freely through the protection placements, thus minimizing friction, or "rope drag".

(2) The other member of the climbing team, the belayer (sometimes referred to as the "second"), is responsible for belaying the leader, removing the belay anchor, and retrieving the protection placed by the leader between belay positions (also called "cleaning the pitch").

(3) Before the climb starts, the second will normally set up the first belay while the leader is arranging his rack. When the belay is ready, the belayer signals, "BELAY ON", affirming that the belay is "on" and the rope will be managed as necessary. When the leader is ready, he double checks the belay. The leader can then signal, "CLIMBING", only as a courtesy, to let the belayer know he is ready to move. The belayer can reply with "CLIMB", again, only as a courtesy, reaffirming that the belay is "on" and the rope will be managed as necessary. The leader then begins climbing.

(4) While belaying, the second must pay close attention to the climber's every move, ensuring that the rope runs free and does not inhibit the climber's movements. If he cannot see the climber, he must "feel" the climber through the rope. Unless told otherwise by the climber, the belayer can slowly give slack on the rope as the climber proceeds on the route. The belayer should keep just enough slack in the rope so the climber does not have to pull it through the belay. If the climber wants a tighter rope, it can be called for. If the belayer notices too much slack developing in the rope, the excess

rope should be taken in quickly. It is the belayer's responsibility to manage the rope, whether by sight or feel, until the climber tells him otherwise.

(5) As the leader protects the climb, slack will sometimes be needed to place the rope through the carabiner (clipping), in a piece of protection above the tie-in point on the leaders harness. In this situation, the leader gives the command "SLACK" and the belayer gives slack, (if more slack is needed the command will be repeated). The leader is able to pull a bight of rope above the tie-in point and clip it into the carabiner in the protection above. When the leader has completed the connection, or the clip, the command "TAKE ROPE" is given by the leader and the belayer takes in the remaining slack.

(6) The leader continues on the route until either a designated belay location is reached or he is at the end of or near the end of the rope. At this position, the leader sets an anchor, connects to the anchor and signals "OFF BELAY". The belayer prepares to climb by removing all but at least one of his anchors and secures the remaining equipment. The belayer remains attached to at least one anchor until the command "BELAY ON" is given.

d. When the leader selects a particular route, he must also determine how much, and what types, of equipment might be required to safely negotiate the route. The selected equipment must be carried by the leader. The leader must carry enough equipment to safely protect the route, additional anchors for the next belay, and any other items to be carried individually such as rucksacks or individual weapons.

(1) The leader will assemble, or "rack," the necessary equipment onto his harness or onto slings around the head and shoulder. A typical leader "rack" consists of:

- Six to eight small wired stoppers on a carabiner.
- Four to six medium to large wired stoppers on a carabiner.
- Assorted hexentrics, each on a separate carabiner.
- SLCDs of required size, each on a separate carabiner.
- Five to ten standard length runners, with two carabiners on each.
- Two to three double length runners, with two carabiners on each.
- Extra carabiners.
- Nut tool.

Note: The route chosen will dictate, to some degree, the necessary equipment. Members of a climbing team may need to consolidate gear to climb a particular route.

(2) The belayer and the leader both should carry many duplicate items while climbing.

- Short Prusik sling.
- Long Prusik sling.
- Cordellette.
- 10 feet of 1-inch webbing.
- 20 feet of 1-inch webbing.
- Belay device (a combination belay/rappel device is multifunctional).
- Rappel device (a combination belay/rappel device is multifunctional).
- Large locking carabiner (pear shape carabiners are multifunctional).

- Extra carabiners.
- Nut tool (if stoppers are carried).

Note: If using an over the shoulder gear sling, place the items in order from smallest to the front and largest to the rear.

e. Leading a difficult pitch is the most hazardous task in roped climbing. The lead climber may be exposed to potentially long, hard falls and must exercise keen judgment in route selection, placement of protection, and routing of the climbing rope through the protection. The leader should try to keep the climbing line as direct as possible to the next belay to allow the rope to run smoothly through the protection with minimal friction. Protection should be placed whenever the leader feels he needs it, and BEFORE moving past a difficult section.

CAUTION

The climber must remember he will fall twice the distance from his last piece of protection before the rope can even begin to stop him.

(1) ***Placing Protection.*** Generally, protection is placed from one stable position to the next. The anchor should be placed as high as possible to reduce the potential fall distance between placements. If the climbing is difficult, protection should be placed more frequently. If the climbing becomes easier, protection can be placed farther apart, saving hardware for difficult sections. On some routes an extended diagonal or horizontal movement, known as a traverse, is required. As the leader begins this type of move, he must consider the second's safety as well as his own. The potential fall of the second will result in a pendulum swing if protection is not adequate to prevent this. The danger comes from any objects in the swinging path of the second.

CAUTION

Leader should place protection prior to, during, and upon completion of any traverse. This will minimize the potential swing, or pendulum, for both the leader and second if either should fall.

(2) ***Correct Clipping Technique.*** Once an anchor is placed, the climber "clips" the rope into the carabiner (Figure 6-29, page 6-50). As a carabiner hangs from the protection, the rope can be routed through the carabiner in two possible ways. One way will allow the rope to run smoothly as the climber moves past the placement; the other way will often create a dangerous situation in which the rope could become "unclipped" from the carabiner if the leader were to fall on this piece of protection. In addition, a

series of incorrectly clipped carabiners may contribute to rope drag. When placing protection, the leader must ensure the carabiner on the protection does not hang with the carabiner gate facing the rock; when placing protection in a crack ensure the carabiner gate is not facing into the crack.

- Grasp the rope with either hand with the thumb pointing down the rope towards the belayer
- Pull enough rope to reach the carabiner with a bight
- Note the direction the carabiner is hanging from the protection
- Place the bight into the carabiner so that, when released, the rope does not cause the carabiner to twist.

(a) If the route changes direction, clipping the carabiner will require a little more thought. Once leaving that piece of protection, the rope may force the carabiner to twist if not correctly clipped. If the clip is made correctly, a rotation of the clipped carabiner to ensure that the gate is not resting against the rock may be all that is necessary.

CAUTION
Ensure the carabiner gate is not resting against a protrusion or crack edge in the rock surface; the rock may cause the gate to open.

(b) Once the rope is clipped into the carabiner, the climber should check to see that it is routed correctly by pulling on the rope in the direction it will travel when the climber moves past that position.

(c) Another potential hazard peculiar to leading should be eliminated before the climber continues. The carabiner is attached to the anchor or runner with the gate facing away from the rock and opening down for easy insertion of the rope. However, in a leader fall, it is possible for the rope to run back over the carabiner as the climber falls below the placement. If the carabiner is left with the gate facing the direction of the route there is a chance that the rope will open the gate and unclip itself entirely from the placement. To prevent this possibility, the climber should ensure that after the clip has been made, the gate is facing away from the direction of the route. There are two ways to accomplish this: determine which direction the gate will face before the protection or runner is placed or once clipped, rotate the carabiner upwards 180 degrees. This problem is more apt to occur if bent gate carabiners are used. Straight gate ovals or "Ds" are less likely to have this problem and are stronger and are highly recommended. Bent gate carabiners are easier to clip the rope into and are used mostly on routes with bolts preplaced for protection. Bent gate carabiners are not recommended for many climbing situations.

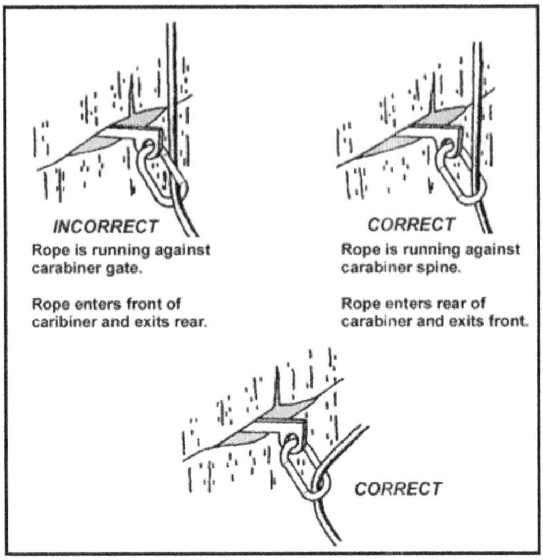

Figure 6-29. Clipping on to protection.

(3) ***Reducing Rope Drag; Using Runners***. No matter how direct the route, the climber will often encounter problems with "rope drag" through the protection positions. The friction created by rope drag will increase to some degree every time the rope passes through a carabiner, or anchor. It will increase dramatically if the rope begins to "zigzag" as it travels through the carabiners. To prevent this, the placements should be positioned so the rope creates a smooth, almost straight line as it passes through the carabiners (Figure 6-30). Minimal rope drag is an inconvenience; severe rope drag may actually pull the climber off balance, inducing a fall.

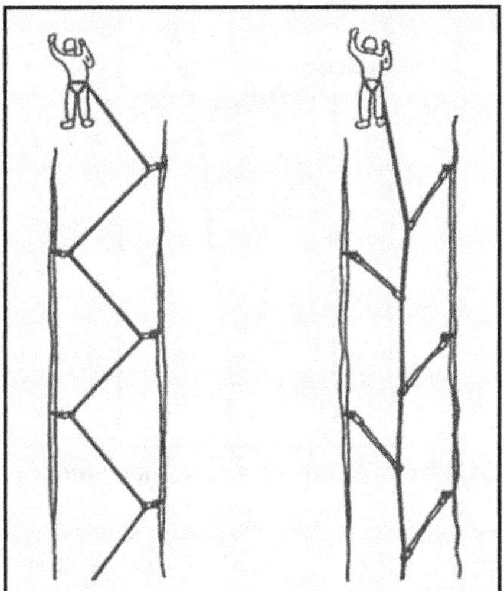

Figure 6-30. Use of slings on protection.

CAUTION

Rope drag can cause confusion when belaying the second or follower up to a new belay position. Rope drag can be mistaken for the climber, causing the belayer to not take in the necessary slack in the rope and possibly resulting in a serious fall.

(a) If it is not possible to place all the protection so the carabiners form a straight line as the rope moves through, you should "extend" the protection (Figure 6-31, page 6-52). Do this by attaching an appropriate length sling, or runner, to the protection to extend the rope connection in the necessary direction. The runner is attached to the protection's carabiner while the rope is clipped into a carabiner at the other end of the runner. Extending placements with runners will allow the climber to vary the route slightly while the rope continues to run in a relatively straight line.

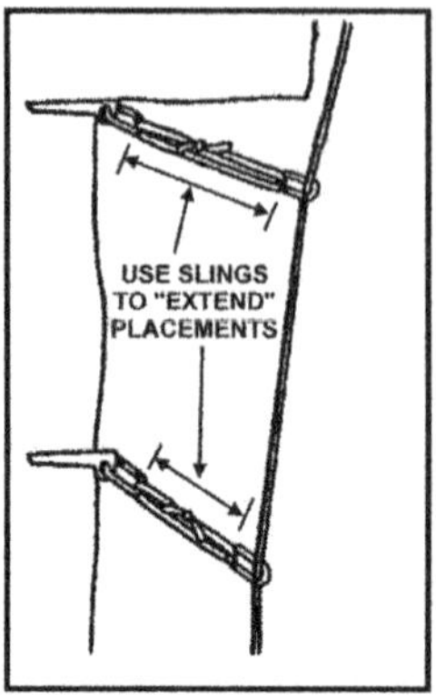

Figure 6-31. Use of slings to extend placement positions.

(b) Not only is rope drag a hindrance, it can cause undue movement of protection as the rope tightens between any "out of line" placements. Rope drag through chock placements can be dangerous. As the climber moves above the placements, an outward or upward pull from rope drag may cause correctly set chocks to pop out, even when used "actively". Most all chocks placed for leader protection should be extended with a runner, even if the line is direct to eliminate the possibility of movement.

(c) Wired chocks are especially prone to wiggling loose as the rope pulls on the stiff cable attachment. All wired chocks used for leader protection should be extended to reduce the chance of the rope pulling them out (Figure 6-32). Some of the larger chocks, such as roped Hexentrics and Tri-Cams, have longer slings pre-attached that will normally serve as an adequate runner for the placement. Chocks with smaller sling attachments must often be extended with a runner. Many of today's chocks are manufactured with pre-sewn webbing installed instead of cable.

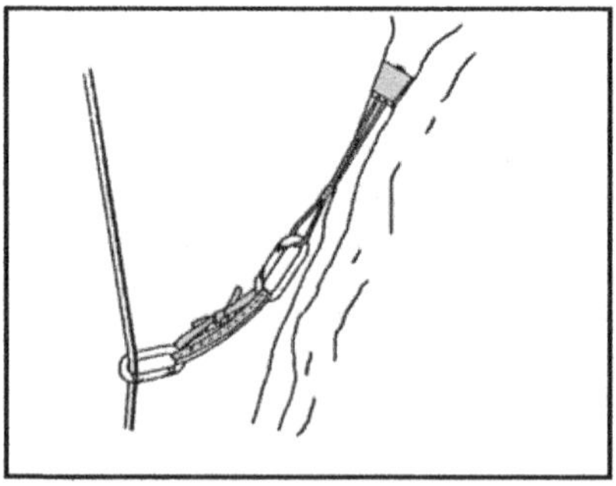

Figure 6-32. Use of sling on a wired stopper.

(d) When a correctly placed piton is used for protection, it will normally not be affected by rope drag. A correctly placed piton is generally a multi-directional anchor, therefore, rope drag through pitons will usually only affect the leader's movements but will continue to protect as expected.

(e) Rope drag will quite often move SLCDs out of position, or "walk" them deeper into the crack than initially placed, resulting in difficult removal or inability to remove them at all. Furthermore, most cases of SLCD movement result in the SLCD moving to a position that does not provide protection in the correct direction or no protection at all due to the lobes being at different angles from those at the original position.

Note: Any placement extended with a runner will increase the distance of a potential fall by the actual length of the sling. Try to use the shortest runners possible, ensuring they are long enough to function properly.

f. Belaying the follower is similar to belaying a top-roped climb in that the follower is not able to fall any farther than rope stretch will allow. This does not imply there is no danger in following. Sharp rocks, rock fall, and inadequately protected traverses can result in damage to equipment or injury to the second.

g. Following, or seconding, a leader has a variety of responsibilities. The second has to issue commands to the leader, as well as follow the leader's commands. Once the lead climber reaches a good belay position, he immediately establishes an anchor and connects to it. When this is completed he can signal "OFF BELAY" to the belayer. The second can now remove the leader's belay and prepare to climb. The second must remain attached to at least one of the original anchors while the leader is preparing the next belay position. The removed materials and hardware can be organized and secured on the second's rack in preparation to climb.

(1) When the leader has established the new belay position and is ready to belay the follower, the "new" belayer signals "BELAY ON." The second, now the climber, removes any remaining anchor hardware/materials and completes any final preparations. The belayer maintains tension on the rope, unless otherwise directed, while the final preparations are taking place, since removal of these remaining anchors can introduce slack into the rope. When the second is ready, he can, as a courtesy, signal "CLIMBING," and the leader can, again as a courtesy, reply with "CLIMB."

(2) Upon signaling "BELAY ON," the belayer must remove and keep all slack from the rope. (This is especially important as in many situations the belayer cannot see the follower. A long pitch induces weight and sometimes "drag" on the rope and the belayer above will have difficulty distinguishing these from a rope with no slack.)

h. When removing protection, the man cleaning the pitch should rack it properly to facilitate the exchange and or arrangement of equipment at the end of the pitch. When removing the protection, or "cleaning the pitch", SLCDs or chocks may be left attached to the rope to prevent loss if they are accidentally dropped during removal. If necessary, the hardware can remain on the rope until the second reaches a more secure stance. If removing a piton, the rope should be unclipped from the piton to avoid the possibility of damaging the rope with a hammer strike.

(1) The second may need to place full body weight on the rope to facilitate use of both hands for protection removal by giving the command "TENSION." The second

must also ensure that he does not climb faster than the rope is being taken in by the belayer. If too much slack develops, he should signal "TAKE ROPE" and wait until the excess is removed before continuing the climb. Once the second completes the pitch, he should immediately connect to the anchor. Once secured, he can signal "OFF BELAY." The leader removes the belay, while remaining attached to an anchor. The equipment is exchanged or organized in preparation for the next pitch or climb.

(2) When the difficulty of the climbing is within the "leading ability" of both climbers, valuable time can be saved by "swinging leads." This is normally the most efficient method for climbing multi-pitch routes. The second finishes cleaning the first pitch and continues climbing, taking on the role of lead climber. Unless he requires equipment from the belayer or desires a break, he can climb past the belay and immediately begin leading. The belayer simply adjusts his position, re-aiming the belay once the new leader begins placing protection. Swinging leads, or "leap frogging," should be planned before starting the climb so the leader knows to anchor the upper belay for both upward and downward pulls during the setup.

6-23. AID CLIMBING

When a route is too difficult to free climb and is unavoidable, if the correct equipment is available you might aid climb the route. Aid climbing consists of placing protection and putting full body weight on the piece. This allows you to hang solely on the protection you place, giving you the ability to ascend more difficult routes than you can free climb. Clean aid consists of using SLCDs and chocks, and is the simplest form of aid climbing.

a. **Equipment**. Aid climbing can be accomplished with various types of protection. Regardless of the type of protection used, the method of aid climbing is the same. In addition to the equipment for free climbing, other specialized equipment will be needed.

(1) ***Pitons***. Pitons are used the same as for free climbing. Most piton placements will require the use of both hands. Piton usage will usually leave a scar in the rock just by virtue of the hardness of the piton and the force required to set it with a hammer. Swinging a hammer to place pitons will lead to climber fatigue sooner than clean aid. Since pitons are multidirectional, the strength of a well-placed piton is more secure than most clean aid protection. Consider other forms of protection when noise could be hazardous to tactics.

(2) ***Bolts***. Bolts are used when no other protection will work. They are a more permanent form of protection and more time is needed to place them. Placing bolts creates more noise whether drilled by hand or by motorized drill. Bolts used in climbing are a multi-part expanding system pounded into predrilled holes and then tightened to the desired torque with a wrench or other tool. Bolts are used in many ways in climbing today. The most common use is with a hanger attached and placed for anchors in face climbing. However, bolts can be used for aid climbing, with or without the hanger.

(a) Placing bolts for aid climbing takes much more time than using pitons or clean aid. Bolting for aid climbing consists of consecutive bolts about 2 feet apart. Drilling a deep enough hole takes approximately thirty minutes with a hand drill and up to two minutes with a powered hammer drill. A lot of time and work is expended in a short distance no matter how the hole is drilled. (The weight of a powered hammer drill becomes an issue in itself.) Noise will also be a factor in both applications. A constant pounding with a hammer on the hand drill or the motorized pounding of the powered drill may alert the enemy to the position. The typical climbing bolt/hanger combination normally is left in the hole where it was placed.

(b) Other items that can be used instead of the bolt/hanger combination are the removable and reusable "spring-loaded removable bolts" such as rivets (hex head threaded bolts sized to fit tightly into the hole and pounded in with a hammer), split-shaft rivets, and some piton sizes that can be pounded into the holes. When using rivets or bolts without a hanger, place a loop of cable over the head and onto the shaft of the rivet or bolt and attach a carabiner to the other end of the loop (a stopper with the chock slid back will suffice). Rivet hangers are available that slide onto the rivet or bolt after it is placed and are easily removed for reuse. Easy removal means a slight loss of security while in use.

(3) ***SLCDs.*** SLCDs are used the same as for free climbing, although in aid climbing, full body weight is applied to the SLCD as soon as it is placed.

(4) ***Chocks.*** Chocks are used the same as for free climbing, although in aid climbing, weight is applied to the chock as soon as it is placed.

(5) ***Daisy Chains.*** Daisy chains are tied or presewn loops of webbing with small tied or presewn loops approximately every two inches. The small loops are just large enough for two or three carabiners. Two daisy chains should be girth-hitched to the tie-in point in the harness.

(6) ***Etriers (or Aiders).*** Etriers (aiders) are tied or presewn webbing loops with four to six tied or presewn internal loops, or steps, approximately every 12 inches. The internal loops are large enough to easily place one booted foot into. At least two etriers (aiders) should be connected by carabiner to the free ends of the daisy chains.

(7) ***Fifi Hook.*** A fifi hook is a small, smooth-surfaced hook strong enough for body weight. The fifi hook should be girth-hitched to the tie-in point in the harness and is used in the small loops of the daisy chain. A carabiner can be used in place of the fifi hook, although the fifi hook is simpler and adequate.

(8) ***Ascenders.*** Ascenders are mechanical devices that will move easily in one direction on the rope, but will lock in place if pushed or pulled the other direction. (Prusiks can be used but are more difficult than ascenders.)

b. **Technique.** The belay will be the same as in normal lead climbing and the rope will be routed through the protection the same way also. The big difference is the movement up the rock. With the daisy chains, aiders, and fifi hook attached to the rope tie-in point of the harness as stated above, and secured temporarily to a gear loop or gear sling, the climb continues as follows:

(1) The leader places the first piece of protection as high as can safely be reached and attaches the appropriate sling/carabiner

(2) Attach one daisy chain/aider group to the newly placed protection

(3) Clip the rope into the protection, (the same as for normal lead climbing)

(4) Insure the protection is sound by weighting it gradually; place both feet, one at a time, into the steps in the aider, secure your balance by grasping the top of the aider with your hands.

(5) When both feet are in the aider, move up the steps until your waist is no higher than the top of the aider.

(6) Place the fifi hook (or substituted carabiner) into the loop of the daisy chain closest to the daisy chain/aider carabiner, this effectively shortens the daisy chain;

maintain tension on the daisy chain as the hook can fall out of the daisy chain loop if it is unweighted.

Note: Moving the waist higher than the top of the aider is possible, but this creates a potential for a fall to occur even though you are on the aider and "hooked" close to the protection with the daisy chain. As the daisy chain tie-in point on the harness moves above the top of the aider, you are no longer supported from above by the daisy chain, you are now standing above your support. From this height, the fifi hook can easily fall out of the daisy chain loop if it is unweighted. If this happens, you could fall the full length of the daisy chain resulting in a static fall on the last piece of protection placed.

(7) Release one hand from the aider and place the next piece of protection, again, as high as you can comfortably reach; if using pitons or bolts you may need both hands free- "lean" backwards slowly, and rest your upper body on the daisy chain that you have "shortened" with the fifi hook

(8) Clip the rope into the protection

(9) Attach the other daisy chain/aider group to the next piece of protection

(10) Repeat entire process until climb is finished

c. **Seconding.** When the pitch is completed, the belayer will need to ascend the route. To ascend the route, use ascenders instead of Prusiks, ascenders are much faster and safer than Prusiks. Attach each ascender to a daisy chain/aider group with carabiners. To adjust the maximum reach/height of the ascenders on the rope, adjust the effective length of the daisy chains with a carabiner the same as with the fifi hook; the typical height will be enough to hold the attached ascender in the hand at nose level. When adjusted to the correct height, the arms need not support much body weight. If the ascender is too high, you will have difficulty reaching and maintaining a grip on the handle.

(1) Unlike lead climbing, there will be a continuous load on the rope during the cleaning of the route, this would normally increase the difficulty of removing protection. To make this easier, as you approach the protection on the ascenders, move the ascenders, one at a time, above the piece. When your weight is on the rope above the piece, you can easily unclip and remove the protection.

CAUTION

If both ascenders should fail while ascending the pitch, a serious fall could result. To prevent this possibility, *tie-in short* on the rope every 10-20 feet by tying a figure eight loop and clipping it into the harness with a separate locking carabiner as soon as the ascent is started. After ascending another 20 feet, repeat this procedure. Do not unclip the previous figure eight until the new knot is attached to another locking carabiner. Clear each knot as you unclip it.

Notes: 1. Ensure the loops formed by the short tie-ins do not catch on anything below as you ascend.

2. If the nature of the rock will cause the "hanging loop" of rope, formed by tying in at the end of the rope, to get caught as you move upward, do not tie into the end of the rope.

(2) Seconding an aid pitch can be done in a similar fashion as seconding free-climbed pitches. The second can be belayed from above as the second "climbs" the protection. However, the rope is unclipped from the protection before the aider/daisy chain is attached.

d. **Seconding Through a Traverse.** While leading an aid traverse, the climber is hanging on the protection placed in front of the current position. If the second were to clean the section by hanging on the rope while cleaning, the protection will be pulled in more than one direction, possibly resulting in the protection failing. To make this safer and easier, the second should hang on the protection just as the leader did. As the second moves to the beginning of the traverse, one ascender/daisy chain/aider group is removed from the rope and clipped to the protection with a carabiner, (keep the ascenders attached to the daisy chain/aider group for convenience when the traverse ends). The second will negotiate the traverse by leapfrogging the daisy chain/aider groups on the next protection just as the leader did. Cleaning is accomplished by removing the protection as it is passed when all weight is removed from it. This is in effect a self-belay. The second maintains a shorter safety tie-in on the rope than for vertical movement to reduce the possibility of a lengthy pendulum if the protection should pull before intended.

e. **Clean Aid Climbing.** Clean aid climbing consists of using protection placed without a hammer or drill involvement: chocks, SLCDs, hooks, and other protection placed easily by hand. This type of aid climbing will normally leave no trace of the climb when completed. When climbing the aiders on clean aid protection, ensure the protection does not "move" from it's original position.

(1) Hooks are any device that rests on the rock surface without a camming or gripping action. Hooks are just what the name implies, a curved piece of hard steel with a hole in one end for webbing attachment. The hook blade shape will vary from one model to another, some have curved or notched "blades" to better fit a certain crystal shape on a face placement. These types of devices due to their passive application, are only secure while weighted by the climber.

(2) Some featureless sections of rock can be negotiated with hook use, although bolts can be used. Hook usage is faster and quieter but the margin of safety is not there unless hooks are alternated with more active forms of protection. If the last twenty foot section of a route is negotiated with hooks, a forty foot fall could result.

6-24. THREE-MAN CLIMBING TEAM

Often times a movement on steep terrain will require a team of more than two climbers, which involves more difficulties. A four-man team (or more) more than doubles the difficulty found in three men climbing together. A four-man team should be broken down into two groups of two unless prevented by a severe lack of gear.

a. Given one rope, a three-man team is at a disadvantage on a steep, belayed climb. It takes at least twice as long to climb an average length pitch because of the third climber and the extra belaying required. The distance between belay positions will be halved if only one rope is used because one climber must tie in at the middle of the rope. Two ropes are recommended for a team of three climbers.

Note: Time and complications will increase when a three-man team uses only one rope. For example: a 100-foot climb with a 150-foot rope would normally require two belays for two climbers; a 100-foot climb with a 150-foot rope would require six belays for three climbers.

b. At times a three-man climb may be unavoidable and personnel should be familiar with the procedure. Although a team of three may choose from many different methods, only two are described below. If the climb is only one pitch, the methods will vary.

CAUTION

When climbing with a team of three, protected traverses will require additional time. The equipment used to protect the traverse must be left in place to protect both the second and third climbers.

(1) The first method can be used when the belay positions are not large enough for three men. If using one rope, two climbers tie in at each end and the other at the mid point. When using two ropes, the second will tie in at one end of both ropes, and the other two climbers will each tie in to the other ends. The most experienced individual is the leader, or number 1 climber. The second, or number 2 climber, is the stronger of the remaining two and will be the belayer for both number 1 and number 3. Number 3 will be the last to climb. Although the number 3 climber does no belaying in this method, each climber should be skilled in the belay techniques required. The sequence for this method (in one pitch increments) is as follows (repeated until the climb is complete):

(a) Number 1 ascends belayed by number 2. Number 2 belays the leader up the first pitch while number 3 is simply anchored to the rock for security (unless starting off at ground level) and manages the rope between himself and number 2. When the leader completes the pitch, he sets up the next belay and belays number 2 up.

(b) Number 2 ascends belayed by number 1, and cleans the route (except for traverses). Number 2 returns the hardware to the leader and belays him up the next pitch. When the leader completes this pitch, he again sets up a new belay. When number 2 receives "OFF BELAY" from the leader, he changes ropes and puts number 3 on belay. He should not have to change anchor attachments because the position was already aimed for a downward as well as an upward pull when he belayed the leader.

(c) Number 3 ascends belayed by number 2. When number 3 receives "BELAY ON," he removes his anchor and climbs to number 2's position. When the pitch is completed he secures himself to one of number 2's belay anchors. When number 1's belay is ready, he brings up number 2 while number 3 remains anchored for security. Number 2 again cleans the pitch and the procedure is continued until the climb is completed.

(d) In this method, number 3 performs no belay function. He climbs when told to do so by number 2. When number 3 is not climbing, he remains anchored to the rock for security. The standard rope commands are used; however, the number 2 climber may include the trailing climber's name or number in the commands to avoid confusion as to who should be climbing.

(d) Normally, only one climber would be climbing at a time; however, the number 3 climber could ascend a fixed rope to number 2's belay position using proper ascending technique, with no effect on the other two members of the team. This would save time for a team of three, since number 2 would not have to belay number 3 and could be either belaying number 1 to the next belay or climbing to number 1. If number 3 is to ascend a fixed rope to the next belay position, the rope will be loaded with number 3's weight, and positioned directly off the anchors established for the belay. The rope should be located so it does not contact any sharp edges. The rope to the ascending number 3 could be secured to a separate anchor, but this would require additional time and gear.

(2) The second method uses either two ropes or a doubled rope, and number 2 and number 3 climb simultaneously. This requires either a special belay device that accepts two ropes, such as the tuber type, or with two Munter hitches. The ropes must travel through the belay device(s) without affecting each other.

(a) As the leader climbs the pitch, he will trail a second rope or will be tied in with a figure eight in the middle of a doubled rope. The leader reaches the next belay position and establishes the anchor and then places both remaining climbers on belay. One remaining climber will start the ascent toward the leader and the other will start when a gap of at least 10 feet is created between the two climbers. The belayer will have to remain alert for differences in rope movement and the climbers will have to climb at the same speed. One of the "second" climbers also cleans the pitch.

(b) Having at least two experienced climbers in this team will also save time. The belayer will have additional requirements to meet as opposed to having just one second. The possible force on the anchor will be twice that of one second. The second that is not cleaning the pitch can climb off route, but staying on route will usually prevent a possible swing if stance is not maintained.

CHAPTER 7
ROPE INSTALLATIONS

Obstacles on the battlefield today are inevitable. They can limit the battlefield and, even worse, prevent a unit from accomplishing its mission. However, with highly skilled personnel trained on rope installations, leaders can be assured that even a unit with limited mountain skills and experience will be able to successfully move and operate in terrain that would otherwise have been impassable.

Section I. FIXED ROPE

A fixed rope is a rope anchored in place to assist soldiers in movement over difficult terrain. Its simplest form is a rope tied off on the top of steep terrain. As terrain becomes steeper or more difficult, fixed rope systems may require intermediate anchors along the route. Moving on a fixed rope requires minimal equipment. The use of harnesses, ascenders, and other technical gear makes fixed rope movement easier, faster, and safer, but adds to total mission weight.

7-1. INSTALLATION

To install a fixed rope, two experienced climbers rope up for a roped climb. The leader must have the necessary equipment to rig the anchor at the top of the pitch. Although leader protection is usually not needed on a typical slope, additional hardware can be brought along and placed at the leader's discretion. The second will establish a belay if protection is being placed. Otherwise, he will stack and manage the rope. He ensures the rope runs smoothly up the slope and does not get tangled as the climber ascends. Upon reaching the end of the pitch, the leader will establish the top anchor. Once the anchor is rigged, the leader will take up any remaining slack between himself and the second. He will anchor the installation rope and remain tied into the rope. The second unties from his end of the rope and begins to climb. If the leader placed protection, the second will clean the pitch on his way up.

7-2. UTILIZATION

All personnel using the fixed rope grasp the rope with the palm downward and use it for assistance as they ascend the slope (Figure 7-1, page 7-2). An individual can easily prevent a long fall by attaching himself to the rope with a sling using a friction knot (for example, Prusik, autoblock). The knot is slid along the rope as the individual ascends. If the climber slips and loses control of the rope, the friction knot will grab the rope and arrest the fall. The friction knot used in this manner is referred to as a self-belay (Figure 7-2, page 7-2).

Figure 7-1. Using a fixed rope.

Figure 7-2. Using a self-belay.

7-3. RETRIEVAL

If the fixed rope is to be used on the descent, it can be left in place and recovered after the last rappel. If not, the last climber will tie into the rope and be belayed from above. The climber now can easily free the rope if it gets caught on anything as it is taken up from the belayer.

7-4. FIXED ROPE WITH INTERMEDIATE ANCHORS

Whenever the route varies from the fall line of the slope, the fixed rope must be anchored at intermediate anchor points (Figure 7-3). Intermediate anchor points should also be used on any long routes that exceed the length of a single rope. The use of intermediate anchor points creates independent sections and allows for changes in direction from one section to the next. The independent sections allow for more personnel to move on the fixed rope. This type of fixed rope is commonly used along exposed ridges and narrow mountain passes.

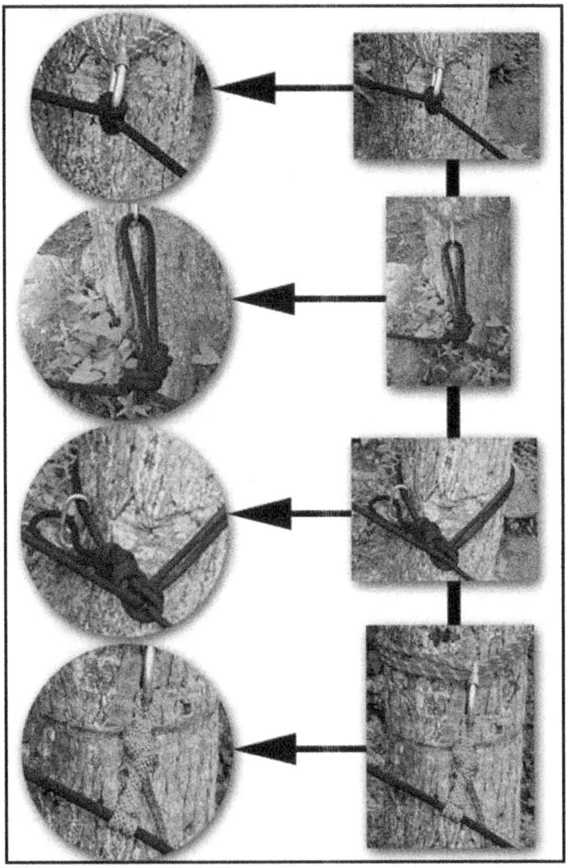

Figure 7-3. Fixed rope with intermediate anchors.

a. **Installation.** Two experienced climbers prepare for a roped climb. The leader will carry a typical rack with enough hardware to place an adequate number of intermediate anchor points. The second sets up a standard belay. The route they select must have the following characteristics:

- Most suitable location, ease of negotiation, avoids obstacles.
- Availability of anchors (natural and artificial).
- Area is safe from falling rock or ice.
- Tactical considerations are met.
- A rope routed between knee and chest height (waist high preferred).
- Rope crossovers should be avoided.

As the leader climbs the route, he will place the anchors and route the climbing rope as in a typical roped climb. The leader makes use of any available natural anchors.

Note: Sling attachments should be kept as short and snug as possible to ensure that a load on the fixed rope below the anchor is placed only on that anchor. This will prevent one section from affecting another section.

(1) The leader places an anchor at all points where a change of direction occurs. He also makes every attempt to route the rope so personnel will not have to cross back and forth over the rope between sections.

(2) When the leader reaches the end of the pitch, he temporarily anchors the rope. He should use a sling to anchor himself if there is any chance of slipping and falling. He then takes up any excess slack, and attaches the rope to the anchor.

Note: Enough slack must be left in the rope so the second can tie the knots necessary to fix the rope.

(3) The second unties from the rope and anchors it at the bottom. He attaches himself to the rope with a sling using a friction knot to create a self-belay. The self-belay will protect the second as he climbs and fixes the rope to the intermediate anchor points. When he reaches an anchor point, he unclips the climbing rope so he can advance the self-belay beyond the anchor point. He then takes the slack out of the section below the anchor point. He ensures that the fixed rope will be approximately knee to chest level as climbers negotiate the installation. He then attaches the rope using an anchor knot (for example, middle-of-the-rope clove hitch, double figure-eight). The second then moves to next anchor point and repeats the process.

(4) If a long runner is to be used at any anchor point, the second should adjust the section below it so the runner is oriented in the direction where the load or pull on the anchor will come from. This will help isolate the section.

(5) The sections are normally adjusted fairly snug between anchor points. A slack section may be necessary to move around obstacles in the route or large bulges in the terrain. If clove hitches are used, adjusting the clove hitches at each end of the section can leave any amount of slack.

(6) A middle-of-the-rope Prusik safetied with a figure eight may be used when utility ropes are available. These are used to adjust the rope height (either higher or lower).

(7) In addition to the fixed rope, the second could anchor etriers to be used as footholds.

(8) When the second reaches the end of the pitch, the rope is removed from the top anchor and the remaining slack is removed from the last section. The rope is reattached to the anchor. If additional fixed rope is required the procedure is repeated using another rope. The second will tie the ropes together before anchoring the next section, creating one continuous fixed rope.

b. **Utilization.** Personnel should be attached to the fixed rope during movement for safety reasons.

(1) If a self-belay is desired, a harness should be worn. A friction knot will be tied to the installation rope using a short sling. The sling will then be attached to the harness. Another short sling will be used as a safety line. One end of the sling will be attached to the harness and the other will have a carabiner inserted. This safety line is also attached to the fixed rope during movement. Once the climber reaches an anchor point, he removes his safety line and attaches it to the anchor or attaches it to the next section of rope. He will then untie the friction knot and tie another friction knot beyond the anchor point. The use of a mechanical ascender in the place of the friction knot could greatly speed up movement.

(2) There will be many situations where a self-belay may not be required. In these situations an individual may attach himself to the fixed rope using only a safety line. The individual will tie into the middle of a sling rope approximately 12 feet long. Fixed loops are tied into the running ends and a carabiner is attached into each of the fixed loops. The individual now has two points of attachments to the fixed rope. Upon reaching an anchor point, one safety line is removed and advanced beyond the anchor point onto the next section. Then the next safety is removed and placed on the next section. This way the individual is always secured to the fixed rope at all times.

(3) Personnel will move one at a time per section during the entire movement. Once an individual changes over to the next section he signals the next man to climb. When descending on the fixed rope, personnel can down climb using the installation for assistance. Another option would be to descend using a hasty rappel.

c. **Retrieval.** When the installation is retrieved, the next to last man on the system will untie the knots at the intermediate anchor points and reclips the rope as he ascends. He will be attached to rope using a self-belay. Once he reaches the top of the pitch, the rope should be running the same as when the leader initially placed it. The last man will untie the rope from the bottom anchor and tie into the rope. He will the clean the pitch as he climbs while being belayed from above.

Section II. RAPPELLING

When an individual or group must descend a vertical surface quickly, a rappel may be performed. Rappelling is a quick method of descent but it is extremely dangerous. These dangers include anchor failure, equipment failure, and individual error. Anchors in a mountainous environment should be selected carefully. Great care must be taken to load the anchor slowly and to ensure that no excessive stress is placed on the anchor. To ensure this, bounding rappels should be prohibited, and only walk down rappels used. Constant vigilance to every detail will guarantee a safe descent every time.

7-5. SELECTION OF A RAPPEL POINT

The selection of the rappel point depends on factors such as mission, cover, route, anchor points, and edge composition (loose or jagged rocks). There must be good anchors (primary and secondary). The anchor point should be above the rappeller's departure point. Suitable loading and off-loading platforms should be available.

7-6. INSTALLATION OF THE RAPPEL POINT

A rappel lane should have equal tension between all anchor points by establishing primary and secondary anchor points. The rappel rope should not extend if one anchor point fails. The following methods of establishing an anchor can be performed with a single or double rope. A double rope application should be used when possible for safety purposes.

a. If a rappel lane is less than half the rope length, the climber may apply one of the following techniques:

(1) Double the rope and tie a three-loop bowline around the primary anchor to include the primary anchor inside two loops and enough rope in the third loop to run to the secondary anchor (another three-loop bowline secured with an overhand knot).

(2) Bowline secured with an overhand knot (or any appropriate anchor knot).

(3) Double the rope and establish a self-equalizing anchor system with a three-loop bowline or any other appropriate anchor knot

b. If a rappel lane is greater than half the rope length, the climber may apply one of the following techniques:

(1) Use two ropes. With both ropes, tie a round turn anchor bowline around a primary anchor point. Take the remaining rope (the tail from the primary anchor bowline) and tie another round turn anchor bowline to a secondary anchor point. The secondary anchor point should be in a direct line behind the primary anchor point. The anchor can be either natural or artificial.

(2) Use two ropes. Establish a multi-point anchor system using a bowline on a bight or any other appropriate anchor knot.

c. Situations may arise where, due to the length of the rappel, the rappel rope cannot be tied to the anchor (if the rope is used to tie the knots, it will be too short to accomplish the rappel). The following techniques can be used:

(1) When using a natural anchor, tie a sling rope, piece of webbing, or another rope around the anchor using proper techniques for slinging natural anchors. The rappel rope will have a fixed loop tied in one end, which is attached to the anchor created.

(2) When using an artificial anchor, tie off a sling rope, piece of webbing, runner, or another rope to form a loop. Use this loop to create an equalizing or pre-equalized anchor, to which the rappel rope will be attached.

7-7. OPERATION OF THE RAPPEL POINT

Due to the inherent dangers of rappelling, special care must be taken to ensure a safe and successful descent.

a. **Communication.** Climbers at the top of a rappel point must be able to communicate with those at the bottom. During a tactical rappel, radios, hand signals, and rope signals are considered. For training situations use the commands shown in Table 7-1.

COMMAND	GIVEN BY	MEANING
LANE NUMBER ___, ON RAPPEL	Rappeller	I am ready to begin rappelling.
LANE NUMBER ___, ON BELAY	Belayer	I am on belay and you may begin your rappel.
LANE NUMBER ___, OFF RAPPEL	Rappeller	I have completed the rappel, cleared the rappel lane, and am off the rope.
LANE NUMBER ___, OFF BELAY	Belayer	I am off belay.

Table 7-1. Rappel commands.

Notes: 1. In a training environment, the lane number must be understood.

2. In a tactical situation, a series of tugs on the rope may be substituted for the oral commands to maintain noise discipline. The number of tugs used to indicate each command is IAW the unit SOP.

b. **Duties and Responsibilities.**

(1) Duties of the rappel point commander are as follows:

- Ensures that the anchors are sound and the knots are properly tied.
- Ensures that loose rock and debris are cleared from the loading platform.
- Allows only one man on the loading platform at a time and ensures that the rappel point is run orderly.
- Ensures that each man is properly prepared for the particular rappel: gloves on, sleeves down, helmet with chin strap fastened, gear prepared properly, and rappel seat and knots correct (if required). He also ensures that the rappeller is hooked up to the rope correctly and is aware of the proper braking position.
- Ensures that the proper signals or commands are used.
- Dispatches each man down the rope.
- Is the last man down the rope.

(2) Duties of the first rappeller down are as follows:

- Selects a smooth route, for the rope, that is clear of sharp rocks.
- Conducts a self-belay.
- Clears the route, placing loose rocks far enough back on ledges to be out of the way, which the rope may dislodge.
- Ensures the rope reaches the bottom or is at a place from which additional rappels can be made.
- Ensures that the rope will run freely around the rappel point when pulled from below.
- Clears the rappel lane by straightening all twists and tangles from the ropes.
- Belays subsequent rappellers down the rope or monitors subsequent belayers
- Takes charge of personnel as they arrive at the bottom (off-loading platform).

Note: A rappeller is always belayed from the bottom, except for the first man down. The first man belays himself down the rope using a self-belay attached to his rappel seat, which is hooked to the rappel rope with a friction knot. As the first man rappels down the rope, he "walks" the friction knot down with him.

(3) Each rappeller down clears the ropes, and shouts, "Off rappel," (if the tactical situation permits). After the rope is cleared and the rappeller is off rappel, he acts as the belayer for next rappeller.

(4) Soldiers wear gloves for all types of rappels to protect their hands from rope burns.

(5) Rappellers descend in a smooth, controlled manner.

(6) The body forms an L-shape with the feet shoulder-width apart, legs straight, and buttocks parallel to the ground. When carrying equipment or additional weight, a modified L-shape is used with the legs slightly lower than the buttocks to compensate for the additional weight. The rappeller's back is straight. He looks over the brake shoulder. The guide hand is extended on the rope with the elbow extended and locked. The rope slides freely through the guide hand. The guide hand is used to adjust equipment and assist balance during descent. The rappeller grasps the rope firmly with the brake hand and places it in the brake position. Releasing tension on the rope and moving the brake hand regulates the rate of descent. The rappeller never lets go of the ropes with his brake hand until the rappel is complete.

c. **Tying Off During the Rappel.** It may be necessary to stop during descent. This can be accomplished by passing the rope around the body and placing three or more wraps around the guide-hand-side leg, or by tying off using the appropriate knot for the rappel device.

7-8. RECOVERY OF THE RAPPEL POINT

After almost all personnel have descended, only two personnel will remain at the top of the rappel point. They will be responsible for establishing a retrievable rappel.

a. **Establishing the Retrievable Rappel.** To set up a retrievable rappel point, a climber must apply one of the following methods:

(1) Double the rope when the rappel is less than half the total length of the rope. Place the rope, with the bight formed by the midpoint, around the primary anchor. Join the tails of the rappel rope and throw the rope over the cliff. Tie a clove hitch around a carabiner, just below the anchor point, with the locking bar outside the carabiner away from the gate opening end and facing uphill. Snap the opposite standing portion into the carabiner. When the rappeller reaches the bottom, he pulls on that portion of the rope to which the carabiner is secured to allow the rope to slide around the anchor point.

(2) When the length of the rappel is greater than half the length of the rope used, join two ropes around the anchor point with an appropriate joining knot (except the square knot). Adjust the joining knot so that it is away from the anchor. Tie a clove hitch around a carabiner just below the anchor point with the locking bar outside the carabiner away from the gate opening end and facing uphill. Snap the opposite standing portion into the carabiner. Upon completion of the rappel, pull the rope to which the carabiner is secured to allow the rope to slide around the anchor point.

Notes: 1. When setting up a retrievable rappel, use only a primary point; care is taken in selecting the point.

2. Ensure the soldiers have a safety line when approaching the rappel point, with only the rappeller going near the edge.

b. **Retrieving the Rappel Rope.** The next to last rappeller will descend the lane, removing any twists, and routes the rope for easiest retrieval. Once he reaches the end of the rappel, he tests the rope for retrieval. If the rappel is retrievable, the last man will rappel down. Once he is off rappel, he pulls the lane down.

7-9. TYPES OF RAPPELS

During military mountaineering operations, many types of rappels may be used. The following paragraphs describe some these rappels.

a. **Hasty Rappel** (Figure 7-4). The hasty rappel is used only on moderate pitches. Its main advantage is that it is easier and faster than other methods. Gloves are worn to prevent rope burns.

(1) Facing slightly sideways to the anchor, the rappeller places the ropes horizontally across his back. The hand nearest to the anchor is his guide hand, and the other is the brake hand.

(2) To stop, the rappeller brings his brake hand across in front of his body locking the rope. At the same time, he turns to face up toward the anchor point.

Figure 7-4. Hasty rappel.

b. **Body Rappel** (Figure 7-5). The rappeller faces the anchor point and straddles the rope. He then pulls the rope from behind, and runs it around either hip, diagonally across the chest, and back over the opposite shoulder. From there, the rope runs to the brake hand, which is on the same side of the hip that the rope crosses (for example, the right hip to the left shoulder to the right hand). The rappeller leads with the brake hand down and faces slightly sideways. The foot corresponding to the brake hand precedes the guide hand at all times. The rappeller keeps the guide hand on the rope above him to guide himself--not to brake himself. He must lean out at a sharp angle to the rock. He keeps his legs spread well apart and relatively straight for lateral stability, and his back straight to reduce friction. The BDU collar is turned up to prevent rope burns on the neck. Gloves are worn, and other clothing may be used to pad the shoulders and buttocks. To brake, the rappeller leans back and faces directly toward the rock area so his feet are horizontal to the ground.

Figure 7-5. Body rappel.

Notes: 1. Hasty rappels and body rappels are not used on pitches that have overhangs; feet must maintain surface contact.

2. Hasty rappels and body rappels are not belayed from below.

c. **Seat-Hip Rappel** (Figure 7-6). The seat rappel differs from the body rappel in that the friction is absorbed by a carabiner that is inserted in a sling rope seat and fastened to the rappeller. This method provides a faster and more frictional descent than other methods. Gloves can be worn to prevent rope burns.

Figure 7-6. Seat-hip rappel.

(1) An alternate technique is to insert two carabiners opposite and opposed. Then insert a locking carabiner into the two carabiners with opening gate on brake hand side. Then run the rope through the single carabiner. This helps to keep the rappel rope away from the harness.

(2) To hook up for the seat-hip method, stand to one side of the rope. If using a right-hand brake, stand to the left of the rappel rope facing the anchor; if using a left-hand brake, stand to the right of the rappel rope. Place the rappel rope(s) into the locking carabiner; slack is taken between the locking carabiner and anchor point and wrapped around the shaft of the locking carabiner and placed into the gate so that a round turn is made around the shaft of the locking carabiner (Figure 7-7, page 7-12). Any remaining slack is pulled toward the uphill anchor point. If a single rope is used, repeat this process to place two round turns around the shaft of the locking carabiner. Face the anchor point and descend using the upper hand as the guide and the lower hand as the brake. This method has minimal friction, and is fast and safe. However, care is taken that the rope is hooked correctly into the carabiner to avoid the gate being opened by the rope. Loose clothing or equipment around the waist may be accidentally pulled into the locking carabiner and lock (stop) the rappel. For this reason, the rappeller must tuck in his shirt and keep his equipment out of the way during his descent.

Figure 7-7. Proper hookup using carabiner wrap.

d. **Figure-Eight Descender**. The figure-eight descender puts less kinks in the rope, and it can be used with one or two ropes (Figure 7-8).

(1) To use the figure-eight descender, pass a bight through the large eye and then over the small eye onto the neck. Place the small eye into a locking carabiner. To reduce the amount of friction on the figure-eight, place the original bight into the carabiner and not around the neck of the descender. (Less friction requires more braking force from the rappeller.)

(2) The guide hand goes on the rope that is running from the anchor. The brake hand goes on the slack rope. The brake is applied by moving the brake hand to the rear or downward.

Figure 7-8. Figure-eight descender.

d. **Other Devices**. Many different types of devices are similar in design and operation to the basic plate. These include slots or plates and tubers. Most of these devices can accommodate two ropes not greater than 7/16 of an inch in size. Follow manufacturer's directions for using these devices for rappelling.

e. **Extending the Rappel Device**. The rappel device can be extended using either a piece of webbing or cordage to move the device away from the body and the harness, preventing accidental damage (Figure 7-9, page 7-14). It also allows for easier self-belay.

f. **Self-Belay Techniques.** A friction knot can be used as a belay for a rappeller (Figure 7-9, page 7-14). The knot acts as the brake hand when the rappeller must work or negotiate an obstacle requiring the use of both hands. The knot acts as a belay if the rappeller loses control of the rope.

Figure 7-9. Extended hookup with self-belay.

Section III. ONE-ROPE BRIDGE

The one-rope bridge is constructed using a static rope. The rope is anchored with an anchor knot on the far side of the obstacle and is tied off at the near end with a tightening system. A one-rope bridge may be built many ways, depending upon the tactical situation and area to be crossed (crossing a gorge above the tree line may require constructing artificial anchors). However, they all share common elements to safely construct and use the bridge: two suitable anchors; good loading and unloading platforms; a rope about 1-meter (waist) high for loading and unloading; a tightening system; and a rope tight enough for ease of crossing. Which side the tightening system is utilized, or whether an anchor knot or retrievable bowline is used, depends on the technique.

7-10. SITE SELECTION

A suitable crossing site must have "bombproof" anchors on both the near side and far side. These anchors must be extremely strong due to the amount of tension that will be placed upon them. Natural anchors, such as large trees and solid rock formations, are always preferred. The site must also have suitable loading and off-loading platforms to facilitate safe personnel movement.

7-11. INSTALLATION USING TRANSPORT TIGHTENING SYSTEM

The transport tightening system provides a mechanical advantage without requiring additional equipment.

a. The rope must first be anchored on the far side of the obstacle. If crossing a stream, the swimmer must be belayed across. If crossing a ravine or gorge, crossing may involve rappelling and a roped climb. Once across, the swimmer/climber will temporarily anchor the installation rope.

b. One man on the near side ties a fixed-loop knot (for example, wireman's, figure-eight slip knot) approximately 3 feet from the near side anchor and places the carabiner into the loop of the knot. The opening gate must be up and away from the loop. If two carabiners are used, the gates will be opposing. At that time, soldiers route the remainder of the rope around the near side anchor point and hook the rope into the carabiner. This system is known as a transport-tightening system (Figure 7-10). The man on the far side pulls the knot out four to six feet from the near anchor.

c. Once the knot has been pulled out, the far side man anchors the rope using a tensionless anchor. The anchor should be waist high.

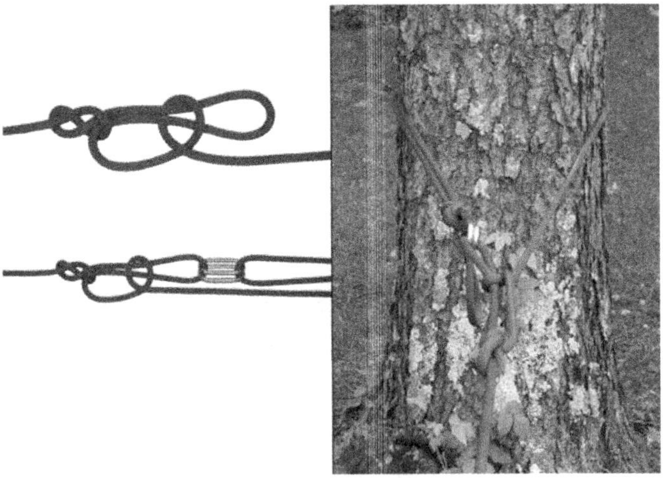

Figure 7-10. Transport tightening system.

d. A three-man pull team on the near side pulls the slack out of the installation rope. The knot should be close enough to the near side anchor to allow personnel to easily load the installation.

Note: No more than three personnel should be used to tighten the rope. Using more personnel can over-tighten the rope and bring the rope critically close to failure.

e. The rope the can be secured using one of three methods: transport knot (Figure 7-11), round turn around anchor and two half hitches on a bight (Figure 7-12), or a tensionless anchor knot (Figure 7-13).

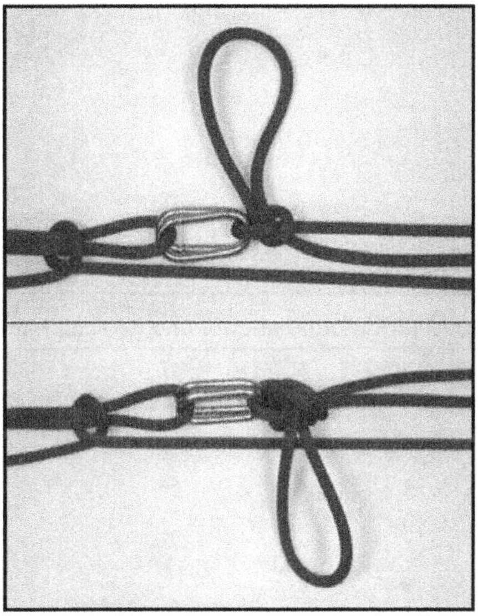

Figure 7-11. Transport knot.

Figure 7-12. Round turn around anchor and two half hitches on a bight.

Figure 7-13. Tensionless anchor knot.

Note: During training, a second static rope may be installed under less tension and alongside the tight rope to increase safety. An individual would clip into both ropes when crossing, thus having a backup in case of failure of the tighter rope.

7-12. INSTALLATION USING Z-PULLEY TIGHTENING SYSTEM

The Z-pulley tightening system (Figure 7-14) is another method for gaining a mechanical advantage.

a. The rope is brought across the obstacle the same way as discussed in paragraph 7-10.

b. Once across, the far side man anchors the rope.

c. One soldier ties a friction knot (autoblock, web wrap, Kleimheist) with a sling rope onto the bridging rope on the near side bank. Two steel carabiners are inserted with opposing gates into the friction knot.

d. The rope is routed around the near side anchor and through the carabiners, from inside to outside, and is run back to the near side anchor.

e. A second sling rope is tied to the bridge rope and then anchored to the near side anchor. This knot will be used as a progress capture device.

Figure 7-14. Z-pulley tightening system.

f. The three-man pull team on the near side then pulls on the rope, creating a pulley effect that tightens the system. As the rope is pulled tight, one man pushes the friction knot back toward the far side.

g. When the rope is tight, it is tied off with a tensionless anchor knot, transport knot, or round turn around anchor and two half hitches on a bight.

7-13. UTILIZATION

The rope bridge can be used to move personnel and equipment over obstacles. There are several methods of accomplishing this.

a. **Method of Crossing**. If dry crossing is impossible, soldiers will use the rope bridge as a hand line. Preferably, all soldiers will tie a safety line and attach it to the rope installation as they cross. If the soldier must cross with his rucksack, he may wear it over both shoulders, although the preferred method is to place another carabiner into the top of the rucksack frame, attach it to the bridge, and pull the rucksack across. Soldiers will always cross on the downstream side of the installation. If a dry crossing is possible soldiers will use one of three methods: commando crawl, monkey crawl, and Tyrolean traverse.

(1) ***Commando Crawl*** (Figure 7-15). The soldier lies on top of the rope with the upstream foot hooked on the rope and the knee bent close to the buttocks; the downstream leg hangs straight to maintain balance. He progresses by pulling with his hands and arms. To recover if he falls over, the soldier hooks one leg and the opposite arm over the rope, and then pushes down with the other hand to regain position.

Figure 7-15. Commando crawl.

Note: Only one man at a time is allowed on the bridge while conducting a commando crawl.

(2) ***Monkey Crawl*** (Figure 7-16, page 7-20). The soldier hangs below the rope suspended by his hands with both heels crossed over the rope. He pulls with his hands and arms, and pushes with his feet to make progress.

Figure 7-16. Monkey crawl.

(3) ***Rappel Seat Method*** (Figure 7-17). The soldier ties a rappel seat (or dons a seat harness) with the carabiner facing up and away from his body. He then faces the rope and clips into the rope bridge. He rotates under the rope and pulls with his hands and arms to make progress. The rappel seat method is the preferred method. If crossing with rucksacks, a carabiner is inserted into the frame and attached to the rope bridge. The soldier the places one or both legs through the shoulder carrying straps and pulls the rucksack across.

Figure 7-17. Rappel seat method.

b. **Rigging Special Equipment**. Any special equipment, such as crew-served weapons, ammunition, or supplies, must be rigged for movement across the rope bridge. A unit SOP may dictate the rigging of these items, but many expedient methods exist. The rigging should use various items that would be readily available to a deployed unit. Some of these items include tubular nylon webbing, cordage (various sizes), and carabiners.

(1) ***Machine Guns***. To rig machine guns, use a sling rope and tie a rerouted figure-eight around the spine of the front sight post. Then tie two evenly spaced fixed loops. Finally, anchor the sling rope to the buttstock of the machine gun. Additional tie downs may be necessary to prevent accidental disassembly of the weapon.

(2) ***ALICE Packs.*** ALICE packs can be joined together with a sling to facilitate moving more than one rucksack at one time.

7-14. HAULING LINE

A hauling line may be used to move rucksacks or casualties across the rope bridge (Figure 7-18).

a. **Construction**. An additional rope is brought across the rope bridge and anchored to the far side. The other end is anchored on the near side. All the slack is pulled to the near side, and a figure-eight slip knot is tied at the loading platform. A carabiner is inserted into the loop and clipped onto the rope bridge.

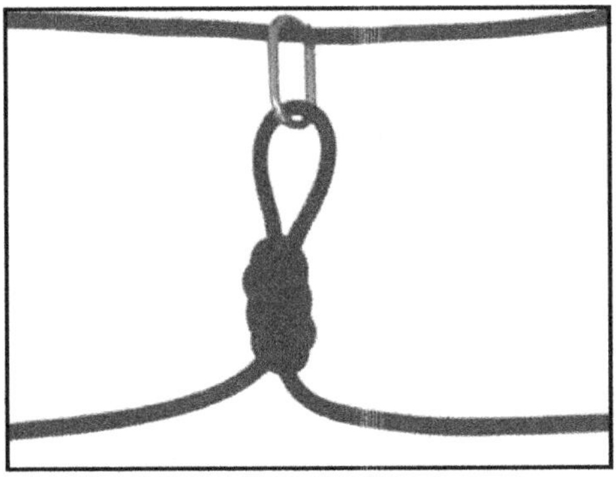

Figure 7-18. Hauling line.

b. **Moving Rucksacks**. Use carabiners to attach the rucksack frames to the rope bridge. Then clip the carabiner of the hauling line into the carabiner of the rucksack closest to the far side. Personnel on the far side pull the rucksacks across using the hauling line while personnel on the near side manages the slack at all times.

c. **Moving Litters**. The carabiner of the hauling line will remain on the rope bridge. On each side of this carabiner, using the hauling line tie a middle-of-the-rope clove hitch around both of the horizontal lift straps of the litter. Remove the slack between the carabiners. Then place the carabiners in each of the lift straps onto the rope bridge. The same technique used for the rucksacks is used to pull the litter across.

7-15. RETRIEVAL

Once all except two troops have crossed the rope bridge, the bridge team commander (BTC) chooses either the wet or dry method to dismantle the rope bridge.

a. If the BTC chooses the dry method, he should have anchored his tightening system with the transport knot.

(1) The BTC back-stacks all of the slack coming out of the transport knot, then ties a fixed loop and places a carabiner into the fixed loop.

(2) The next to last man to cross attaches the carabiner to his rappel seat or harness, and then moves across the bridge using the Tyrolean traverse method.

(3) The BTC then removes all knots from the system. The far side remains anchored. The rope should now only pass around the near side anchor.

(4) A three-man pull team, assembled on the far side, takes the end brought across by the next to last man and pulls the rope tight again and holds it.

(5) The BTC then attaches himself to the rope bridge and moves across.

(6) Once across, the BTC breaks down the far side anchor, removes the knots, and then pulls the rope across.

(b) If the BTC chooses a wet crossing, any method can be used to anchor the tightening system.

(1) All personnel cross except the BTC or the strongest swimmer.

(2) The BTC then removes all knots from the system.

(3) The BTC ties a fixed loop, inserts a carabiner, and attaches it to his rappel seat or harness. He then manages the rope as the slack is pulled to the far side.

(4) The BTC then moves across the obstacle while being belayed from the far side.

Section IV. SUSPENSION TRAVERSE

The suspension traverse is used to move personnel and equipment over rivers, ravines, chasms, and up or down a vertical rock face (Figure 7-19). The system may be established on a plane, varying from horizontal to near vertical.

Figure 7-19. Suspension traverse.

7-16. SITE SELECTION

The crossing site must have bombproof anchors at the near side and the far side, and suitable loading and off-loading platforms. If the anchors do not provide sufficient height to allow clearance, an A-frame must be used.

7-17. INSTALLATION

Installation of a suspension traverse can be time-consuming and equipment-intensive. All personnel must be well trained and well rehearsed in the procedures.

a. **A-frames**. Even in wooded mountainous terrain constructing an A-frame may be necessary due to the lack of height where the installation is needed. Site selection determines whether more height is needed; mission requirements determine site selection. The two main installations that use A-frames are the suspension traverse and vertical hauling line.

b. **Equipment**. Two sturdy poles are needed. The exact size of the poles depends on the type of load and location of the installation. The average size A-frame pole should be at least 3 inches in diameter and 9 to 12 feet long. Three to five 14-foot sling ropes are needed, depending on the size of the poles used for the A-frame.

c. **Construction**. Place two poles with the butt ends flush, and mark the apex on both poles.

(1) Ensure that proper height is attained and that the installation runs in a straight line between the two anchors. An A-frame placed out of proper alignment can cause the system to collapse. Try to find natural pockets in which to place the base of the A-frame poles.

(2) With a sling rope, tie a clove hitch around the left pole (standing at the base of the poles and facing the top) 3 inches above the apex marking, leaving about 18 inches of the sling rope free on top of the clove hitch. Place the locking bar on the outside edge of the pole. Make sure the rope end is pointing down as it is tied. (See Figure 7-20A, page 7-24.)

(3) Place the poles side by side and wrap the sling rope horizontally around both poles six to eight times, wrapping down from the clove hitch (Figure 7-20B, page 7-24). It may be necessary to join another sling rope to the first by using a square knot secured with overhand knots. Position this knot on the outside of one of the poles so as not to interfere with the vertical wraps. Make at least two additional wraps below the joining square knot. (See Figure 7-20C, page 7-24.)

(4) On the last horizontal wrap (ensure there are at least two wraps below the joining knot) to which the clove hitch is not tied, pass the rope between the poles below the wraps, and make four to six tight vertical wraps around the horizontal wraps (Figure 7-20D, page 7-24). Make the wraps as tight as possible. The vertical wraps must be as flat as possible next to each other. When starting the first vertical wrap, ensure it is in the same direction as the 18-inch tail on the top of the clove hitch. Insert a carabiner into the last two vertical wraps (Figure 7-20E, page 7-24).

(5) On the last vertical wrap, pass the rope between the poles above the horizontal wraps. Tie it off with a square knot in the section of rope coming from the clove hitch. Secure with overhand knots tied in the tails. (See Figure 7-20F, page 7-24.)

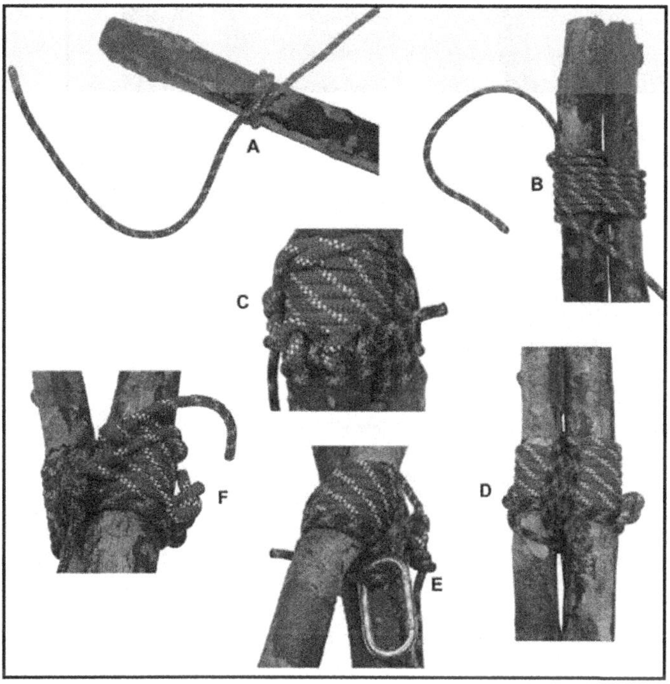

Figure 7-20. A-frame horizontal and vertical wraps.

(6) Use a spreader rope to prevent the A-frame from collapsing from pressure applied at the apex (Figure 7-21). If the ground is soft, dig the legs in about 6 inches. Tie a sling rope between the legs with a round turn with two half hitches around each leg. Remove all slack in the rope between the legs.

(7) If the ground is a hard surface, tie end-of-the-rope clove hitches with the locking portions facing to the rear, the direction of kick. Tie the tails off at a 45-degree angle with a round turn and two half hitches to a secondary anchor point. The spreader rope should be no more than 6 inches above ground level. The use of clove hitches and half hitches permits easy adjustment of the spreader rope. If more than one sling rope is needed, tie the two ropes together with a square knot and secure with half hitches or overhand knots.

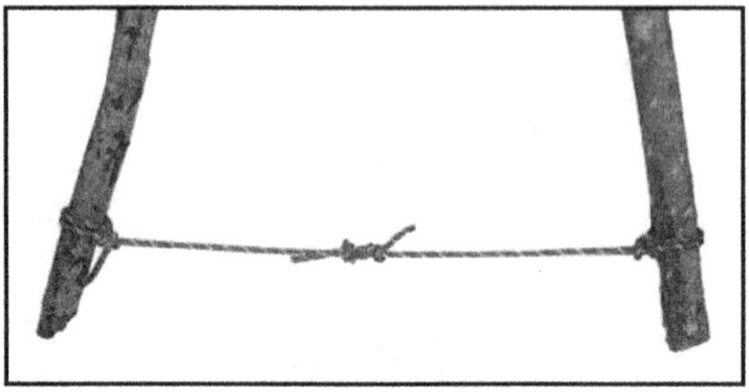

Figure 7-21. A-frame spreader.

d. **Installation Construction**. One man rappels down the pitch and secures two installation (traverse) ropes to the far anchor with an anchor knot. Place a transport tightening system in each installation rope at the near (upper) anchor. Run the installation ropes through or around the anchor in opposite directions and tie off. Anchor the traverse ropes as close together as possible so that the ropes do not cross.

(1) Place the A-frame (if needed) so that both traverse ropes run over the apex and the A-frame splits the angle formed between the near (upper) and far (lower) anchors, with the legs firmly emplaced or anchored with pitons. Ensure that the A-frame is in line with the anchors. Adjust the A-frame under the traverse ropes after tightening to firmly implant the A-frame.

(2) Tighten the installation ropes using either the transport tightening system (paragraph 7-11) or the z-pulley tightening system (paragraph 7-12).

(3) Anchor the A-frame to the traverse rope. Tie a clove hitch at the center of a sling rope. Place it over one of the poles above the apex and move down to the apex so that the locking bar of the clove hitch is to the inside of the A-frame. Secure each end of the sling rope to one of the tightened static lines with two Prusik knots-one forward and one to the rear of the A-frame on the same static line rope (Figure 7-22).

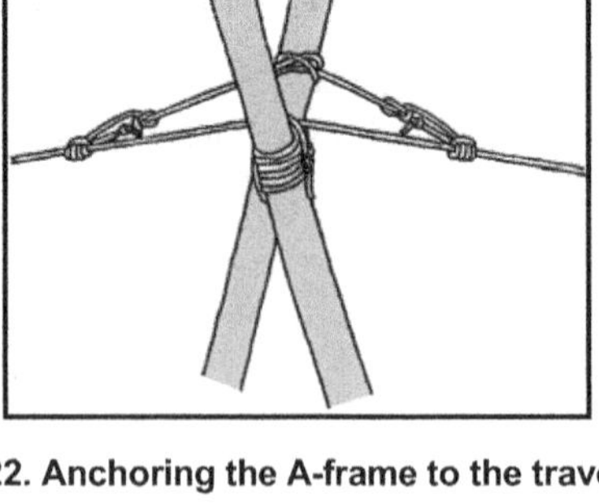

Figure 7-22. Anchoring the A-frame to the traverse rope.

Note: The A-frame should be positioned so that the angles created by the A-frame bisecting the installation rope are approximately equal on both sides. This creates downward pressure holding the A-frame in position, not forcing it in a lateral direction. It must also be placed in a straight line between the upper and lower anchor points.

(4) Use a carrying rope to attach loads to the traverse ropes (Figure 7-23). Join the ends of a 14-foot sling rope with a square knot and two overhand knots. Displace the knot one-third of the distance down the loop and tie an overhand knot both above and below the square knot. This forms two small loops and one large loop that is longer than the two small loops combined.

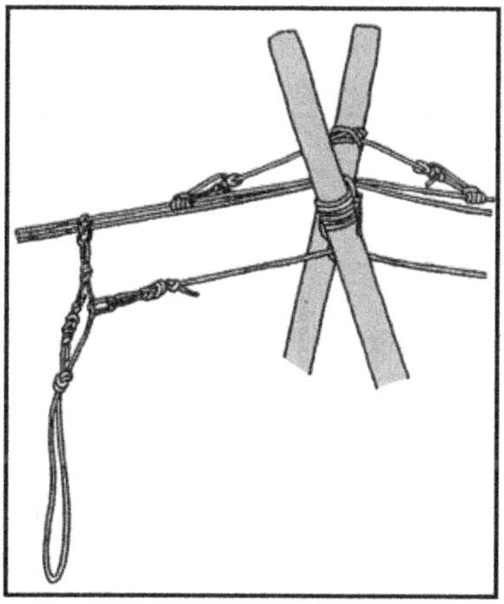

Figure 7-23. Carrying rope for use on a traverse.

(5) Attach the carrying rope to the traverse ropes with carabiners (or a pulley) that have the gates reversed and opening in opposite directions. Attach a belay rope to the center loop of the carrying rope using a fixed loop or locking carabiner on the side opposite the joining knot (Figure 7-23). When the suspension traverse is near horizontal, a second rope may be needed to pull the load across and should be attached to the carrying rope the same as the first.

(6) Insert second carabiner into the one placed into the wraps of the A-frame. This is where a belay rope will be attached

(7) With a sling rope, tie a six wrap middle-of-the-rope Prusik knot to both static ropes near the far side off-loading point. This acts as a stopper knot for the man descending, preventing him from hitting the lower anchor.

(8) Attach the load by running the long loop of the carrying rope through the load or through the soldier's harness and attaching the bottom loop to the traverse rope carabiner.

Descent must be belayed slowly and be controlled. Soldiers descending should hold onto the carrying rope and keep their feet high to avoid contact with the ground. Due to the constant tension maintained on the belayer, use a mechanical belay. If the belayer cannot view the entire descent route, use a relay man.

7-18. RETRIEVAL

The suspension traverse is not as readily retrievable as the one-rope bridge. Therefore, the installing unit should dismantle it after it is no longer needed.

Section V. VERTICAL HAULING LINE

The vertical hauling line is an installation used to move men and equipment up vertical or near-vertical slopes (Figure 7-24). It is often used with a fixed rope for personnel movement. The hauling line is used to move equipment, such as mortars or other crew-served weapons, rucksacks, or supplies

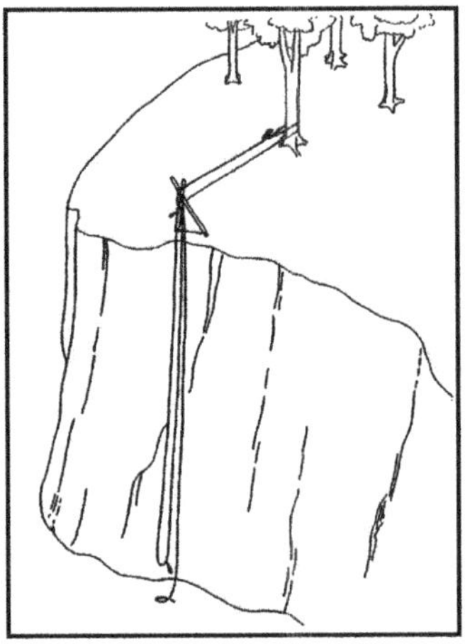

Figure 7-24. Vertical hauling line.

7-19. SITE SELECTION

The first and most important task is to determine where to construct the vertical hauling line. The site must have an appropriate top anchor that is secure enough to hold the system and load. Loading and unloading platforms should be easily accessible natural platforms that provide a safe working area. The ideal platform at the top allows construction of the vertical hauling line without the use of an A-frame. The site should also have sufficient clearance to allow for space between the slope and pulley rope for easy hauling of troops or equipment.

7-20. INSTALLATION

Construct an A-frame, if necessary, and anchor it. Double one installation rope, find the middle, and lay the middle of the installation rope over the apex of the A-frame; a 30-centimeter (12-inch) bight should hang below the apex.

a. To maintain the 12-inch bight, tie clove hitches above the A-frame lashing on each side of the apex with the installation rope, ensuring that the locking bars of the clove hitches are on the inside. Ensure that the portion of the rope that forms the bight comes out of the bottom of the clove hitch. (See Figure 7-25.)

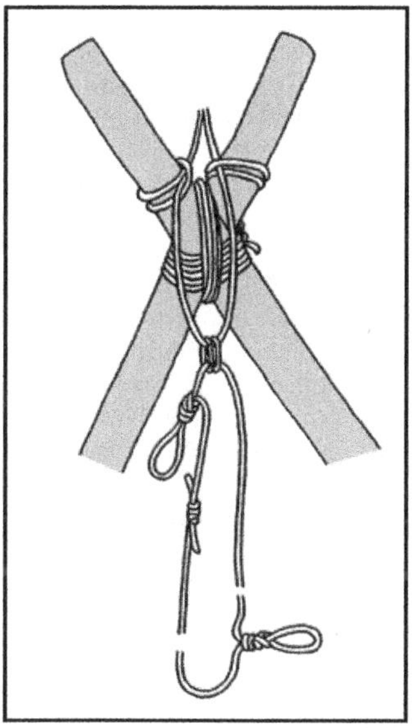

Figure 7-25. Attaching the anchor rope to the A-frame.

b. To anchor the A-frame, use a transport tightening system with the doubled rope, which is tied to the A-frame. Tie this off at an anchor point to the rear of the A-frame installation and adjust the angle of the A-frame so it leans out over the cliff edge. The angle should be 15 to 25 degrees unloaded. The A-frame should not lean outward more than 45 degrees once loaded since the legs can lose their position.

c. Tie the ends of another installation rope together with a joining knot to form the hauling line. Attach the rope to the system by two carabiners with gates up and opposed or one mountain rescue pulley with a locking steel carabiner in the 12-inch bight hanging from the apex of the A-frame. Tie fixed loops (wireman's, directional figure-eight, or single butterfly) on opposite sides of the endless rope at the loading and unloading platforms.

d. Attach equipment to the hauling line 12 inches above the joining knot by a carabiner in the fixed loop.

e. Additional fixed loops may be tied in the hauling line for more control over the object when moving large loads. Attach personnel to the hauling line by use of a rappel seat or seat harness.

Note: Mortar tubes and similar objects are attached to the line by two knots so that the tube stays parallel and as close to the hauling line as possible.

f. When personnel are moved using a vertical hauling line, make a knotted hand line; anchor it in line with, or to, the primary anchor (round turn with a bowline); and place it over the spreader on the legs of the A-frame. Space the overhand knots in the knotted hand line 12 inches apart, with about 20 feet of rope without knots at one end for the anchor. Throw the knotted hand line over the A-frame spreader rope and down the side of the cliff. Personnel ascending the vertical hauling line use this as a simple fixed rope.

g. Use as many men as needed to pull the load to the top by pulling on the rope opposite the load. If equipment and personnel are only being lowered, belay from the top using the hauling line. Station two climbers at the unloading platform to retrieve loads.

h. If only equipment is being hauled up, it is not necessary to use the knotted hand line rope, but it may be necessary to use a belay rope. To move materials or troops up on one side of the hauling line, pull the other side from below.

Note: Personnel using the hauling line for movement must apply all related principles of climbing. Always station two operators at the top of the vertical hauling line to aid men or to retrieve loads when they reach the top. They will always be safetied while working near the edge. When in use, the A-frame should lean slightly over the edge of the cliff to prevent excessive wear on the ropes that pass over sharp rocks. Reduce excessive friction on the system. Remove all obstacles and any loose objects that could be dislodged by personnel and equipment.

7-21. RETRIEVAL

The vertical hauling line is used along a main supply route. When it is no longer needed, the installing unit will return and dismantle the system.

Section VI. SIMPLE RAISING SYSTEMS

Moving heavy objects with limited manpower may be necessary in mountainous terrain. To reduce fatigue of those personnel moving the load, simple rigging techniques can be used to increase the mechanical advantage of the hauling system.

7-22. Z-PULLEY SYSTEM

The Z-pulley system is a simple, easily constructed hauling system (Figure 7-26, page 7-30).

a. **Considerations**. Anchors must be sturdy and able to support the weight of the load. Site selection is governed by different factors: tactical situation, weather, terrain, equipment, load weight, and availability of anchors.

b. **Theory.** Use carabiners as a substitute if pulleys are not available. The mechanical advantage obtained in theory is 3:1. The less friction involved the greater the mechanical advantage. Friction is caused by the rope running through carabiners, the load rubbing against the rock wall, and the rope condition.

c. **Construction**. Use the following procedures to construct a Z-pulley system.

(1) Establish an anchor (anchor pulley system [APS]). Place a carabiner on the runner at the anchor point, place a pulley into the carabiner, and run the hauling rope through the pulley.

(2) With a sling rope (preferably 7 millimeter), tie a middle-of-rope Prusik knot secured with a figure-eight knot on the load side of the pulley. This will be used as a progress capture device (PCD). A mechanical descender may be used in place of the Prusik knot. Take the tails exiting the figure-eight and tie a Munter hitch secured by a mule knot. Ensure the Munter hitch is loaded properly before tying the mule knot.

(3) At an angle away from the APS, establish a moveable pulley system (MPS) to create a "Z" in the hauling rope. Tie another Prusik knot on the load side of the hauling rope. Secure it with a figure-eight knot. Using the tails tie a double-double figure-eight knot. Insert a locking carabiner into the two loops formed, then place the working end into the carabiner. Mechanical ascenders should not be used as an MPS. Move the working end back on a parallel axis with the APS. Provide a pulling team on the working end with extra personnel to monitor the Prusik knots.

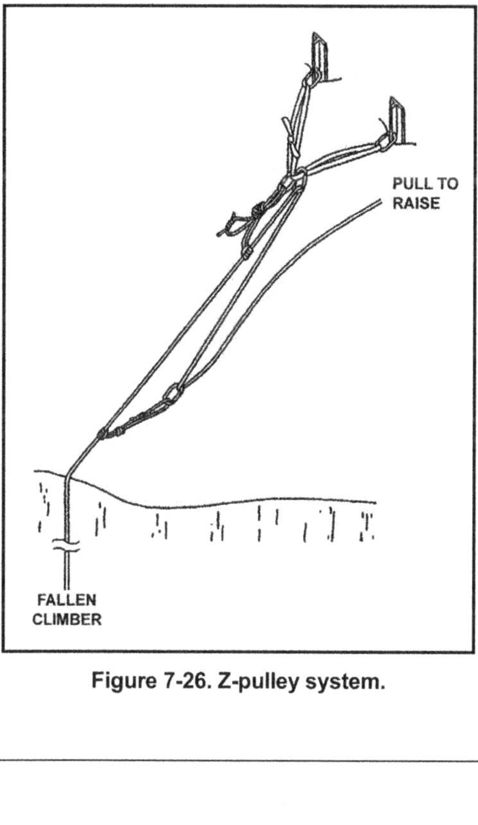

Figure 7-26. Z-pulley system.

d. **Other Factors**. If the two pulleys touch, the "Z" is lost along with the mechanical advantage. For greater efficiency, the main anchor should be well back from the edge and all ropes should pull parallel to the load.

Note: Avoid the possibility of overstressing the anchors. Be aware of reduced sensitivity to the load due to the mechanical advantage. Use belays and backup safeties. Protect the rope from edges and other abrasive parts of the rock.

7-23. U-PULLEY SYSTEM

The U-pulley system is another simple, easily-constructed hauling system (Figure 7-27, page 7-32).

a. **Considerations**. Anchors must be sturdy and able to support the weight of the load. Site selection is governed by different factors: tactical situation, weather, terrain, equipment, load weight, and availability of anchors.

b. **Theory.** Use carabiners as a substitute if pulleys are not available. The mechanical advantage obtained in theory is 2:1. The less friction involved the greater the mechanical advantage. Friction is caused by the rope running through carabiners, the load rubbing against the rock wall, and the rope condition.

c. **Construction.** Use the following procedures construct a U-pulley system.

(1) Anchor the hauling rope.

(2) Prepare the load or casualty for hauling. Place a locking carabiner the on to the harness or the rigged load.

(3) Lower a bight to the casualty or the load.

(4) Place the bight into the carabiner; or place the bight on to a pulley and then place pulley into the carabiner.

(5) Construct a second anchor. Attach a locking carabiner to the anchor.

(6) Tie a middle of the rope Prusik onto the haul rope exiting the pulley. Secure the Prusik with a double-double figure eight. This is the PCD. Place the fixed loops into the locking carabiner of the second anchor.

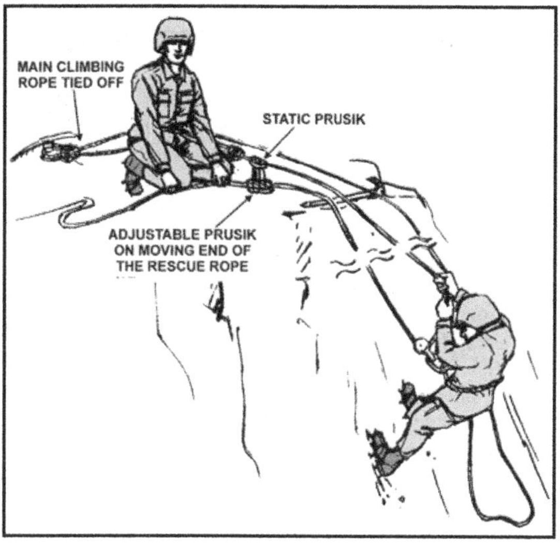

Figure 7-27. U-pulley system.

CHAPTER 8
MOUNTAIN WALKING TECHNIQUES

Mountain travel encompasses the full spectrum of techniques used to negotiate steep, rugged terrain. Mountain walking on rock and snow, technical rock and ice climbing, skiing or snow shoeing, rappelling, and stream crossing are the key travel skills a military mountaineer must possess.

8-1. BASIC PRINCIPLES

Up scree or talus, through boulder fields or steep wooded mountainsides, over snow or grass-covered slopes, the basic principles of mountain walking remain the same.

a. The soldier's weight is centered directly over the feet at all times. He places his foot flat on the ground to obtain as much (boot) sole-ground contact as possible. Then, he places his foot on the uphill side of grass tussocks, small talus and other level spots to avoid twisting the ankle and straining the Achilles tendon. He straightens the knee after each step to allow for rest between steps, and takes moderate steps at a steady pace. An angle of ascent or descent that is too steep is avoided, and any indentations in the slope are used to advantage.

b. In addition to proper technique, pace is adapted to conditions. The mountaineer sets a tempo, or number of steps per minute, according to the pace of the unit in which he is moving. (Physical differences mean that the tempos of two people moving at the same speed will not always be the same.) The soldier maintains tempo and compensates for changes of slope or terrain by adjusting the length of his stride. Tempo, pace, and rhythm are enhanced when an interval of three to five paces is kept between individuals. This interval helps lessen the "accordion" effect of people at the end of the file who must constantly stop and start.

c. The terrain, weather, and light conditions affect the rate of climb. The more adverse the conditions, the slower the pace. Moving too fast, even under ideal conditions, produces early fatigue, requires more rest halts, and results in loss of climbing time. A soldier can only move as fast as his lungs and legs will allow. The trained, conditioned and acclimatized soldier has greater endurance and moves more efficiently. Rest, good nutrition and hydration, conditioning, acclimatization, proper training, and the will to climb are key to successful mountain operations.

d. Breaks are kept to a minimum. When a moderate pace is set, the need for rest halts decreases, the chance of personnel overheating is lessened, and a unit can cover a given distance in a minimal time. If possible, rests should be taken on level ground avoiding steeper inclines.

(1) During the first half-hour of movement an adjustment halt should be taken. Soldiers will loosen or tighten bootlaces as needed, adjust packs and add or remove appropriate layers of clothing.

(2) Following the first halt, a well-conditioned party may take a short rest every 1 to 1.5 hours. If possible, soldiers lean against a tree, rock, or hillside to relieve the shoulders of pack weight, breathe deeply, hydrate, and snack on trail food. These halts are kept short enough to avoid muscles stiffening (one to two minutes).

(3) Later in the march longer halts may be necessary due to fatigue or mission requirements. At these halts soldiers should immediately put on additional clothing to avoid chilling—it is much easier to keep a warm body warm than to warm up a cold one.

(4) After a climb, a good rest is needed to revive tired muscles.

e. The rest step is used for steep slopes, snowfields, and higher elevations. It controls pace and limits fatigue by giving the lungs and legs a moment to recuperate between steps. Pace is kept slow and rhythmic.

(1) After each step forward, the soldier pauses briefly, relaxing the muscles of the forward leg while resting his entire bodyweight on the rear leg. The rear leg is kept straight with the knee locked so that bone, not muscle, supports the weight.

(2) Breathing is synchronized with the rest step. The number of breaths per step will change depending on the difficulty of the climb. Steeper slopes or higher elevations may require several breaths per step. When the air thins at altitude it is especially important to breathe deeply, using the "pressure breathing" technique. The soldier exhales strongly, enabling an easier, deeper inhale.

(3) This slow, steady, halting rest step is more efficient than spurts of speed, which are rapidly exhausting and require longer recovery.

f. Downhill walking uses less energy than uphill but is much harder on the body. Stepping down can hammer the full bodyweight onto the feet and legs. Blisters and blackened toenails, knee damage, and back pain may follow. To avoid these problems the soldier should start by tightening bootlaces to ensure a snug fit (also keep toenails trimmed). A ski pole, ice ax, or walking stick will help take some of the load and give additional stability. (Refer to Chapter 11 for techniques and use of the ice ax.) Keep a moderate pace and walk with knees flexed to absorb shock.

g. Side hill travel on any surface should be avoided whenever possible. Weighted down with a rucksack, the soldier is vulnerable to twisted ankles, back injury, and loss of balance. If side hill travel is necessary, try to switchback periodically, and use any lower angle flat areas such as rocks, animal trails, and the ground above grass or brush clumps to level off the route.

8-2. TECHNIQUES

Mountain walking techniques can be divided according to the general formation, surface, and ground cover such as walking on hard ground, on snow slopes and grassy slopes, through thick brush, and on scree and talus slopes.

a. **Hard Ground.** Hard ground is firmly compacted, rocky soil that does not give way under the weight of a soldier's step. It is most commonly found under mature forest canopy, in low brush or heather, and areas where animals have beaten out multiple trails.

(1) When ascending, employ the rest step to rest the leg muscles. Steep slopes can be traversed rather than climbed straight up. To turn at the end of each traverse, the soldier should step off in the new direction with the uphill foot. This prevents crossing the feet and possible loss of balance. While traversing, the full sole-to-ground principle is accomplished by rolling the ankle downhill on each step. For small stretches the herringbone step may be used—ascending straight up a slope with toes pointed out. A normal progression, as the slope steepens, would be from walking straight up, to a herringbone step, and then to a traverse on the steeper areas.

(2) Descending is best done by walking straight down the slope without traversing. The soldier keeps his back straight and bends at the knees to absorb the shock of each step. Body weight is kept directly over the feet and the full boot sole is placed on the ground with each step. Walking with a slight forward lean and with the feet in a normal position make the descent easier.

b. **Snow Slopes.** Snow-covered terrain can be encountered throughout the year above 1,500 meters in many mountainous areas. Talus and brush may be covered by hardened snowfields, streams made crossable with snowbridges. The techniques for ascending and descending moderate snow slopes are similar to walking on hard ground with some exceptions.

(1) ***Diagonal Traverse Technique.*** The diagonal traverse is the most efficient means to ascend snow. In conjunction with the ice ax it provides balance and safety for the soldier. This technique is a two-step sequence. The soldier performs a basic rest step, placing the leading (uphill) foot above and in front of the trailing (downhill) foot, and weighting the trail leg. This is the in-balance position. The ice ax, held in the uphill hand, is placed in the snow above and to the front. The soldier shifts his weight to the leading (uphill) leg and brings the unweighted trail (downhill) foot ahead of the uphill foot. He shifts weight to the forward (downhill) leg and then moves the uphill foot up and places it out ahead of the trail foot, returning to the in-balance position. At this point the ax is moved forward in preparation for the next step.

(2) ***Step Kicking.*** Step kicking is a basic technique used when crampons are not worn. It is best used on moderate slopes when the snow is soft enough to leave clear footprints. On softer snow the soldier swings his foot into the snow, allowing the leg's weight and momentum to carve the step. Fully laden soldiers will need to kick steps, which take half of the boot. The steps should be angled slightly into the slope for added security. Succeeding climbers will follow directly in the steps of the trailbreaker, each one improving the step as he ascends. Harder snow requires more effort to kick steps, and they will not be as secure. The soldier may need to slice the step with the side of his boot and use the diagonal technique to ascend.

(3) ***Descending Snow.*** If the snow is soft and the slope gentle, simply walk straight down. Harder snow or steeper slopes call for the plunge step, which must be done in a positive, aggressive manner. The soldier faces out, steps off, and plants his foot solidly, driving the heel into the snow while keeping his leg straight. He shifts his weight to the new foot plant and continues down with the other foot. On steeper terrain it may be necessary to squat on the weighted leg when setting the plunge step. The upper body should be kept erect or canted slightly forward.

(4) ***Tips on Snow Travel.*** The following are tips for travelling on snow.

(a) Often the best descent is on a different route than the ascent. When looking for a firmer travel surface, watch for dirty snow—this absorbs more heat and thus hardens faster than clean snow.

(b) In the Northern Hemisphere, slopes with southern and western exposures set up earlier in the season and quicker after storms, but are more prone to avalanches in the spring. These slopes generally provide firm surfaces while northern and eastern exposures remain unconsolidated.

(c) Travel late at night or early in the morning is best if daytime temperatures are above freezing and the sun heats the slopes. The night's cold hardens the snow surface.

(d) Avoid walking on snow next to logs, trees, and rocks as the subsurface snow has melted away creating hidden traps.

c. **Grassy Slopes.** Grassy slopes are usually composed of small tussocks of growth rather than one continuous field.

(1) When ascending, step on the upper side of each hummock or tussock, where the ground is more level.

(2) When descending a grassy slope, the traverse technique should be used because of the uneven nature of the ground. A climber can easily build up too much speed and fall if a direct descent is tried. The hop-skip step can be useful on this type of slope. In this technique, the lower leg takes all of the weight, and the upper leg is used only for balance. When traversing, the climber's uphill foot points in the direction of travel. The downhill foot points about 45 degrees off the direction of travel (downhill). This maintains maximum sole contact and prevents possible downhill ankle roll-out.

Note: Wet grass can be extremely slippery; the soldier must be aware of ground cover conditions.

d. **Thick Brush.** For the military mountaineer, brush is both a help and a hindrance. Brush-filled gullies can provide routes and rally points concealed from observation; on the other hand steep brushy terrain is hazardous to negotiate. Cliffs and steep ravines are hidden traps, and blow downs and thickets can obstruct travel as much as manmade obstacles. When brush must be negotiated take the most direct route across the obstacle; look for downed timber to use as raised paths through the obstacle; or create a tunnel through the obstacle by prying the brush apart, standing on lower branches and using upper limbs for support.

e. **Scree Slopes.** Slopes composed of the smallest rocks are called scree slopes. Scree varies in size from the smallest gravel to about the size of a man's fist.

(1) Ascending scree slopes is difficult and tiring and should be avoided, if possible. All principles of ascending hard ground and snow apply, but each step is carefully chosen so that the foot does not slide down when weighted. This is done by kicking in with the toe of the upper foot (similar to step-kicking in snow) so that a step is formed in the loose scree. After determining that the step is stable, weight is transferred to the upper leg, the soldier then steps up and repeats the process with the lower foot.

(2) The best method for descending scree slopes is to come straight down the slope using a short shuffling step with the knees bent, back straight, feet pointed downhill, and heels dug in. When several climbers descend a scree slope together, they should be as close together as possible (one behind the other at single arm interval) to prevent injury from dislodged rocks. Avoid running down scree as this can cause a loss of control. When the bottom of the slope (or run out zone) cannot be seen, use caution because drop-offs may be encountered.

(3) Scree slopes can be traversed using the ice ax as a third point of contact. Always keep the ax on the uphill side. When the herringbone or diagonal method is used to ascend scree, the ax can be used placing both hands on the top and driving the spike into the scree slope above the climber. The climber uses the ax for balance as he moves up to it, and then repeats the process.

f. **Talus Slopes.** Talus slopes are composed of rocks larger than a man's fist. When walking in talus, ascending or descending, climbers should always step on the uphill side of rocks and stay alert for movement underfoot. Disturbing unstable talus can cause rockslides. Climbers must stay in close columns while walking through talus so that dislodged rocks do not reach dangerous speeds before reaching lower soldiers. To prevent rock fall injuries, avoid traversing below other climbers. All other basics of mountain walking apply.

8-3. SAFETY CONSIDERATIONS

The mountain walking techniques presented here are designed to reduce the hazards of rock fall and loss of control leading to a fall. Carelessness can cause the failure of the best-planned missions.

a. Whenever a rock is kicked loose, the warning, "Rock!" is shouted immediately. Personnel near the bottom of the cliff immediately lean into the cliff to reduce their exposure, and do not look up. Personnel more than 3 meters away from the bottom of the cliff may look up to determine where the rock is heading and seek cover behind an obstacle. Lacking cover, personnel should anticipate which way the rock is falling and move out of its path to the left or right.

b. If a soldier slips or stumbles on sloping terrain (hard ground, grass, snow, or scree) he must immediately self-arrest, digging into the slope with hands, elbows, knees and toes. If he falls backwards and rolls over he must immediately try to turn over onto his stomach with his legs downhill and self-arrest with hands and toes.

c. When traveling through steep terrain, soldiers should be trained in the use of the ice ax for self-arrest. The ax can be used to arrest a fall on solid ground, grass and scree as well as snow. It may also be used as a third point of contact on difficult terrain. If not in use the ice ax is carried in or on the rucksack with its head down and secured.

8-4. NAVIGATION

Navigation is the process of determining one's present position, the location of a target objective, and selecting and following a route between these two points. Navigation consists of three distinct stages: orientation, navigation, and route finding.

- Orientation is simply figuring out exactly where one is. The use of the map, compass and identifiable terrain features, assisted by an altimeter and GPS, is the foundation of good navigation.
- Navigation includes the determination of the objective's location and the direction from the soldier's starting point to the objective. The same skills and equipment used in orientation are essential for good navigation.
- Route finding is picking the best line of travel that matches the equipment and capabilities of the team. Good route finding incorporates a comprehensive awareness of terrain, a solid base of mountaineering experience, good judgement and sound tactical instincts.

a. **Compasses.** The magnetic compass is the simplest and most widely used instrument for measuring directions and angles in the mountains. The lensatic compass is most commonly used in the military and can be employed in a variety of ways for either day or night navigation.

b. **Altimeters.** The altimeter is a vital piece of navigational equipment that can save valuable time in determining position through elevation.

(1) The standard altimeter is a modified barometer. A barometer is an instrument that measures the weight of a column of air above itself and displays the result on a scale marked in units of pressure, usually inches of mercury, millimeters of mercury, or millibars. Since air pressure drops uniformly as elevation is gained, it can be used to read altitude by means of the altimeter's scale, marked in feet or meters of elevation above sea level. By measuring air pressure, the altimeter/barometer gives the navigator new techniques for position finding, route planning, checking progress and terrain identification. It also gives the navigator valuable weather information specific to his immediate location.

(2) Changes in the weather are usually accompanied by air pressure changes, which are reflected in the altimeter. As the air pressure drops due to the approach of inclement weather for instance, the displayed elevation will rise by a corresponding amount. This means that a barometric pressure change of one inch of mercury equals roughly 1,000 feet of elevation. If the altimeter displays an elevation gain of 300 feet, a loss of barometric pressure of .3 inches has occurred, and bad weather should be expected.

(3) Altimeters come in two types: wrist-mounted digital altimeters and analog altimeters, usually attached to a cord.

(4) Because the altimeter is sensitive to changes in air pressure it must be recalibrated whenever a point of known elevation (summits, saddles, stream-trail intersections, survey monuments, and so forth) is reached. This is especially important when weather fronts are moving rapidly through the area.

(5) The altimeter may expand or contract because of changes in temperature. This can result in faulty elevation readings. Although some altimeters are temperature-compensated, rapid ascents or descents sometime overcome the adjustment, causing them to give poor readings.

(6) Keep the altimeter at a constant temperature. This is best accomplished by storing the altimeter (analog) in a pocket or on a cord around the neck, or on the wrist under the parka and hand gear (digital).

(7) Even though altimeters can be precise they are affected by both pressure and temperature changes and should be monitored carefully. The soldier should become familiar with the specific altimeter he employs and understand its capabilities and limitations.

c. **Global Positioning System.** The GPS is a space-based, global, all-weather, continuously available radio positioning navigation system. It is highly accurate in determining position location derived from a satellite constellation system. It can determine the latitude, longitude and elevation of the individual user. Location information is also displayed in military grid coordinates.

(1) The GPS provides precise steering information as well as position locations. The receiver can accept many checkpoints entered in any coordinate system by the user and convert them to the desired coordinate system. The user then calls up the desired checkpoint and the receiver will display direction and distance to the checkpoint. It can also compute travel time to the next checkpoint.

(2) Because the GPS does not need visible landmarks to operate, it can provide position (accurate up to 16 meters) in whiteouts or on featureless terrain. It also does not compound navigational errors as compass use can.

(3) During route planning, after choosing critical checkpoints, start point and objective, enter their coordinates as way points. The best use of the GPS is to verify these as they are reached, as a backup to terrain association and compass navigation.

(4) Since the 21-satellite constellation is not yet complete, coverage may be limited to specific hours of the day in certain areas of the world. The GPS navigational signals are similar to light rays, so anything that blocks light will reduce or block the effectiveness of the signals. The more unobstructed the view of the sky, the better the system performs. Although the GPS can be used in any terrain, it is performs best in more open areas such as the desert.

(5) Because the GPS requires horizon to horizon views for good satellite reception its use can be limited in the mountains. Canyons, deep valleys, saddles, and steep mountainsides are all problematic spots to use for shots. Ridgelines, spurs, summits, open valleys, or plateaus are better.

(6) When using GPS in regions with questionable surveying and mapping products, operational datum of the local maps must be reconciled with the datum used in navigational and targeting systems. Identify the spheroid and datum information on the pertinent map sheets and then check that the GPS receiver has the compatible datum loaded. If not then you must contact the S2 for updated datum or maps. Otherwise, the GPS will show different locations than those on the map.

(7) Extremely cold temperatures (-4 degrees F and below) and high elevations will adversely affect the operation of the GPS, due to the freezing of the batteries and the LCD screen. Battery life and overall performance can be improved by placing the GPS inside the parka or coat.

d. **Navigation Techniques.** The choice of movement technique often determines the route and navigational technique. For navigation, three techniques can be used: dead reckoning, terrain association, or altimeter navigation. The three are not mutually exclusive and are normally used together, with one chosen as the primary technique. The GPS can be used to supplement these techniques, but due to the problems associated with the restricted line of sight in the mountains, it should not be used as the main technique.

(1) ***Dead Reckoning.*** Because of the complex nature of mountainous terrain, dead reckoning is usually of limited value on most movements. The compass is generally employed more to support terrain association and to orient the map, than as a primary navigational aid. The main exception is during periods of limited visibility on featureless terrain. Heavy fog, snowy or whiteout conditions on a snowfield, glacier, large plateau or valley floor all would call for dead reckoning as a primary navigational technique.

(2) ***Terrain Association.*** The standard terrain association techniques all apply. Handrails, checkpoints, catching features, navigational corridors, boxing-in areas, and attack points are all used. When a small objective lies near or on an easily identifiable feature, that feature becomes an expanded objective. This simplifies the navigational problem by giving a large feature to navigate to first. The altimeter may finalize the search for the objective by identification through elevation. Rough compass headings are used to establish a general direction to the next checkpoint; used when the checkpoint headed toward is a linear feature, and not a precise point. The shape, orientation, size,

elevation, slope (SOSES) strategy is especially valuable in mountain terrain association and should be practiced extensively (FM 3-25.26).

(a) After extensive study of the map and all available sources of information it helps to create a mental image of the route. This will enable the navigator to make the terrain work in his favor. Avoid brush for speed and ease of movement; the military crest of spurs and ridgelines generally provides the best route while providing terrain masking effects. When clear cut, burned-over, or large avalanche slide areas are encountered, it may be necessary to box or contour around them as they may be full of slash or brushy second-growth small trees. Old-growth forest provides the easiest travel.

(b) The following situations will result in objects appearing closer than they actually are:

- When most of the object is visible and offers a clear outline.
- When you are looking across a partially cleared depression.
- When looking down a straight, open road or track.
- When looking over a smooth, uniform surface, such as snow, water, or desert.
- When the light is bright and the sun is shining from behind the observer.
- When the object is in sharp contrast to the background.
- When seen in the clear air of high altitude.
- When looking down from high ground to low ground.

(c) The following situations will result in objects appearing farther away than they actually are:

- When only part of the object is seen or it is small in relation to its surroundings.
- When you are looking across an exposed depression.
- When looking up from low ground to high ground.
- When your vision is narrowly confined.
- When the light is poor, such as dawn, dusk, or low visibility weather; or when the sun is in your eyes, but not behind the object being viewed.
- When the object blends into the background.

(2) ***Altimeter Navigation.*** Altimeters provide assistance to the navigator in several ways. They aid in orientation, in computing rates of ascent or descent, in resection, and in weather prediction.

(a) When moving along any linear feature such as a ridge, watercourse, or trail which is shown on the map, check the altimeter. The point where the indicated elevation contour crosses that feature is your location.

(b) The navigator frequently finds it necessary to determine his position through the use of resection. A modified resection can be performed by shooting an azimuth to a known, clearly visible summit or similar feature and then plotting the back azimuth on the map. By determining your present elevation and finding where that particular contour crosses the back azimuth you should locate your position. This can be difficult when in low ground, as mountain summits can rarely be clearly seen from valley floors. In addition, most mountaintops are so large that there is usually no specific point to shoot at. In this case, the soldier should take multiple azimuths to known features. If he is located on a good linear feature he will have a decent idea of where he is. The altimeter can be used to verify elevation and establish a notional linear feature—a contour line. The point where the resecting back azimuths cross the contour line is the navigator's location.

(c) Using the altimeter to calculate rates of ascent can help in sound decision-making. Rates of travel, along with weather conditions, light conditions (time of day), and the physical condition of the team, are all key variables that can influence the success or failure of the mission.

(d) Altimeters can be used as barometers to assist in weather prediction.

e. **Approach Observations.** Watch the mountain during the approach march, studying it for climbing routes. Distant views can reveal large-scale patterns of ridges, cliffs, snowfields and glaciers. General angles of the large rock masses can be seen from afar.

(1) Closer viewing displays these patterns and angles on a smaller scale. Fault lines, gross bedding planes of rock, cliff bands, and crevasse zones become visible. Snowy or vegetated ledge systems appear. Weaknesses in the mountain walls, such as couloirs or gullies, may present themselves.

(2) Most of these features repeat themselves at increasingly finer levels, as they are generally derived from the overall structure of the particular mountain group. A basic knowledge of mountain geology, combined with the specific geological background of the operational area, pays off in more efficient travel.

f. **Natural Indicators of Direction in the Northern Hemisphere.** Southern slopes are sunnier and drier than northern slopes, with sparser or different types of vegetation. Northern slopes can be snowier and, because of more intense glaciation in past ages, are often steeper.

Note: Opposite rules apply in the Southern Hemisphere.

g. **Winter Route Selection.** The following must be considered when selecting a route in the winter.

(1) Conduct a thorough map reconnaissance considering the weather, individual ski abilities, avalanche danger, vegetation, water features, terrain relief, and the size of the unit.

(2) Weather conditions will affect the chosen route. During calm weather, your rate of movement will be significantly faster than during periods of inclement weather.

(3) Individual ski abilities will affect your rate of movement, constrain your choice of terrain, and impact on your route choices.

(4) Avalanche danger zones must be identified by map review and data gathered during route planning. During movement, snow pits, shovel tests, and ski shear tests must be conducted prior to crossing an avalanche danger zone. Bottom line: avoid avalanche danger areas. If you must cross one, cross above the starting zone or below the run-out zone.

(5) Vegetation can work for you or against you. Thickly forested areas usually have a deep snow pack. For weaker skiers, forested areas are full of potentially dangerous obstacles. On slopes with an angle of 30 to 45 degrees that are sparsely vegetated an avalanche danger is still present. If the weather turns bad, forested areas provide welcome relief from wind and blowing snow.

(6) Water features provide valuable navigation aids. Under deep snow pack small creeks and ponds may be hard to locate. Large frozen lakes and rivers can provide excellent means of increasing your rate of march.

(7) During ski movements, efficient use of terrain will greatly improve morale and reduce fatigue. While traveling in mountainous terrain, do not needlessly give up elevation gained. Maintain a steady climb rate and avoid over exertion. Avoid north, east, and south facing slopes when the avalanche danger is high. Avoid cornices and be aware of their probable and improbable fracture lines. Weather and tactical situation permitting, travel on the windward side of ridgelines. If weak skiers are in the group, stay away from restrictive terrain with sheer drop-offs. When touring use climbing skins to maintain control and lessen lost time per hour due to individuals falling.

(8) The following are additional hints for navigation in snowy conditions:

- Keep the compass warm.
- If no terrain features exist for steering marks, use your back azimuth and tracks to maintain course.
- Limit steering marks to shorter distances since visibility can change quickly.
- Never take azimuths near metallic objects. Hold the compass far enough from your weapon, ice ax, and so on to get accurate readings.
- Make frequent compass checks.
- Preset azimuths on your compass.
- Use a steady, unshifting wind to aid you in maintaining course.

h. **Problems.** The following conditions and characteristics of cold weather and mountainous regions make accurate navigation difficult.

(1) In winter, short hours of daylight, fog, snowfall, blizzards, whiteouts, and drifting snow, especially above tree line, drastically limit visibility. At times, an overcast sky and snow-covered terrain create a phenomenon called flat light, which makes recognition of irregularities in the terrain extremely difficult.

(2) Heavy snow may completely cover existing tracks, trails, outlines of small lakes and similar landmarks. Because the appearance of the terrain is quite different in winter from that in summer, particular attention must be paid to identifying landmarks, both on the ground and from aerial photographs.

(3) Magnetic disturbances, caused by large ore deposits, are frequently encountered and make magnetic compass readings difficult and sometimes unreliable.

(4) Handling maps, compasses, and other navigation instruments in low temperatures with bare hands is difficult. Removing hand wear may only be possible for short periods.

(5) Keeping count of pace is extremely difficult in winter and mountain environments. Thick vegetation and rough, steep slopes hamper attempts at accurate pace counts. The most reliable method is the use of a 50-meter long piece of field wire or rope.

8-5. ROUTE PLANNING

Proper route planning can make the difference between success and failure on long mountain movements. Careful map reconnaissance, knowledge of the enemy situation, terrain analysis of the operational area, and an accurate assessment of the unit's capabilities are all key parts of the planning process.

a. **Map Reconnaissance.** Topographic maps provide the primary source of information concerning the area of operations. A 1:25,000 map depicts greater detail than a 1:50,000 map and should be used whenever possible. Because examination of the micro-terrain is so important for mountain operations, even larger scale maps are desirable. Civilian 1:12,000 maps can be used if available. Aerial, oblique angle, photographs give details not always shown on maps (crags and overhangs). Sketch maps supplement other sources of information but should not be relied on for accuracy since they are seldom drawn to scale. Along with sketch maps, verbal descriptions, documented information gathered from units previously in the area, or published sources such as alpine journals or climbing guides may help. Forest service and logging and mining company maps provide additional information, often showing the most recent changes to logging trails and mining access roads. Standard military topographic maps are generally accurate graphic depictions of the operational area.

(1) When conducting a map reconnaissance, pay close attention to the marginal information. Mountain-specific terrain features may be directly addressed in the legend. In addition, such facilities as ski lifts, cable and tramways are often found. Check the datum descriptor (for foreign maps) to ensure compatibility with entered datum in GPS units. Along with the standard topographic map color scheme, there are some commonly seen applications for mountainous terrain. White with blue contours indicates glaciers or permanent snowfields. The outline of the snow or ice is shown by dashed blue lines while their contour lines are solid blue. High ice cliffs which are equal to or exceed the contour interval will be shown. Low ice cliffs and ice caves may be indicated if they provide local landmarks. Brown contour lines on white mean dry areas without significant forest cover. Areas above tree line, clear cuts, rock or avalanche slide paths and meadows are all possible. Study the surrounding terrain and the legend for other clues. An important point to remember is that thick brush in small gullies and streambeds may not be depicted by green, but should still be expected.

(2) Obstacles, such as rivers and gorges, will require technical equipment to cross if bridges are not present. Fords and river crossing sites should be identified. Due to the potential for hazardous weather conditions, potential bivouac sites are noted on the map. Ruins, barns, sheds and terrain-protected hollows are all possible bivouac sites. Danger areas in the mountains; isolated farms and hamlets, bridges, roads, trails, and large open areas above tree line, are factored in, and plans made to avoid them. Use of terrain-masking becomes essential because of the extended visibility offered by enemy observation points on the dominant high ground.

(3) Helicopter support, weather permitting, requires identification of tentative landing zones for insertions, extractions, resupply and medevac. The confined nature of mountain travel means that crucial passes become significant chokepoints and planners should designate overwatches/surveillance positions beforehand. Alternate routes should be chosen with weather imposed obstacles in mind: spring flood or afternoon snowmelt turns small streams into turbulent, impassable torrents. Avalanche danger prohibits travel on certain slopes or valley floors.

b. **Enemy Situation.** Route selection should only be done after reviewing all available information about the friendly and enemy situation.—Is the enemy force on his own ground? Are they accustomed to the terrain and the weather? Are they trained mountain troops with specialized equipment?—Only after answering these and other questions can an effective route plan be completed. If the enemy force is better prepared to maneuver in the mountains, they have a marked advantage, and route selection must be scrutinized.

c. **Analysis of the Operational Area.** Not all mountainous terrain is created equal and not all movement plans have the same expectation of success. Planners must undertake a thorough analysis of the general terrain to be crossed, including the geology, mountain structure and forms, and ground cover.

(1) Heavily glaciated granite mountains pose different problems than does river-carved terrain. The U-shaped valley bulldozed out by a glacier forces maneuver elements down to the valley floor or up to the ridge tops, while the water-cut V-shape of river valleys allows movement throughout the compartment.

(2) Routes through granite rock (long cracks, good friction; use of pitons, chocks and camming units) will call for different equipment and technique than that used for steep limestone (pockets, smooth rock; bolts, camming units).

(3) Operations above tree line in temperate climates or in the brushy zone of arid mountains means that material for suspension traverse A-frames must be packed. The thick brush and krummholtz mats of subalpine zones and temperate forested mountains can create obstacles that must be bypassed.

(4) Heavy spruce/fir tangles slow progress to a crawl, therefore planners should ensure routes do not blindly traverse these zones.

d. **Unit Assessment.** When assessing unit movement capabilities the key indices are training and conditioning levels. Soldiers who have received basic military mountaineer training, who know how to move through rough terrain, and who have been hardened with training hikes through the mountains, will perform better than troops without this background.

e. **Time-Distance Formulas.** Computing march rates in the mountains is extremely difficult, especially when there is snow cover. The following rates are listed as a guide (Table 8-1). Rates are given for movement over flat or gently rolling terrain for individuals carrying a rifle and loaded rucksack.

	UNBROKEN TRAIL	BROKEN TRAIL
On foot (no snow cover)	2 to 3 kph (cross-country)	3 to 4 kph (trail walking)
On foot (less than 1 foot of snow)	1.6 to 3.2 kph	2 to 3.2 kph
On foot (more than 1 foot of snow)	.4 to 1.2 kph	2 to 3.2 kph
Snowshoeing	1.6 to 3.2 kph	3.2 to 4 kph
Skiing	1. to 5.6 kph	4.8 to 5.6 kph
Skijoring	N/A	3 to 24 kph

Table 8-1. Time-distance formulas.

(1) March distances in mountainous terrain are often measured in time rather than distance units. In order to do this, first measure the map distance. This distance plus 1/3 is a good estimate of actual ground distance. Add one hour for each 1,000 feet of ascent or 2,000 feet of descent to the time required for marching a map distance.

(2) As Table 8-1 indicates, snow cover will significantly affect rates of march. Since snow can be expected in the mountains most months of the year, units should have some experience at basic snow travel.

(3) Individual loads also affect march rates. Combined soldier loads that exceed 50 pounds per man can be expected to slow movement significantly in mountainous terrain. Given the increased weight of extra ammunition for crew-served weapons, basic mountaineering gear, and clothing for mountain travel, it becomes obvious that soldiers will be carrying weights well in excess of that 50-pound limit. Units should conduct cross-country movements in the mountains with the expected rucksack and LCE weights in order to obtain accurate, realistic rates of march.

(4) In the harsh environment of the mountains, helicopter support cannot be relied on. The process of transporting extra equipment and sustainment supplies will result in vastly increased movement times. The heavier loads will exhaust soldiers mentally and physically. Tactical movements, such as patrolling or deliberate assaults, should take this into account.

8-6. ROUTE SELECTION

Many variables affect the selection of the proper route. The following guidelines apply to all situations.

a. **Select a Current Map.** Check the date of the map for an indication of the reliability of the map in depicting vegetation, clearings, roads, and trails accurately. The leader should use all the latest topographic data he can find.

b. **Gather Intelligence Information.** The most important consideration in every leader's mind when plotting a movement is "where is the enemy?" The latest intelligence reports are essential. Additionally, weather reports, snow condition reports, avalanche probability, aerial photos, and any past or recent history of operations in the area may be of help.

c. **Select a Route.** Identify the starting point and determine the movement objective. Plot start and end points. Carefully scrutinize the area in between and begin to select the route. Consider the following:

(1) ***Trafficability***. This includes degree of slopes, vegetation, width of trails, snow depth, avalanche probability, and the likelihood of crevasses.

(2) ***Time-Distance Formula.*** Time allotted and distance to be covered must be considered.

(3) ***Required Equipment.*** Carry enough equipment to move along the route and to survive if an extended stay becomes necessary. Do not plan a route beyond the means of your equipment.

(4) ***Location of Enemy.*** Plan a route that allows maximum use of the masking effect of the terrain. Avoid danger areas or areas of recent enemy activity unless required by the mission. Use vegetation to mask your movement if possible (especially coniferous forests). Avoid silhouetting on ridgelines.

(5) ***Communications.*** Communications will be severely limited in the mountains. Dead spaces or communications holes are common. Use all available information and plan accordingly.

(6) ***Conditions/Capabilities of Unit.*** The unit must be able to negotiate the route chosen. Take into consideration their present health, as well as their training level when selecting your intended route.

(7) ***Checkpoints/Control Points.*** When plotting a route on the map, utilize prominent terrain features on either side of the route as checkpoints. Ensure that when you select

your checkpoints they are visually significant (elevation) and that they are easily identifiable. Avoid the use of manmade features as checkpoints due to their unreliability and lack of permanence. Select features that are unique to the area.

CHAPTER 9
MOUNTAIN STREAM CROSSINGS

Operations conducted in mountainous terrain may often require the crossing of swift flowing rivers or streams. Such crossings should not be taken lightly. The force of the flowing water may be extremely great and is most often underestimated. All rivers and streams are obstacles to movement. They should be treated as danger areas and avoided whenever possible. When rivers or streams must be crossed, there are a variety of techniques the small-unit leader may choose from, depending upon the type of stream, its width, speed of the current, and depth of the water.

There are limits on the safe use of these techniques. Not all mountain rivers or streams will be fordable with these techniques. If a water obstacle is too wide, swift, or deep, an alternate route should be used, or the crossing will require major bridging by engineers. It may require the use of rafts or boats. Reconnaissance of questionable crossing sites is essential. This chapter covers the techniques for crossing mountain streams that have a depth generally not exceeding waist deep.

9-1. RECONNAISSANCE

Reconnaissance of the route (map, photo, and or aerial) may not always reveal that a water obstacle exists. In a swamp, for example, unfordable sloughs may not show on the map, and they may be concealed from aerial observation by a canopy of vegetation. Whenever it is possible that a unit will be required to cross a water obstacle, its commander must plan some type of crossing capability.

a. Site selection is extremely important once you determine that you must make a crossing (Figure 9-1). Look for a high place from which you can get a good view of the obstacle and possible crossing sites. A distant view, perhaps from a ridge, is sometimes better than a hundred close views from a riverbank. Site selection must be made before the arrival of the main body.

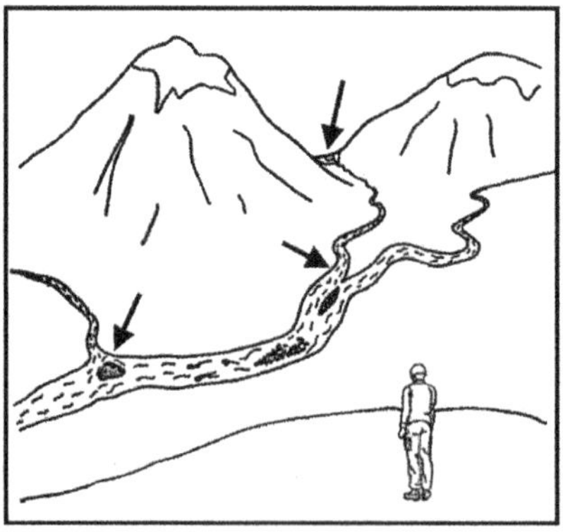

Figure 9-1. Normal locations of shallowest water and safest crossing sites.

b. A dry crossing on fallen timber or log jams is preferable to attempting a wet crossing. Depending upon the time of year, the closer you are to the source, or headwaters, the better your chances are of finding a natural snow or ice bridge for crossing. If a dry crossing is unavailable, the following considerations should be made:

(1) The time of day the crossing can be an important factor. Although early morning is generally best because the water level is normally lower during this period, recent weather is a big factor; there may have been heavy rain in the last eight hours. As glaciers, snow, or ice melt during the day, the rivers rise, reaching their maximum height between mid afternoon and late evening, depending on the distance from the source. Crossings, if made during the early morning, will also allow clothing to dry more quickly during the heat of the day.

(2) A crossing point should normally be chosen at the widest, and thus shallowest, point of the river or stream. Sharp bends in the river should be avoided since the water is likely to be deep and have a strong current on the outside of the bend. Crossings will be easiest on a smooth, firm bottom. Large rocks and boulders provide poor footing and cause a great deal of turbulence in the water.

(3) Many mountain streams, especially those which are fed by glacier run-off, contain sections with numerous channels. It is often easier to select a route through these braided sections rather than trying to cross one main channel. A drawback to crossing these braided channels, however, is the greater distance to the far bank may increase exposure time and often the sand and gravel bars between the channels will offer little cover or concealment, if any.

(4) The crossing site should have low enough banks on the near and far side to allow a man carrying equipment to enter and exit the stream with relative ease. If a handline or rope bridge is to be constructed, the crossing site should have suitable anchors on the near and far bank, along with safe loading and unloading areas. Natural anchors are not a necessity, however the time required to find a site with solid natural anchors will probably be less than the time required to construct artificial anchors. In some areas, above the tree line for example, artificial anchors may have to be constructed. Deadman anchors buried in the ground, or under a large pile of boulders work well.

(5) Log jams and other large obstructions present their own hazards. Logs floating downstream will generally get hung up in shallower sections creating the jam. Once a log jam is formed, however, the water forced to flow around it will erode the stream bottom. Eventually deep drop-offs or holes may develop, especially around the sides and off the downstream end of the log jam. A log jam that totally bridges a section of the stream may be the best way to cross. A wet crossing in the vicinity of a log jam should be performed a good distance below or above it. Some things to consider when crossing near log jams are:

- Cross well to the downstream side when possible.
- Keep a sharp lookout for floating timber that could knock you off your feet.
- If you must cross on the upstream side, stay well upstream from the log jam. If a person is swept off his feet and caught in the debris of the jam, he could easily drown. A handline will greatly increase safety here.

(6) When possible, select a crossing site that has enough natural protection on the near and far banks so that security teams may be placed out and enough cover and concealment is available for the size of the element making the crossing. When cover and

concealment is minimal, as in the higher alpine zones, the crossings must be conducted as efficiently as possible to minimize exposure to enemy observation.

9-2. PREPARATION OF TROOPS AND EQUIPMENT

Prepare men and equipment for a crossing as far in advance as feasible. Final preparation should be completed in a security perimeter on the near side just before crossing. Preparation includes the following.

a. Waterproof water-sensitive items. Wrap radios, binoculars, SOI, papers, maps and any extra clothing in waterproof bags (trash bags also work well), if available. These bags also provide additional buoyancy in case of a fall.

b. Trousers are unbloused and shirts are pulled out of the trousers. All pockets are buttoned. This allows water to escape through the clothing. Otherwise the clothing would fill up and retain water, which would weigh the body down. This is especially critical if an individual must swim to shore. Depending on the circumstances of the crossing (for example, tactical situation, temperature of the air and water), the crossing can be made in minimal clothing so that dry clothing is available after the crossing. Boots should be worn to protect feet from rocks; however, socks and inner soles should be removed. On the far side, the boots can be drained and dry socks replaced.

c. Load-carrying equipment harness and load-bearing vest (LBV) is unbuckled and worn loosely. It is extremely difficult to remove a buckled harness in the water in an emergency.

d. Helmets are normally removed and placed in the rucksack in slow moving streams with sandy or gravel bottoms. If you have to resort to swimming it is easier done without the helmet. However, when crossing swift flowing streams, especially those with large rocks and other debris, the risk of head injury if a person slips is high. In this case the helmet should be worn with the chinstrap fastened.

e. The rucksack should be worn well up on the shoulders and snug enough so it does not flop around and cause you to lose your balance. The waist strap MUST be unbuckled so you can get rid of the pack quickly if you are swept off your feet and have to resort to swimming. If a pack has a chest strap it must also be unbuckled. Secure everything well within your pack. It is easier to find one large pack than to find several smaller items.

f. Individual weapons should be attached to the pack or slung over the shoulder.

9-3. INDIVIDUAL CROSSINGS

Whenever possible, and when the degree of experience permits, streams should be forded individually for a speedier crossing. The average soldier should be able to cross most streams with mild to moderate currents and water depths of not much more than knee deep using proper techniques.

a. The individual should generally face upstream and slightly sideways, leaning slightly into the current to help maintain balance. At times, he may choose to face more sideways as this will reduce the surface area of the body against the current, thus reducing the current's overall force on the individual.

b. The feet should be shuffled along the bottom rather than lifted, with the downstream foot normally in the lead. He should take short, deliberate steps. Lunging steps and crossing the feet result in a momentary loss of balance and greatly increase the chance of a slip.

c. The individual should normally cross at a slight downstream angle so as not to fight the current. There is normally less chance of a slip when stepping off with the current as opposed to stepping off against the current.

d. The individual must constantly feel for obstacles, holes and drop-offs with the lead foot and adjust his route accordingly. If an obstacle is encountered, the feet should be placed on the upstream side of it where the turbulence is less severe and the water normally shallower.

e. To increase balance, and if available, a long ice ax, sturdy tree limb, or other staff can be used to give the individual a third point of contact (Figure 9-2). The staff should be used on the upstream side of the individual and slightly leaned upon for support. The staff should be moved first, then the feet shuffled forward to it. This allows two points of contact to be maintained with the streambed at all times. The individual still moves at a downstream angle with the downstream foot in the lead.

Figure 9-2. Individual crossing with staff.

9-4. TEAM CROSSING

When the water level begins to reach thigh deep, or anytime the current is too swift for personnel to safely perform an individual crossing, a team crossing may be used. For chain crossing, two or more individuals cross arms with each other and lock their hands in front of themselves (Figure 9-3). The line formed faces the far bank. The largest individual should be on the upstream end of the line to break the current for the group. The line formed will then move across the stream using the same principles as for individual crossings, but with the added support of each other. The line should cross parallel to the direction of the current. The team still moves at a slight downstream angle, stepping off with the downstream foot in the lead.

Figure 9-3. Chain method for stream crossing.

9-5. ROPE INSTALLATIONS

When the water level begins to reach waist deep or the current is too swift for even a team crossing, the chosen site must be closely examined. The stream at this point may be impassable. Many times though, a crossing site which may be unsafe for individual or team crossings can be made safe with the installation of a handline or rope bridge. Crossing on a handline will still require each individual to enter the water and get wet. If a one-rope bridge can be constructed, it may require only a couple of individuals to enter the water. Deciding whether to install a handline or a rope bridge will depend on the anchors available, height of the anchors above the water, and the distance from the near and far anchors. The maximum distance a one-rope bridge is capable of spanning is approximately 1/2 to 2/3 the length of the rope in use.

a. **Establishing the Far Anchor.** Whether a handline or rope bridge is to be installed, someone must cross the stream with one end of the rope and anchor it on the far side. This duty should be performed by the most capable and strongest swimmer in the party. The swimmer should be belayed across for his own safety. The belay position should be placed as far above the crossing as possible. In the event that the current is too strong for the individual, he will pendulum back to the near bank. Rescuers should be poised on the near bank at points where the individual will pendulum back, should he fail to reach the far bank. The initial crossing site should be free of obstacles that would snag the rope and prevent the pendulum back to the bank for an easy recovery.

(1) The individual may attach the belay rope to his seat harness or a swami belt with a carabiner. He should NEVER tie directly into the rope when being belayed for a stream crossing. If the swimmer should be swept away and become tangled, he must be able to release himself quickly from the rope and swim to shore as best he can. The individual may also choose to tie a fixed loop into the end of the belay rope and hang on to it, where he can immediately release it in an emergency.

(2) Anytime a crossing site must be used where the swimmer may encounter problems getting to the far bank, he should have on a life vest or other personal flotation

device (PFD). If the swimmer must release the rope at any time, he will have to rely on his own water survival skills and swimming ability to get to shore. A PFD will greatly increase his own personal safety. A PFD may also be used by the last man across, as he will release the rope from the anchor and be belayed across as the first man

b. **Installation of a Handline.** If it is possible to use a rope high enough above the water to enable soldiers to perform a dry crossing, then a rope bridge should be installed as such. If this is impossible, and the rope must be installed to assist in a wet crossing, then it should be installed as a handline (Figure 9-4).

(1) The far anchor should be downstream from the near anchor so that the rope will run at an angle downstream from the near anchor, approximately thirty to forty-five degrees, rather than straight across the stream. Here again, it is easier to move with the current as opposed to directly across or against it.

(2) The rope may be anchored immediately on the far bank, pulled tight, and anchored on the near bank, or it may be installed with a transport tightening system if a tighter rope is required.

(3) Crossing will always be performed on the downstream side of the handline, shuffling the feet with the downstream foot in the lead.

(4) A second climbing rope is used as a belay (Figure 9-5). One end of the belay rope will be on the near bank and the other end on the far bank. It should be sent across with the strong-swimmer. An appropriate knot is tied into the middle of the belay rope to form two fixed loops with each loop being approximately 6 inches long. One loop is connected to the handline with carabiner(s) and the individual crossing connects one loop to himself. The loops are short enough so the individual is always within arms reach of the handline should he slip and let go. The individuals are belayed from both the near and far banks. If a mishap should occur the individual can be retrieved from either shore, whichever appears easiest. The belay on the opposite shore can be released allowing the individual to pendulum to the bank. It is important that the belay rope NOT be anchored or tied to the belayer so that it may be quickly released if necessary.

Figure 9-4. Stream crossing using a handline.

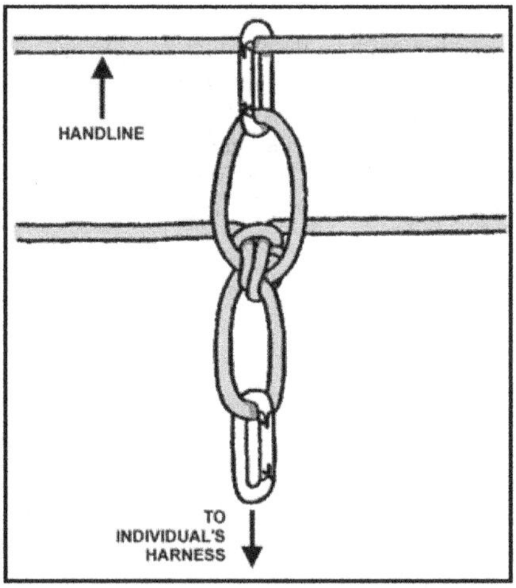

Figure 9-5. Belay rope for crossing using a handline.

(5) Under most circumstances, the handline should be crossed one person at a time. This keeps rope stretch and load on the anchors to a minimum.

(6) Rucksacks can be either carried on the back the same way as for individual crossings, or they can be attached to the handline and pulled along behind the individual.

(7) If a large amount of equipment must be moved across the stream, especially heavier weapons, such as mortars, recoilless rifles, and so on, then a site should be selected to install a rope bridge.

9-6. SAFETY

River and stream crossings present one of the most hazardous situations faced by the military mountaineer. The following safety procedures are minimum guidelines that should be followed when conducting a river or stream crossing.

a. All weak and nonswimmers should be identified before a crossing so that stronger swimmers may give assistance in crossing.

b. Not every river or stream can be crossed safely. It is always possible to cross at a different time or place, use a different technique, or choose another route.

c. The technique used is directly dependent upon water depth, speed of the current, stream bottom configuration, width of the stream, and individual experience.

d. The safest methods of crossing are always with the use of a handline or one-rope bridge.

e. If the installation of a handline or rope bridge becomes too difficult at a given crossing site, then that site should be considered too hazardous and another site selected.

f. A lookout should be posted 50 to 100 meters upstream to watch for any obstacles that may be carried downstream and interfere with the crossing.

g. When conducting individual crossings (those without a handline or rope bridge), lifeguards should be posted downstream with poles or ropes prepared to throw, for assistance or rescue.

h. When the unit knows a rope installation will be required for crossing, at least two life vests or other PFDs should be on hand to provide additional safety for the strong swimmer who must establish the far anchor, and the last man across who retrieves the system.

9-7. SWIMMING

There are times when you might be alone and have no choice but to swim across, or there may be a time that you find yourself suddenly plunged into a swift river or rapids. In either case, the following techniques could save your life.

a. Immediately jettison any equipment or clothing that restricts movement.

b. Do not try to fight the current. Maneuver towards shore in a position with the feet downstream, facing downstream, and fanning the hands alongside the body to add buoyancy and to fend off submerged rocks. Use the feet to protect the rest of the body and to fend off submerged rocks.

c. Keep the head above water to observe for obstacles and attempt to maneuver away from them.

d. Try to avoid backwater eddies and converging currents as they often contain dangerous swirls. Avoid bubbly water under falls as it has little buoyancy. Breath between the wave troughs.

e. If the shore is too difficult to reach, seek out the closest and safest spot, such as a sandbar, to get yourself out of the water as quickly as possible. Hypothermia will set in quickly in colder waters.

CHAPTER 10

MOVEMENT OVER SNOW AND ICE

Movement over snow- and ice-covered slopes presents its own unique problems. Movement on steeper slopes requires an ice ax, crampons, and the necessary training for this equipment. Personnel will also have to learn how to place solid anchors in snow and ice to protect themselves during these movements if roped. Snow-covered glaciers present crevasse fall hazards even when the slope is relatively flat, requiring personnel to learn unique glacier travel and crevasse rescue techniques.

All the principles of rock climbing, anchor placement, belays, and rope usage discussed throughout the previous chapters apply to snow and ice climbing as well. This chapter will focus on the additional skills and techniques required to move safely through snow-covered mountains and over glaciated terrain.

10-1. MOVEMENT OVER SNOW

The military mountaineer must be equally adept on both snow and ice due to route necessity and rapidly changing conditions. On steep slopes in deep snow, the climber may climb straight up facing the slope. The ice ax shaft, driven directly into the snow, provides a quick and effective self-belay in case of a slip—the deeper the shaft penetrates the snow, the better the anchor (Figure 10-1). It is usually best, however, to climb snow-covered slopes in a traversing fashion in order to conserve energy, unless there is significant avalanche danger.

Figure 10-1. Self-belay on snow.

a. The progression from walking on flat terrain to moving on steep terrain is the same as for moving over snow-free terrain. If the snow is packed the sole of the boot will generally hold by kicking steps, even on steep slopes. Where it is difficult to make an effective step with the boot, a cut made with the adze of the ice ax creates an effective step. In these situations crampons should be used for faster and easier movement.

b. When descending on snow, one can usually come straight downhill, even on steep terrain. Movement downhill should be slow and deliberate with the climber using an even pace. The heels should be kicked vigorously into the snow. The body may be kept erect with the aid of an ice ax, which may be jammed into the snow at each step for additional safety. Here again, crampons or step cutting may be necessary. A technique known as glissading may also be used as an easy method of descent and is covered in detail later in this chapter.

10-2. MOVEMENT OVER ICE

Ice is found in many areas of mountains when snow is present, and during the summer months also where perennial snowpack exists. Many times an ice area will be downslope of a snowfield and sometimes the ice pack itself will be lightly covered with snow. Even if using an ice ax and or crampons, movement will still be difficult without proper training.

10-3. USE OF ICE AX AND CRAMPONS

Movement over snow and ice is almost impossible without an ice ax and or crampons.

a. **Ice Ax.** When walking on snow or ice, the ice ax can be used as a third point of contact. When the terrain steepens, there are a number of ways to use the ice ax for snow or ice climbing. Some positions are more effective than others, depending on the intended result. You may find other ways to hold and use the ax, as long the security remains in effect.

(1) ***Cane Position.*** The ice ax can be used on gentle slopes as a walking stick or cane (Figure 10-2). The ax is held by the head with the spike down and the pick facing to the rear in preparation for self-arrest. When moving up or down gentle slopes the ice ax is placed in front as the third point of contact, and the climber moves toward it. When traversing, the ax is held on the uphill side, in preparation for a self-arrest.

Figure 10-2. Using the ice ax in the cane position.

(2) ***Cross Body Position or Port Arms Position.*** On steeper slopes the ax can be used in the port arms position, or cross body position (Figure 10-3). It is carried across the chest, upslope hand on the shaft, spike towards the slope. The head of the ax is held away from the slope with the pick to the rear in preparation for self-arrest. Ensure the leash is connected to the upslope hand, which allows the ax to be used in the hammer position on the upslope side of the climber. The spike, in this case, is used as an aid for maintaining balance.

Figure 10-3. Ice ax in the cross body or port arms position.

(3) ***Anchor Position.*** As the slope continues to steepen, the ax may be used in the anchor position (Figure 10-4). The head is held in the upslope hand and the pick is driven into the slope. The spike is held in the downhill hand and pulled slightly away from the slope to increase the "bite" of the pick into the ice. If the climber is wearing a harness, the pick can be deeply inserted in the ice or hard snow and the ax leash could be connected to the tie-in point on the harness for an anchor (ensure the ax is placed for the intended direction of pull).

Figure 10-4. Ice ax in the anchor position.

(4) ***Push-Hold Position.*** Another variation on steep slopes is the push-hold position (Figure 10-5). The hand is placed on the shaft of the ax just below the head with the pick forward. The pick is driven into the slope at shoulder height. The hand is then placed on the top of the ax head for use as a handhold.

Figure 10-5. Ice ax in the push-hold position.

(5) ***Dagger Position.*** The dagger position is used on steep slopes to place a handhold above shoulder height (Figure 10-6). The hand grasps the head of the ax with the pick forward and the shaft hanging down. The ax is driven into the surface in a stabbing action. The hand is then placed on the ax head for use as a handhold.

Figure 10-6. Ice ax in the dagger position.

(6) ***Hammer Position.*** The hammer position will set the pick deepest in any snow or ice condition (Figure 10-7, page 10-6). The ax is used like a hammer with the pick being driven into the slope. On vertical or near-vertical sections, two axes used in the hammer position will often be required.

Figure 10-7. Ice ax in the hammer position.

b. **Crampons.** Walking in crampons is not complicated but it does present difficulties. When walking in crampons, the same principles are used as in mountain walking, except that when a leg is advanced it is swung in a slight arc around the fixed foot to avoid locking the crampons or catching them in clothing or flesh. The trousers should be bloused to prevent catching on crampons. All straps should be secured to prevent stepping on them and, potentially, causing a fall. The buckles should be located on the outside of each foot when the crampons are secured to prevent snagging. Remember, when the crampon snags on the pants or boots, a tear or cut usually results, and sometimes involves the skin on your leg and or a serious fall.

(1) Two methods of ascent are used on slopes: traversing and straight up.

(a) A traverse on ice or snow looks much like any mountain walking traverse, except that the ankles are rolled so that the crampons are placed flat on the surface (Figure 10-8). On snow the points penetrate easily; on ice the foot must be pressed or stamped firmly to obtain maximum penetration. At the turning points of a traverse, direction is changed with the uphill foot as in mountain walking.

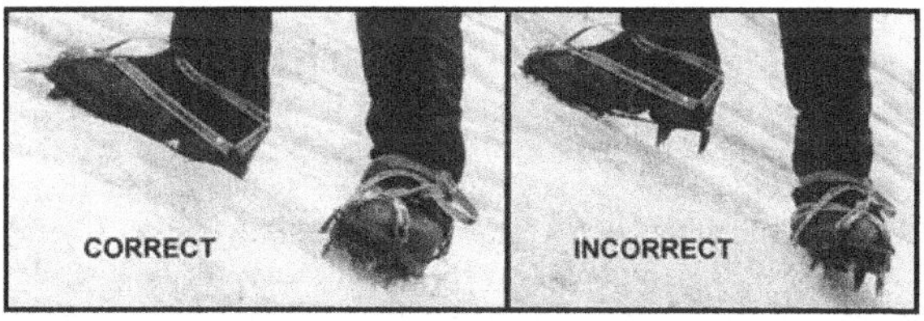

Figure 10-8. Correct and incorrect crampon technique.

(b) A straight up method is for relatively short pitches, since it is more tiring than a traverse. The climber faces directly up the slope and walks straight uphill. As the slope steepens, the herringbone step is used to maintain the flatfoot technique. For short steep pitches, the climber may also face downslope, squatting so the legs almost form a

90-degree angle at the knees, driving the spike of the ice ax into the slope at hip level, and then moving the feet up to the ax. By repeating these steps, the ax and crampon combination can be used to climb short, steep pitches without resorting to step cutting. This method can be tiring. The technique is similar to the crab position used for climbing on slab rock and can also be used for short descents.

(2) A technique known as "front-pointing" may be used for moving straight uphill (Figure 10-9). It is especially useful on steep terrain, in combination with the ice ax in the push-hold, dagger, or hammer position. Front-pointing is easiest with the use of more rigid mountain boots and rigid crampons. The technique is similar to doing calf raises on the tips of the toes and is much more tiring than flat-footing.

(a) The technique starts with the feet approximately shoulder width apart. When a step is taken the climber places the front points of the crampons into the ice with the toe of the boot pointing straight into the slope.

(b) When the front points have bitten into the ice the heel of the boot is lowered slightly so that the first set of vertical points can also bite. The body is kept erect, with the weight centered over the feet as in climbing on rock.

Figure 10-9. Front-pointing with crampons.

c. **Vertical Ice.** When a climb on ice reaches the 60- to 70-degree angle, two ice axes may be helpful, and will become necessary as the angle approaches 90 degrees. The same basic climbing techniques described in Chapter 6 should be applied. If leashes of the correct length and fit are attached to both axes, it may be possible to hang completely from the axes while moving the feet.

d. **Descending with Crampons and Ice Ax.** Whenever possible, descend straight down the fall line. As the slope steepens, gradually turn sideways; on steeper slopes, bend at the waist and knees as if sitting, keeping the feet flat to engage all vertical crampon points and keep the weight over the feet as in descending rock slab (Figures 10-10 and 10-11, page 10-8). On steep terrain, assume a cross body or port arms position with the ax, and traverse. The crab position or front-pointing may also be used for descending. Regardless of the technique used, always ensure the points of the crampons are inserted in the snow or ice and take short, deliberate steps to minimize the chance of tripping and falling down the slope.

Figure 10-10. Flat-footing in the crab position.

Figure 10-11. Use of ice ax in descent.

e. **Normal Progression.** The use of the ice ax and crampons follows a simple, logical progression. The techniques can be used in any combination, dictated by the terrain and skill of the individual. A typical progression could be as follows:

(1) ***Crampons.*** Use crampons in the following situations:

- Walking as on flat ground.

- Herringbone step straight up the slope.
- Traverse with feet flat.
- Backing up the slope (crab position).
- Front-pointing.

(2) ***Ice Ax.*** Use the ice ax in these situations:

- Cane position on flat ground.
- Cane position on uphill side as slope steepens.
- Port arms position with spike on uphill side.
- Anchor position with pick on uphill side.
- Push-hold position using front-pointing technique.
- Dagger position using front-pointing technique.
- Hammer position using front-pointing technique.

e. **Climbing Sequence.** Using most of these positions, a single ax can be "climbed" in steps to move upslope on low-angle to near vertical terrain (Figure 10-12). Begin by positioning the feet in a secure stance and placing the ax in the hammer position as high as possible. Slowly and carefully move the feet to higher positions alternately, and move the hand up the ax shaft. Repeat this until the hand is on top of the head of the ax. Remove the ax and place it at a higher position and begin again.

Figure 10-12. Climbing sequence.

f. **Step Cutting.** Step cutting is an extremely valuable technique that is a required skill for any military mountaineer (Figure 10-13). Using cut steps can save valuable time that would be spent in donning crampons for short stretches of ice and can, in some cases, save the weight of the crampons altogether. Steps may also have to be cut by the lead team to enable a unit without proper equipment to negotiate snow- or ice-covered terrain. As units continue to move up areas where steps have been cut they should continue to improve each step. In ascending, steps may be cut straight up the slope, although a traverse will normally be adopted. In descending, a traverse is also the preferred method. When changing direction, a step large enough for both feet and crampons must be made. Once the step is formed, the adze is best used to further shape and clean the step.

(1) ***Snow.*** On slopes of firm snow and soft ice, steps may be cut by swinging the ax in a near-vertical plane, using the inside corner of the adze for cutting. The step should be fashioned so that it slopes slightly inward and is big enough to admit the entire foot. Steps used for resting or for turning must be larger.

(2) ***Ice.*** Hard ice requires that the pick of the ax be used. Begin by directing a line of blows at right angles to the slope to make a fracture line along the base of the intended step. This technique will reduce the chance of an unwanted fracture in the ice breaking out the entire step. Next, chop above the fracture line to fashion the step. When using the pick it should be given an outward jerk as it is placed to prevent it from sticking in the ice.

(3) ***Step Cutting in a Traverse.*** When cutting steps in a traverse, the preferred cutting sequence is to cut one step at an arm's length from the highest step already cut, then cut one between those two. Cutting ahead one step then cutting an intermediate step keeps all of the steps relatively close to one another and maintains a suitable interval that all personnel can use.

(4) ***Handholds.*** If handholds are cut, they should be smaller than footholds, and angled more.

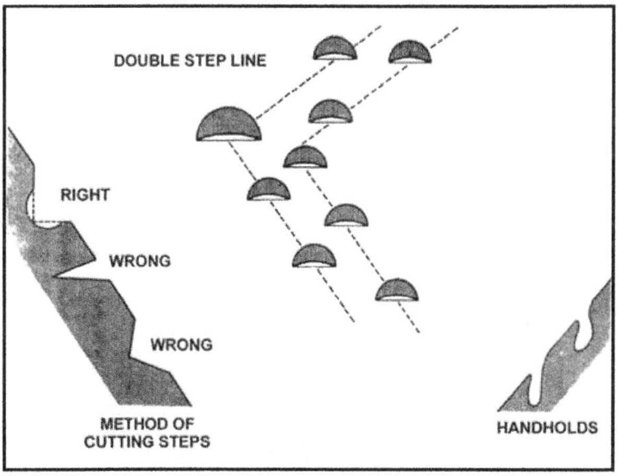

Figure 10-13. Step cutting and handhold cutting.

g. **Self-Arrest.** The large number of climbers injured or killed while climbing on snow and ice can be attributed to two major failings on the part of the climber: climbing unroped, and a lack of knowledge and experience in the techniques necessary to stop, or arrest, a fall (Figure 10-14). A climber should always carry an ice ax when climbing on steep snow or ice; if a fall occurs, he must retain possession and control of his ice ax if he is to successfully arrest the fall. During movement on steep ice, the ax pick will be in the ice solidly before the body is moved, which should prevent a fall of any significance (this is a self belay not a self-arrest).

CAUTION

Self-arrest requires the ax pick to gradually dig in to slow the descent. Self-arrest is difficult on steep ice because the ice ax pick instantly "bites" into the ice, possibly resulting in either arm or shoulder injury, or the ax is deflected immediately upon contact.

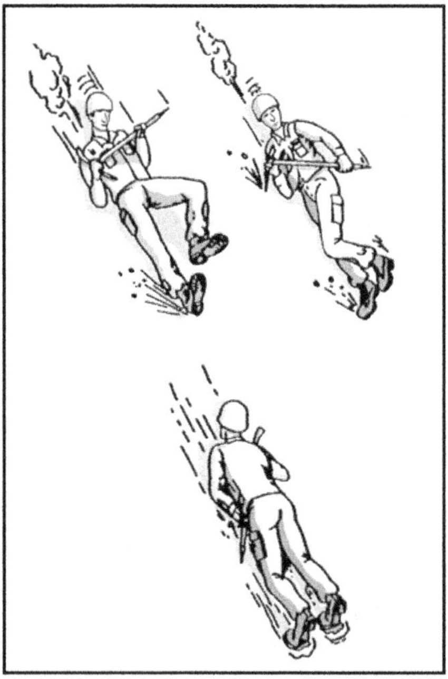

Figure 10-14. Self-arrest technique.

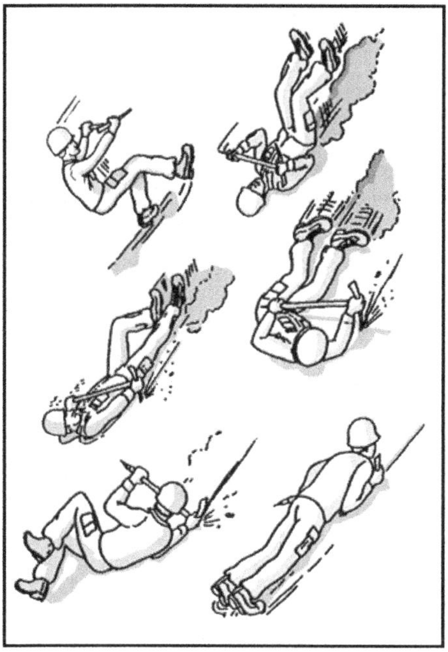

Figure 10-14. Self-arrest technique (continued).

(1) A climber who has fallen may roll or spin; if this happens, the climber must first gain control of his body, whether it is with his ice ax or simply by brute force. Once the roll or spin has been controlled, the climber will find himself in one of four positions.

- Head upslope, stomach on the slope, and feet pointed downslope.
- Head upslope, back to the slope, and feet pointed downslope.
- Head downslope, stomach on the slope, and feet pointed upslope.
- Head downslope, back to the slope, and feet pointed upslope.

(2) To place the body in position to arrest from the four basic fall positions the following must be accomplished.

(a) In the first position, the body is in proper relation to the slope for an arrest.

(b) In the second position, the body must first be rotated from face up to face down on the slope. This is accomplished by rolling the body toward the head of the ax.

(c) In the third position, the pick of the ice ax is placed upslope and used as a pivot to bring the body into proper position.

(d) In the fourth position, the head of the ax must be driven into the snow to the climber's side. This will cause the body to rotate into a head up, stomach down position.

(3) The final position when the arrest of the fall is completed should be with the head upslope, stomach on the slope, with the feet pointed downslope. If crampons are not

worn, the toe of the boots may be dug into the slope to help arrest the fall. The ax is held diagonally across the chest, with the head of the ax by one shoulder and the spike near the opposite hip. One hand grasps the head of the ax, with the pick pointed into the slope, while the other hand is on the shaft near the spike, lifting up on it to prevent the spike from digging into the slope.

Note: If crampons are worn, the feet must be raised to prevent the crampons from digging into the snow or ice too quickly. This could cause the climber to tumble and also, could severely injure his ankles and legs.

(4) When a fall occurs, the climber should immediately grasp the ax with both hands and hold it firmly as described above. Once sufficient control of the body is attained, the climber drives the pick of the ice ax into the slope, increasing the pressure until the fall is arrested. Raising the spike end of the shaft increases the biting action of the pick. It is critical that control of the ice ax be maintained at all times.

10-4. GLISSADING

Glissading is the intentional, controlled, rapid descent, or slide of a mountaineer down a steep slope covered with snow (Figure 10-15, page 10-14). Glissading is similar to skiing, except skis are not used. The same balance and control are necessary, but instead of skis the soles of the feet or the buttocks are used. The only piece of equipment required is the standard ice ax, which serves as the rudder, brake, and guide for the glissade. The two basic methods of glissading are:

a. **Squatting Glissade.** The squatting glissade is accomplished by placing the body in a semi-crouched position with both knees bent and the body weight directly over the feet. The ice ax is grasped with one hand on the head, pick, and adze outboard (away from the body), and the other hand on the shaft. The hand on the shaft grips it firmly in a position that allows control as well as the application of downward pressure on the spike of the ax.

b. **Sitting Glissade.** Using this method the glissader sits on the snow with the legs flat, and the heels and feet raised and pointed downslope. The ice ax is firmly grasped in the same manner as the squatting glissade, with the exception that the hand on the shaft must be locked against the hip for control. The sitting glissade is slower but easier to control than the squatting glissade.

c. **Safety.** A glissade should never be attempted on a slope where the bottom cannot be seen, since drop-offs may exist out of view. Also, a sitting glissade should not be used if the snow cover is thin, as painful injury could result.

Figure 10-15. Glissading techniques.

10-5. SNOW AND ICE ANCHORS

Ice and snow anchors consist of snow pickets, flukes, deadman-type anchors, ice screws, and ice pitons. Deadman anchors can be constructed from snowshoes, skis, backpacks, sleds, or any large items.

a. **Ice Pitons.** The ice piton is used to establish anchor points. The ice piton is not seen in modern ice climbing but may still be available to the military. The standard ice piton is made of tubular steel and is 10 inches in length. Ice pitons installed in pairs are a bombproof anchor; however, ice pitons have no threads for friction to hold them in the ice once placed and are removed easily. Safe use of ice pitons requires placement in pairs. Used singularly, ice pitons are a strong anchor but are easily removed, decreasing the perceived security of the anchor. Follow the instructions below for placing ice pitons in pairs.

(1) Cut a horizontal recess into the ice, and also create a vertical surface (two clean surfaces at right angles to each other).

(2) Drive one piton into the horizontal surface and another into the vertical surface so that the two pitons intersect at the necessary point (Figure 10-16).

(3) Connect the two rings with a single carabiner, ensuring the carabiner is not cross-loaded. Webbing or rope can be used if the rings are turned to the inside of the intersection.

(4) Test the piton pair to ensure it is secure. If it pulls out or appears weak, move to another spot and replace it. The pair of pitons, when placed correctly, are multidirectional.

Figure 10-16. Ice piton pair.

(5) The effective time and or strength for an ice piton placement is limited. The piton will heat from solar radiation, or the ice may crack or soften. Solar radiation can be nearly eliminated by covering the pitons with ice chips once they have been placed. If repeated use is necessary for one installation, such as top roping, the pitons should be inspected frequently and relocated when necessary. When an ice piton is removed, the ice that has accumulated in the tube must be removed before it freezes in position, making further use difficult.

c. **Ice Screws**. The ice screw is the most common type of ice protection and has replaced the ice piton for the most part (Figure 10-17). Some screws have longer "hangers" or handles, which allow them to be easily twisted into position by hand. Place ice screws as follows:

(1) Clear away all rotten ice from the surface and make a small hole with the ax pick to start the ice screw in.

(2) Force the ice screw in until the threads catch.

Figure 10-17. Placement of ice screw using the pick.

(3) Turn the screw until the eye or the hanger of the ice screw is flush with the ice and pointing down. The screw should be placed at an angle 90 to 100 degrees from the

lower surface. Use either your hand or the pick of the ice ax to screw it in. If you have a short ax (70 centimeters or less), you may be able to use the spike in the eye or hanger to ease the turning. (Remember that you may only have use of one hand at this point depending on your stance and the angle of the terrain.)

(4) As with ice pitons, melting of the ice around a screw over a period of time must be considered. The effective time and or strength of an ice screw placement is limited. The screw will heat from solar radiation, or the ice may crack or soften. Solar radiation can be nearly eliminated by covering the screw with ice chips once it has been emplaced. If repeated use is necessary for one installation, such as top roping, the screws should be inspected frequently and relocated when necessary. When an ice screw is removed, the ice that has accumulated in the tube, must be removed before further use.

d. **Horseshoe or Bollard Anchor.** This is an artificial anchor shaped generally like a horseshoe (Figure 10-18). It is formed from either ice or snow and constructed by either cutting with the ice ax or stamping with the boots. When constructed of snow, the width should not be less than 10 feet. In ice, this width may be narrowed to 2 feet, depending on the strength of the ice. The length of the bollard should be at least twice the width. The trench around the horseshoe should be stamped as deeply as possible in the snow and should be cut not less than 6 inches into the ice after all rotten ice is removed. The backside of the anchor must always be undercut to prevent the rope from sliding off and over the anchor.

(1) This type of anchor is usually available and may be used for fixed ropes or rappels. It must be inspected frequently to ensure that the rope is not cutting through the snow or ice more than one-third the length of the anchor; if it is, a new anchor must be constructed in a different location.

(2) A horseshoe anchor constructed in snow is always precarious, its strength depending upon the prevailing texture of the snow. For dry or wind-packed snow, the reliability of the anchor should always be suspect. The backside of the bollard can be reinforced with ice axes, pickets, or other equipment for added strength.

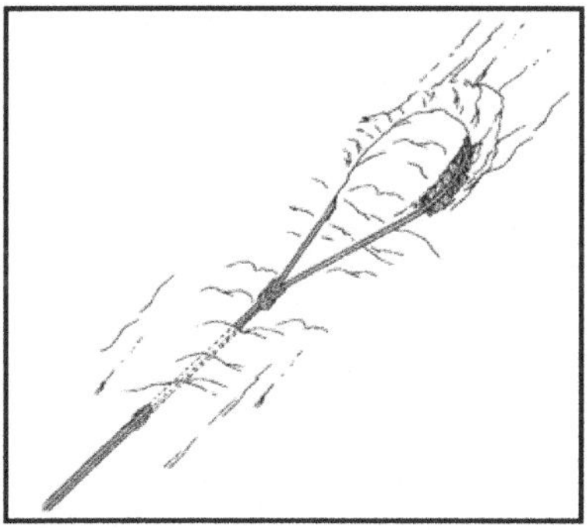

Figure 10-18. Horseshoe or bollard anchor.

e. **Pickets and Ice Axes.** Pickets and ice axes may be used as snow anchors as follows.

(1) The picket should be driven into the snow at 5 to 15 degrees off perpendicular from the lower surface. If the picket cannot be driven in all the way to the top hole, the carabiner should be placed in the hole closest to the snow surface to reduce leverage. The picket may also be tied off with a short loop of webbing or rope as in tying off pitons.

(2) An ice ax can be used in place of a picket. When using an ice ax as a snow anchor, it should be inserted with the widest portion of the ax shaft facing the direction of pull. The simplest connection to the ax is to use a sling or rope directly around the shaft just under the head. If using the leash ensure it is not worn, frayed, or cut from general use; is strong enough; and does not twist the ax when loaded. A carabiner can be clipped through the hole in the head, also.

(3) Whenever the strength of the snow anchor is suspect, especially when a picket or ax cannot be driven in all the way, the anchor may be buried in the snow and used as a "dead man" anchor. Other items suitable for dead man anchor construction are backpacks, skis, snowshoes, ski poles, or any other item large enough or shaped correctly to achieve the design. A similar anchor, sometimes referred to as a "dead guy," can be made with a large sack either stuffed with noncompressible items or filled with snow and buried. Ensure the attaching point is accessible before burying. The direction of pull on long items, such as a picket or ax, should be at a right angle to its length. The construction is identical to that of the dead man anchor used in earth.

f. **Equalized Anchors.** Snow and ice anchors must be constantly checked due to melting and changing snow or ice conditions.

(1) Whenever possible, two or more anchors should be used. While this is not always practical for intermediate anchor points on lead climbs or fixed ropes, it should be mandatory for main anchors at all belay positions, rappel points, or other fixed rope installations. (Figure 10-19, page 10-18, shows an example of three snow pickets configured to an equalized anchor.)

(2) As with multipoint anchors on rock, two or more snow or ice anchors can be joined together with a sling rope or webbing to construct one solid, equalized anchor. A bowline on a bight tied into the climbing rope can also be used instead of sling ropes or webbing.

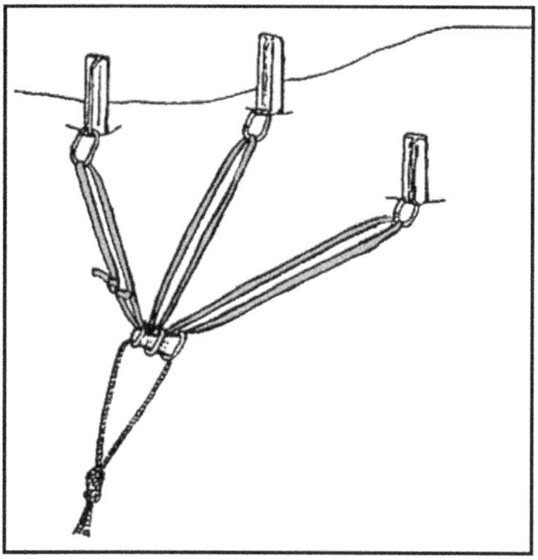

Figure 10-19. Equalized anchor using pickets.

10-6. ROPED CLIMBING ON ICE AND SNOW

When climbing on ice or snow team members tie into a climbing rope the same as when they climb on rock. When crevasses are expected, a three-man rope team is recommended.

a. **Tie-In Method.** For climbing on snow and ice, the tie-in procedure is normally the same as for rock climbing; however, when moving over snow-covered glaciers, the tie-in is modified slightly. (See paragraph 10-7, Movement on Glaciers, for more information).

b. **Movement.** For movement on gentle or moderate slopes where there is little chance of a serious fall, all climbers move simultaneously. Normally the climbers move in single file using the steps created by the lead climber and improving them when necessary. The rope between the climbers should be fully extended and kept reasonably tight. Should any member fall, he immediately yells "FALLING." The other rope team members immediately drop into a self-arrest position. The fallen climber also applies the self-arrest procedure. By using this method, called the "team arrest," the entire team as a whole arrests the fall of one member. On steeper slopes, and when crossing snow-covered crevasses where the snow bridges appear weak, the climbers use belayed climbing techniques as in rock climbing.

c. **Belaying on Snow and Ice.** The principles of belaying on ice and snow are the same as on rock. Generally, the high-force falls found in rock climbing are not present on snow and ice unless the pitch being climbed is extremely steep.

(1) ***Boot-Ax Belay.*** This belay can be useful in areas where the full length of the ice ax can penetrate the snow. The holding strength of the boot-ax belay is directly related to the firmness of the snow and to the strength of the ice ax shaft. The shaft of the ax is

tilted slightly uphill and jammed into the snow. The belayer places his uphill foot against the downhill side of the ax for support. A bight formed in the rope is placed over the boot and around the shaft of the ice ax. The brake is applied by wrapping the rope around the heel of the boot (Figure 10-20).

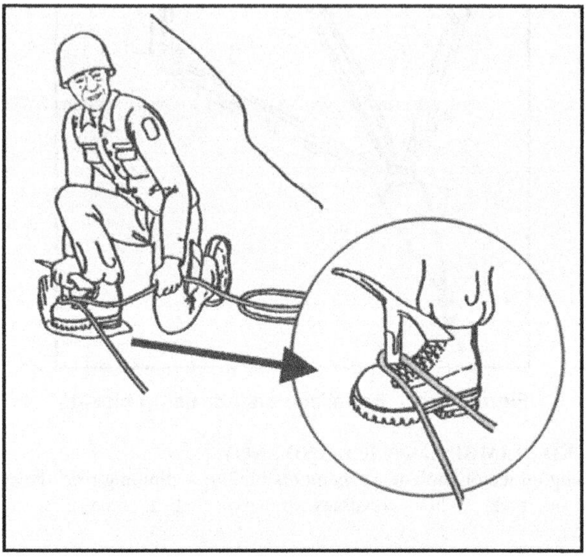

Figure 10-20. Boot-ax belay.

(2) ***Body Belay.*** The body belay can be used on snow and ice, also. The principles are the same as for belays on rock—solid anchors must be used and a well-braced position assumed. The position can be improved by digging depressions into the snow or ice for a seat and footholds. A strong platform should be constructed for the standing body belay.

(3) ***Munter Hitch.*** This belay technique is also used on snow and ice. When using the hitch off of the anchor, a two-point equalized anchor should be constructed as a minimum.

d. **Fixed Ropes.** The use of fixed ropes on ice is recommended for moving units through icefall areas on glaciers or other steep ice conditions. The procedures for emplacing fixed ropes on ice are basically the same as on rock with the exception that anchors need more attention, both in initial placement and in subsequent inspection, and steps may have to be cut to assist personnel.

10-7. MOVEMENT ON GLACIERS

Movement in mountainous terrain may require travel on glaciers. An understanding of glacier formation and characteristics is necessary to plan safe routes. A glacier is formed by the perennial accumulation of snow and other precipitation in a valley or draw. The accumulated snow eventually turns to ice due to metamorphosis. The "flow" or movement of glaciers is caused by gravity. There are a few different types of glaciers identifiable primarily by their location or activity.

- Valley glacier—resides and flows in a valley.
- Cirque glacier—forms and resides in a bowl.
- Hanging glacier—these are a result of valley or cirque glaciers flowing and or deteriorating. As the movement continues, portions separate and are sometimes left hanging on mountains, ridgelines, or cliffs.
- Piedmont glacier—formed by one or more valley glaciers; spreads out into a large area.
- Retreation glacier—a deteriorating glacier; annual melt of entire glacier exceeds the flow of the ice.
- Surging glacier—annual flow of the ice exceeds the melt; the movement is measurable over a period of time.

a. **Characteristics and Definitions.** This paragraph describes the common characteristics of glaciers, and defines common terminology used in reference to glaciers. (Figure 10-21 shows a cross section of a glacier, and Figure 10-22 depicts common glacier features.)

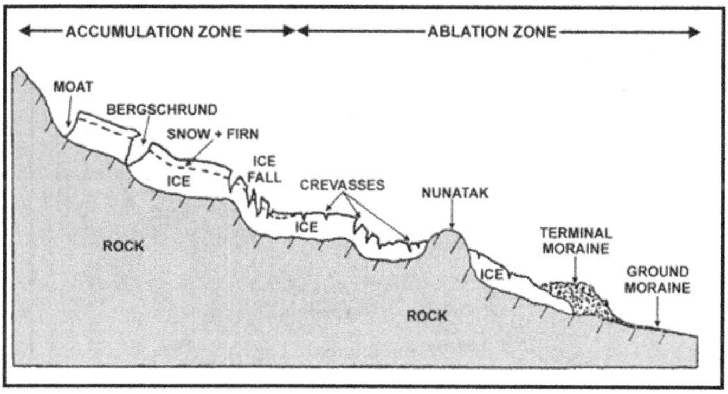

Figure 10-21. Glacier cross section.

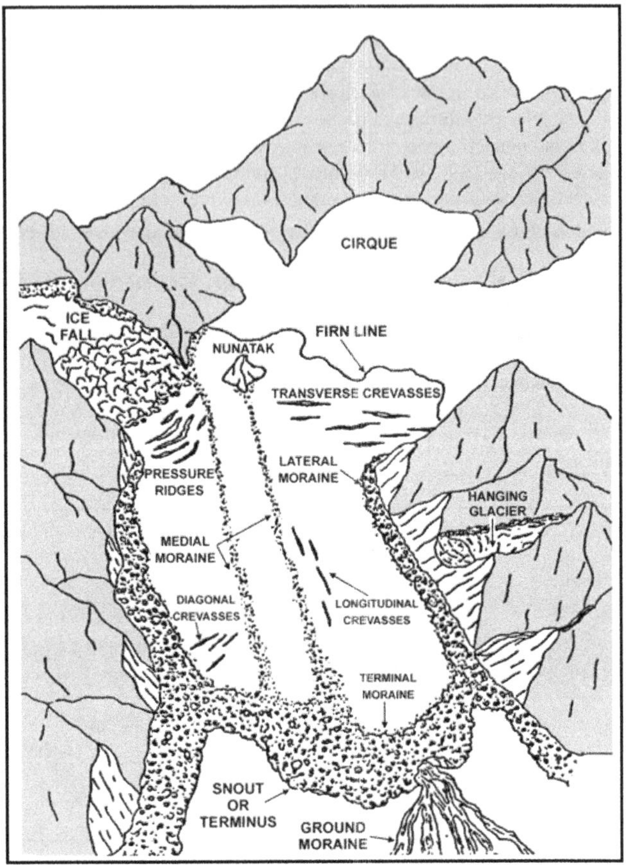

Figure 10-22. Glacier features.

(1) Firn is compacted granular snow that has been on the glacier at least one year. Firn is the building blocks of the ice that makes the glacier.

(2) The accumulation zone is the area that remains snow-covered throughout the year because of year-round snowfall. The snowfall exceeds melt.

(3) The ablation zone is the area where the snow melts off the ice in summer. Melt equals or exceeds snowfall.

(4) The firn line separates the accumulation and ablation zones. As you approach this area, you may see "strips" of snow in the ice. Be cautious, as these could be *snow bridges* remaining over crevasses. Remember that snow bridges will be weakest lower on the glacier as you enter the accumulation zone. The firn line can change annually.

(5) A bergschrund is a large crevasse at the head of a glacier caused by separation of active (flowing) and inactive (stationary) ice. These will usually be seen at the base of a major incline and can make an ascent on that area difficult.

(6) A moat is a wall formed at the head (start) of the glacier. These are formed by heat reflected from valley wall.

(7) A crevasse is a split or crack in the glacier surface. These are formed when the glacier moves over an irregularity in the bed surface.

(8) A transverse crevasse forms perpendicular to the flow of a glacier. These are normally found where a glacier flows over a slope with a gradient change of 30 degrees or more.

(9) Longitudinal crevasses form parallel to the flow of a glacier. These are normally found where a glacier widens.

(10) Diagonal crevasses form at an angle to the flow of a glacier. These are normally found along the edges where a glacier makes a bend.

(11) A snow bridge is a somewhat supportive structure of snow that covers a crevasse. Most of these are formed by the wind. The strength of a snow bridge depends on the snow itself.

(12) Icefalls are a jumble of crisscross crevasses and large ice towers that are normally found where a glacier flows over a slope with a gradient change of 25 degrees or more.

(13) Seracs are large pinnacles or columns of ice that are normally found in icefalls or on hanging glaciers.

(14) Ice avalanches are falling chunks of ice normally occurring near icefalls or hanging glaciers.

(15) The moraine is an accumulation of rock or debris on a glacier caused by rockfall or avalanche of valley walls.

(16) The lateral moraine is formed on sides of glacier.

(17) The medial moraine is in the middle of the glacier. This is also formed as two glaciers come together or as a glacier moves around a central peak.

(18) The terminal moraine is at the base of a glacier and is formed as moraines meet at the snout or terminus of a glacier.

(19) The ground moraine is the rocky debris extending out from the terminus of a glacier. This is formed by the scraping of earth as the glacier grew or surged and exposed as the glacier retreats.

(20) A Nunatak is a rock projection protruding through the glacier as the glacier flows around it.

(21) An ice mill is a hole in the glacier formed by swirling water on the surface. These can be large enough for a human to slip into.

(22) Pressure ridges are wavelike ridges that form on glacier normally after a glacier has flowed over icefalls.

(23) A glacier window is an opening at the snout of the glacier where water runs out of the glacier.

b. **Dangers and Obstacles.** The principle dangers and obstacles to movement in glacial areas are crevasses, icefalls, and ice avalanches. Snow-covered crevasses make movement on a glacier extremely treacherous. In winter, when visibility is poor, the difficulty of recognizing them is increased. Toward the end of the summer, crevasses are widest and covered by the least snow. Crossing snow bridges constitutes the greatest potential danger in movement over glaciers in the summer. On the steep pitch of a glacier, ice flowing over irregularities and cliffs in the underlying valley floor cause the ice to break up into ice blocks and towers, criss-crossed with crevasses. This jumbled cliff of ice is known as an icefall. Icefalls present a major obstacle to safe movement of troops on glaciers.

(1) Moving on glaciers brings about the hazard of falling into a crevasse. Although the crevasses are visible in the ablation zone in the summer (Figure 10-23), the accumulation zone will still have hidden crevasses. The risk of traveling in the accumulation zone can be managed to an acceptable level when ropes are used for connecting the team members (Figure 10-24, page 10-24). Crampons and an ice ax are all that is required to safely travel in the ablation zone in the summer.

Figure 10-23. Ablation zone of glacier in summer.

Figure 10-24. Rope teams moving in the accumulation zone of a glacier.

(2) When conditions warrant, three to four people will tie in to one rope at equal distances from each other. To locate the positions, if three people are on a team, double the rope and one ties into the middle and the other two at the ends. If four people are on a team, form a "z" with the rope and expand the "z" fully, keeping the end and the bight on each "side" of the "z" even. Tie in to the bights and the ends.

(3) Connect to the rope with the appropriate method and attach the Prusik as required. The rope should be kept relatively tight either by Prusik belay or positioning of each person. If the team members need to assemble in one area, use the Prusik to belay each other in.

(4) If a team member falls into a crevasse, the remaining members go into team arrest, assess the situation, and use the necessary technique to remove the person from the crevasse. The simplest and most common method for getting someone out of a crevasse is for the person to climb out while being belayed.

(5) All items should be secured to either the climber or the rope/harness to prevent inadvertent release and loss of necessary items or equipment. Packs should be secured to the rope/harness with webbing or rope. If traveling with a sled in tow, secure it not only to a climber to pull it, but connect it to the rope with webbing or rope also.

(6) If marking the route on the glacier is necessary for backtracking or to prevent disorientation in storms or flat-light conditions, use markers that will be noticeable against the white conditions. The first team member can place a new marker when the last team member reaches the previous marker.

c. **Roped Movement.** The first rule for movement on glaciers is to rope up (Figure 10-25). A roped team of two, while ideal for rock climbing, is at a disadvantage on a snow-covered glacier. The best combination is a three-man rope team. Generally, the rope team members will move at the same time with the rope fully extended and reasonably tight between individuals, their security being the team arrest. If an individual should break through a snow bridge and fall into a crevasse, the other members immediately perform self-arrest, halting the fall. At points of obvious weakness in the snow bridges, the members may decide to belay each other across the crevasse using one of the established belay techniques.

Figure 10-25. Preparation for roped movement.

(1) Even with proper training in crevasse rescue techniques, the probability exists that an individual may remain suspended in a crevasse for a fairly lengthy amount of time while trying to get himself out or while awaiting help from his rope team members. Because of this, it is strongly recommended that all personnel wear a seat/chest combination harness, whether improvised or premanufactured.

(2) Rope team members must be able to quickly remove the climbing rope from the harness(es) during a crevasse rescue. The standard practice for connecting to the rope for glacier travel is with a locking carabiner on a figure-eight loop to the harness. This allows quick detachment of the rope for rescue purposes. The appropriate standing part of the rope is then clipped to the chest harness carabiner.

(3) If a rope team consists of only two people, the rope should be divided into thirds, as for a four-person team. The team members tie into the middle positions on the rope, leaving a third of the rope between each team member and a third on each end of the rope. The remaining "thirds" of the rope should be coiled and either carried in the rucksack, attached to the rucksack, or carried over the head and shoulder. This gives each

climber an additional length of rope that can be used for crevasse rescue, should one of the men fall through and require another rope. If necessary, this excess end rope can be used to connect to another rope team for safer travel.

Note: The self-arrest technique used by one individual will work to halt the fall of his partner on a two-man rope team; however, the chance of it failing is much greater. Crevasse rescue procedures performed by a two-man rope team, by itself, may be extremely difficult. For safety reasons, movement over a snow-covered glacier by a single two-man team should be avoided wherever possible.

d. **Use of Prusik Knots.** Prusik knots are attached to the climbing rope for all glacier travel. The Prusiks are used as a self-belay technique to maintain a tight rope between individuals, to anchor the climbing rope for crevasse rescue, and for self-rescue in a crevasse fall. The Prusik slings are made from the 7-millimeter by 6-foot and 7-millimeter by 12-foot ropes. The ends of the ropes are tied together, forming endless loops or slings, with double fisherman's knots. Form the Prusik knot on the rope in front of the climber. An overhand knot can be tied into the sling just below the Prusik to keep equal tension on all the Prusik wraps. Attach this sling to the locking carabiner at the tie in point on the harness.

Note: An ascender can replace a Prusik sling in most situations. However, the weight of an ascender hanging on the rope during movement will become annoying, and it could be stepped on during movement and or climbing.

e. **Securing the Backpack/Rucksack.** If an individual should fall into a crevasse, it is essential that he be able to rid himself of his backpack. The weight of the average pack will be enough to hinder the climber during crevasse rescue, or possibly force him into an upside down position while suspended in the crevasse. Before movement, the pack should be attached to the climbing rope with a sling rope or webbing and a carabiner. A fallen climber can immediately drop the pack without losing it. The drop cord length should be minimal to allow the fallen individual to reach the pack after releasing it, if warm clothing is needed. When hanging from the drop cord, the pack should be oriented just as when wearing it (ensure the cord pulls from the top of the pack).

f. **Routes.** An individual operating in the mountains must appreciate certain limitations in glacier movement imposed by nature.

(1) Additional obstacles in getting onto a glacier may be swift glacier streams, steep terminal or lateral moraines, and difficult mountain terrain bordering the glacier ice. The same obstacles may also have to be overcome in getting on and off a valley glacier at any place along its course.

(2) Further considerations to movement on a glacier are steep sections, heavily crevassed portions, and icefalls, which may be major obstacles to progress. The use of current aerial photographs in conjunction with aerial reconnaissance is a valuable means of gathering advance information about a particular glacier. However, they only supplement, and do not take the place of, on-the-ground reconnaissance conducted from available vantage points.

g. **Crossing Crevasses.** Open crevasses are obvious, and their presence is an inconvenience rather than a danger to movement. Narrow cracks can be jumped, provided the take off and landing spots are firm and offer good footing. Wider cracks will have to be circumvented unless a solid piece of ice joins into an ice bridge strong enough to support at least the weight of one member of the team. Such ice bridges are often formed in the lower portion of a crevasse, connecting both sides of it.

(1) In the area of the firn line, the zone that divides seasonal melting from permanent falls of snow, large crevasses remain open, though their depths may be clogged with masses of snow. Narrow cracks may be covered. In this zone, the snow, which covers glacier ice, melts more rapidly than that which covers crevasses. The difference between glacier ice and narrow snow-covered cracks is immediately apparent; the covering snow is white, whereas the glacier ice is gray.

(2) Usually the upper part of a glacier is permanently snow covered. The snow surface here will vary in consistency from dry powder to consolidated snow. Below this surface cover are found other snow layers that become more crystalline in texture with depth, and gradually turn into glacier ice. It is in this snow-covered upper part of a glacier that crevasses are most difficult to detect, for even wide crevasses may be completely concealed by snow bridges.

h. **Snow Bridges.** Snow bridges are formed by windblown snow that builds a cornice over the empty interior of the crevasse. As the cornice grows from the windward side, a counter drift is formed on the leeward side. The growth of the leeward portion will be slower than that to the windward so that the juncture of the cornices occurs over the middle of the crevasse only when the contributing winds blow equally from each side. Bridges can also be formed without wind, especially during heavy falls of dry snow. Since cohesion of dry snow depends only on an interlocking of the branches of delicate crystals, such bridges are particularly dangerous during the winter. When warmer weather prevails the snow becomes settled and more compacted, and may form firmer bridges.

(1) Once a crevasse has been completely bridged, its detection is difficult. Bridges are generally slightly concave because of the settling of the snow. This concavity is perceptible in sunshine, but difficult to detect in flat light. If the presence of hidden crevasses is suspected, the leader of a roped team must probe the snow in front of him with the shaft of his ice ax. As long as a firm foundation is encountered, the team may proceed, but should the shaft meet no opposition from an underlying layer of snow, a crevasse is probably present. In such a situation, the prober should probe closer to his position to make sure that he is not standing on the bridge itself. If he is, he should retreat gently from the bridge and determine the width and direction of the crevasse. He should then follow and probe the margin until a more resistant portion of the bridge is reached. When moving parallel to a crevasse, all members of the team should keep well back from the edge and follow parallel but offset courses.

(2) A crevasse should be crossed at right angles to its length. When crossing a bridge that seems sufficiently strong enough to hold a member of the team, the team will generally move at the same time on a tight rope, with each individual prepared to go into self-arrest. If the stability of the snow bridge is under question, they should proceed as follows for a team of three glacier travelers:

(a) The leader and second take up a position at least 10 feet back from the edge. The third goes into a self-belay behind the second and remains on a tight rope.

(b) The second belays the leader across using one of the established belay techniques. The boot-ax belay should be used only if the snow is deep enough for the ax to be inserted up to the head and firm enough to support the possible load. A quick ice ax anchor should be placed for the other belays. Deadman or equalizing anchors should be used when necessary.

(c) The leader should move forward, carefully probing the snow and evaluating the strength of the bridge, until he reaches firm snow on the far side of the crevasse. He then continues as far across as possible so number two will have room to get across without number one having to move.

(d) The third assumes the middle person's belay position. The middle can be belayed across by both the first and last. Once the second is across, he assumes the belay position. Number one moves out on a tight rope and anchors in to a self-belay. Number two belays number three across.

(3) In crossing crevasses, distribute the weight over as wide an area as possible. Do not stamp the snow. Many fragile bridges can be crossed by lying down and crawling to the other side. Skis or snowshoes help distribute the weight nicely.

i. **Arresting and Securing a Fallen Climber.** The simplest and most common method for getting someone out of a crevasse is for the person to climb out while being belayed. Most crevasse falls will be no more than body height into the opening if the rope is kept snug between each person.

(1) To provide a quick means of holding an unexpected breakthrough, the rope is always kept taut. When the leader unexpectedly breaks through, the second and third immediately go into a self-arrest position to arrest the fall. A fall through a snow bridge results either in the person becoming jammed in the surface hole, or in being suspended in the crevasse by the rope. If the leader has fallen only partially through the snow bridge, he is supported by the snow forming the bridge and should not thrash about as this will only enlarge the hole and result in deeper suspension. All movements should be slow and aimed at rolling out of the hole and distributing the weight over the remainder of the bridge. The rope should remain tight at all times and the team arrest positions adjusted to do so. It generally is safer to retain the rucksack, as its bulk often prevents a deeper fall. Should a team member other than the leader experience a partial fall, the rescue procedure will be same as for the leader, only complicated slightly by the position on the rope.

(2) When the person falls into a crevasse, the length of the fall depends upon how quickly the fall is arrested and where in the bridge the break takes place. If the fall occurs close to the near edge of the crevasse, it usually can be checked before the climber has fallen more than 6 feet. However, if the person was almost across, the fall will cause the rope to cut through the bridge, and then even an instantaneous check by the other members will not prevent a deeper fall. The following scenario is an example of the sequence of events that take place after a fall by the leader in a three-person team. (This scenario is for a team of three, each person referred to by position; the leader is number 1.)

(a) Once the fall has been halted by the team arrest, the entire load must be placed on number 2 to allow number 3 to move forward and anchor the rope. Number 3 slowly releases his portion of the load onto number 2, always being prepared to go back into self-arrest should number 2's position begin to fail.

(b) Once number 2 is confident that he can hold the load, number 3 will proceed to number 2's position, using the Prusik as a self belay, to anchor the rope. In this way the rope remains reasonably tight between number 2 and number 3. Number 3 must always be prepared to go back into self-arrest should number 2's position begin to fail.

(c) When number 3 reaches number 2's position he will establish a bombproof anchor 3 to 10 feet in front of number 2 (on the load side), depending on how close number 2 is to the lip of the crevasse. This could be either a deadman or a two-point equalized anchor, as a minimum.

(d) Number 3 connects the rope to the anchor by tying a Prusik with his long Prusik sling onto the rope leading to number 1. An overhand knot should be tied into the long Prusik sling to shorten the distance to the anchor, and attached to the anchor with a carabiner. The Prusik knot is adjusted toward the load.

(e) Number 2 can then release the load of number 1 onto the anchor. Number 2 remains connected to the anchor and monitors the anchor.

(f) A fixed loop can be tied into the slack part of the rope, close to number 2, and attached to the anchor (to back up the Prusik knot).

(g) Number 3 remains tied in, but continues forward using a short Prusik as a self-belay. He must now quickly check on the condition of number 1 and decide which rescue technique will be required to retrieve him.

(3) These preliminary procedures must be performed before retrieving the fallen climber. If number 3 should fall through a crevasse, the procedure is the same except that number 1 assumes the role of number 3. Normally, if the middle person should fall through, number 1 would anchor the rope by himself. Number 3 would place the load on number 1's anchor, then anchor his rope and move forward with a Prusik self-belay to determine the condition of number 2.

j. **Crevasse Rescue Techniques.** Snow bridges are usually strongest at the edge of the crevasse, and a fall is most likely to occur some distance away from the edge. In some situations, a crevasse fall will occur at the edge of the snow bridge, on the edge of the ice. If a fall occurs away from the edge, the rope usually cuts deeply into the snow, thus greatly increasing friction for those pulling from above. In order to reduce friction, place padding, such as an ice ax, ski, ski pole, or backpack/rucksack, under the rope and at right angles to the stress. Push the padding forward as far as possible toward the edge of the crevasse, thus relieving the strain on the snow. Ensure the padding is anchored from falling into the crevasse for safety of the fallen climber.

(1) ***Use of Additional Rope Teams.*** Another rope team can move forward and assist in pulling the victim out of a crevasse. The assisting rope team should move to a point between the fallen climber and the remaining rope team members. The assisting team can attach to the arresting team's rope with a Prusik or ascender and both rope teams' members can all pull simultaneously. If necessary, a belay can be initiated by the fallen climber's team while the assisting team pulls. The arresting team member closest to the fallen climber should attach the long Prusik to themselves and the rope leading to the fallen climber, and the assisting team can attach their Prusik or ascender between this long Prusik and the arresting team member. As the assisting team pulls, the Prusik belay will be managed by the arresting team member at the long Prusik.

Note: Safety in numbers is obvious for efficient crevasse rescue techniques. Additional rope teams have the necessary equipment to improve the main anchor or establish new ones and the strength to pull a person out even if he is deep in the crevasse. Strength of other rope teams should always be used before establishing more time-consuming and elaborate rescue techniques.

(2) ***Fixed Rope.*** If the fallen climber is not injured, he may be able to climb out on a fixed rope. Number 1 clips number 3's rope to himself. He then climbs out using number 3's rope as a simple fixed line while number 2 takes up the slack in number 1's rope through the anchor Prusik for a belay.

(3) ***Prusik Ascending Technique.*** There may be times when the remaining members of a rope team can render little assistance to the person in the crevasse. If poor snow conditions make it impossible to construct a strong anchor, the rope team members on top may have to remain in self-arrest. Other times, it may just be easier for the fallen climber to perform a self-rescue. (Figure 10-26 shows the proper rope configuration.) The technique is performed as follows:

(a) The fallen climber removes his pack and lets it hang below from the drop cord.

(b) The individual slides their short Prusik up the climbing rope as far as possible.

(c) The long Prusik is attached to the rope just below the short Prusik. The double fisherman's knot is spread apart to create a loop large enough for one or both feet. The fallen climber inserts his foot/feet into the loop formed allowing the knot to cinch itself down.

(d) The individual stands in the foot loop, or "stirrup," of the long sling.

(e) With his weight removed from the short Prusik, it is slid up the rope as far as it will go. The individual then hangs from the short Prusik while he moves the long Prusik up underneath the short Prusik again.

(f) The procedure is repeated, alternately moving the Prusiks up the rope, to ascend the rope. Once the crevasse lip is reached, the individual can simply grasp the rope and pull himself over the edge and out of the hole.

(g) Besides being one of the simplest rope ascending techniques, the short Prusik acts as a self-belay and allows the climber to take as long a rest as he wants when sitting in the harness. The rope should be detached from the chest harness carabiner to make the movements less cumbersome. However, it is sometimes desirable to keep the chest harness connected to the rope for additional support. In this case the Prusik knots must be "on top" of the chest harness carabiner so they can be easily slid up the rope without interference from the carabiner. The long Prusik sling can be routed through the chest harness carabiner for additional support when standing up in the stirrup.

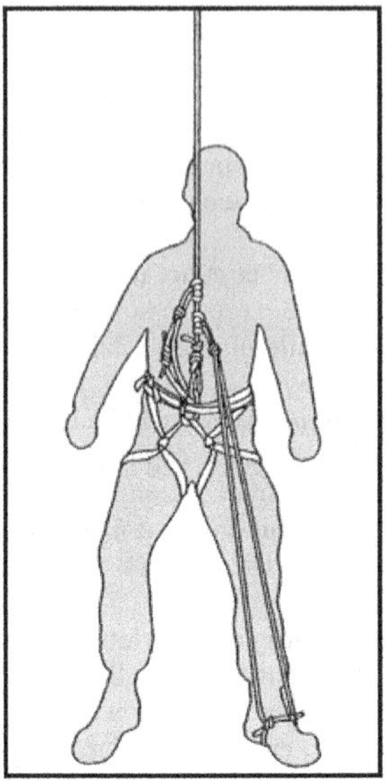

Figure 10-26. Prusik ascending technique.

(4) ***Z-Pulley Hauling System.*** If a fallen climber is injured or unconscious, he will not be able to offer any assistance in the rescue. If additional rope teams are not immediately available, a simple raising system can be rigged to haul the victim out of the crevasse. The Z-pulley hauling system is one of the simplest methods and the one most commonly used in crevasse rescue (Figure 10-27, page 10-32). The basic Z rig is a "3-to-1" system, providing mechanical advantage to reduce the workload on the individuals operating the haul line. In theory, it would only take about 33 pounds of pull on the haul rope to raise a 100-pound load with this system. In actual field use, some of this mechanical advantage is lost to friction as the rope bends sharply around carabiners and over the crevasse lip. The use of mechanical rescue pulleys can help reduce this friction in the system. The following describes rigging of the system. (This scenario is for a team of three, each person referred to by position; the leader is number 1.)

(a) After the rope team members have arrested and secured number 1 to the anchor, and they have decided to install the Z rig, number 2 will attach himself to the anchor without using the rope and clear the connecting knot used. Number 3 remains connected to the rope.

(b) The slack rope exiting the anchor Prusik is clipped into a separate carabiner attached to the anchor. A pulley can be used here if available.

(c) Number 3 will use number 2's short Prusik to rig the haul Prusik. He moves toward the crevasse lip (still on his own self-belay) and ties number 2's short Prusik onto number 1's rope (load rope) as close to the edge as possible.

(d) Another carabiner (and pulley if available) is clipped into the loop of the haul Prusik and the rope between number 3's belay Prusik and the anchor is clipped (or attached through the pulley). Number 3's rope becomes the haul rope.

(e) Number 3 then moves towards the anchor and number 2. Number 2 could help pull if necessary but first would connect to the haul rope with a Prusik just as number 3. If the haul Prusik reaches the anchor before the victim reaches the top, the load is simply placed back on the anchor Prusik and number 3 moves the haul Prusik back toward the edge. The system is now ready for another haul.

CAUTION

The force applied to the fallen climber through use of the Z-pulley system can be enough to destroy the harness-to-rope connection or injure the fallen climber if excess force is applied to the pulling rope.

Notes: 1. The Z-pulley adds more load on the anchor due to the mechanical advantage. The anchor should be monitored for the duration of the rescue.

2. With the "3-to-1" system, the load (fallen climber) will be raised 1 foot for every 3 feet of rope taken up during the haul.

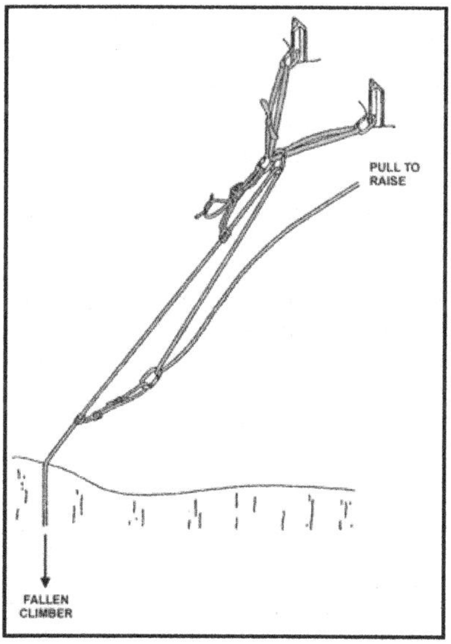

Figure 10-27. Z-pulley hauling system.

10-8. GLACIER BIVOUAC PROCEDURES

When locating a bivouac site or a gathering area where the team might need or want to unrope, at least one person will need to "probe" the area for hidden crevasses. The best type of probe will be the manufactured collapsing probe pole, at least eight feet in length. Other items could be used but the length and strength of the probe is most important. Other rope team members will belay the probers. The prober is "feeling" for a solid platform to place the tent by pushing the probe as hard and deep as possible into the surface. Probing should be in 2-foot intervals in all directions within the site.

a. If the probe suddenly has no resistance while pushing down, a crevasse is present. Attempts to outline the crevasse can be futile if the crevasse is large. Normally, the best decision is to relocate the proposed bivouac area far enough away to avoid that crevasse. (Sometimes only a few feet one way or the other is all that's needed to reach a good platform.) Probe the tent site again after digging to the desired surface. Mark boundaries with wands or other items such as skis, poles, and so on.

b. Occasionally while probing, increased pressure will be noticed without reaching a solid platform. The amount of snowfall may be such that even after digging into the snow, the probe still doesn't contact a hard surface. Try to find a solid platform.

c. There should be no unroped movement outside the probed/marked areas. If a latrine area is needed, probe a route away from the bivouac site and probe the latrine area also. If a dugout latrine is necessary, probe again after digging.

d. Multiple tent sites can be connected, which keeps tents closer together. Probe all areas between the tents if you plan to move in those areas. Closer tents will make communicating between tent groups and rope teams easier.

e. If there is a chance for severe storms with high winds, snow walls may be constructed to protect the tent site from wind. The walls can be constructed from loose snow piled on the perimeter, or blocks can be cut from consolidated snow layers. In deep soft snow, digging three or four feet to find a consolidated layer will result in enough snow moved to build up decent walls around the tent site.

(1) For block construction, move the soft snow from the surface into the wall foundation areas (down to a consolidated layer of snow).

(2) Cut blocks approximately 1 by 1 by 2 feet, and construct the walls by interlocking the blocks with overlapping placements. The walls should be slightly higher than the tent. At a minimum, build walls on the windward side of the tent site.

(3) Snow walls can also provide shelter from wind for food preparation.

CHAPTER 11
MOUNTAIN RESCUE AND EVACUATION

Steep terrain and adverse weather are common in mountainous environments. Under these conditions, relatively minor injuries may require evacuation. The evacuation technique chosen is determined by the type of injury, distance to be moved, terrain, and existing installations. Air evacuation is preferred; however, the weather, tactical situation, or operational ceiling of the aircraft may make this impossible. It is, therefore, imperative that all personnel are trained in mountain evacuation techniques and are self-sufficient. Casualties should be triaged before evacuation. Triage is performed by the most experienced medical personnel available (physician, physician's assistant, medic).

Performing a rescue operation can be a significant emotional event. Rescue scenarios must be practiced and rehearsed until rescue party members are proficient in the many tasks required to execute a rescue. To perform most of the high-angle rescues, Level I and Level II mountaineers are required with a Level III supervising.

11-1. CONSIDERATIONS

The techniques of evacuation are proven techniques. They are, however, all subject to improvement and should be discarded or modified as better methods of handling victims are developed.

a. When evacuating a victim from mountainous areas keep in mind that the purpose of a rescue operation is to save a life, and physical risk to the rescuers must be weighed against this purpose. However, there is no excuse for failing to make the maximum effort within this limitation. Work and expense should be no deterrent when a life is at stake.

b. Rescues will be unplanned (improvised) or planned rescue operations. For a planned rescue, equipment that is especially suited and designed for rescue should be used. For training missions always have a medical plan developed before an emergency arises (plan for the worst and hope for the best). Ensure that the MEDEVAC plan is a comprehensive plan and must be thought out and understood by all that may be involved in a potential rescue.

c. The following actions will be done immediately at the rescue scene.

(1) Assume command. One person, and one person only, is overall in charge at all times.

(2) Prevent further injuries to the victim and to others. Use reasonable care in reaching the victim.

(3) Immediately ensure the victim has an open airway, resume victim's breathing, control serious bleeding, and maintain moderate body warmth. If the victim is unconscious, continually monitor pulse. Protect the patient from environmental hazards.

(4) Do not move the victim until you have ascertained the extent of injuries, unless it is necessary to prevent further injuries or the victim is located in a dangerous location (for example, avalanche run-out zone, hanging glacier, possibility of falling rocks).

(5) Do nothing more until you have thoroughly considered the situation. Resist the urge for action. Speed is less important than correct action.

(6) Decide whether to evacuate with available facilities or to send for help. Speed in getting to a hospital must be balanced against the probability of further injury if working with inexperienced people, lack of equipment or wrong equipment, and terrain at hand.

(7) When the evacuation route is long and arduous, a series of litter relay points or stations should be established. These stations must be staffed with the minimum medical personnel to provide proper emergency treatment. When a victim develops signs of shock or worsens while being evacuated, he should be treated and retained at one of these stations until his condition allows evacuation.

(8) Helicopters or heated vehicles, if available, should be used for evacuation. While the use of aircraft or vehicles is preferred and can expedite a rescue operation, evacuation of a seriously wounded soldier should never be delayed to await aircraft, vehicle, or a change in weather.

11-2. PLANNING RESCUE OPERATIONS

Every commander should have a medical evacuation plan before undertaking an operation. This plan should have contingencies included so as not to rely on a single asset.

a. When rescuing a casualty (victim) threatened by hostile action, environmental hazard, or any other immediate hazard, the rescuer should not take action without first determining the extent of the hazard and his ability to handle the situation. THE RESCUER MUST NOT BECOME A CASUALTY.

b. The rescue team leader must evaluate the situation and analyze the factors involved. This evaluation can be divided into three major steps:

- Identify the task.
- Evaluate the circumstances of the rescue.
- Plan the action.

c. The task must be identified. In planning a rescue, the rescuer tries to obtain the following information:

- Who, what, where, when, why, and how the situation happened.
- Number of casualties by precedence (urgent, priority, routine, tactical immediate),
- number of casualties by type (litter or ambulatory), and the nature of their injuries.
- Terrain features and location of the casualties.
- Tactical situation.
- If adequate assistance is available to aid in security, rescue, treatment, and evacuation.
- If treatment can be provided at the scene; if the victims require movement to a safer location.
- Equipment required for the rescue operation.

d. Circumstances of the rescue are as follows:

(1) After identifying the task, relate it to the circumstances of the situation.

- Are additional personnel, security, medical, or special rescue equipment needed?
- Are there circumstances, such as aircraft accidents (mass casualties), that may require specialized skills?

- What is the weather condition?
- Is the terrain hazardous?
- How much time is available?

(2) The time element may cause a rescuer to compromise planning stages or treatment (beyond first aid). Make a realistic estimate of time available as quickly as possible to determine the action time remaining. The key elements are the casualty's condition and environment.

(3) Mass casualties are to be expected on the modern battlefield. All problems or complexities of rescue are now multiplied by the number of casualties. Time becomes the critical element.

(4) Considerations for the main rescue group for a planned rescue are as follows:

(a) Carry all needed equipment, hot food and drinks, stove, sleeping bags, tents, bivouac sacks, warm clothes, ropes, and stretchers.

(b) Prepare the evacuation route (ground transport to hospital, walking trails, fixed lines, lowering lines, anchor points, and rescue belay points). If the victim is airlifted out, attach a paper with the medical actions that were performed on the ground (for example, blood pressure, pulse rate, drugs started, and so on).

(c) When performing all rescues, the rescuers are always tied in for safety. With all rescue techniques, remember to think things through logically for safety and to prevent the rescuer from accidentally untying himself or the fallen climber.

(d) Constantly inform the casualty (if they are conscious) as to what you are doing and what he must do.

e. The rescue plan should proceed as follows:

(1) In estimating time available, the casualties' ability to endure is of primary importance. Age and physical condition may vary. Time available is a balance of the endurance time of the casualty, the situation, and the personnel and equipment available.

(2) Consider altitude and visibility. Maximum use of secure, reliable trails or roads is essential.

(3) Ensure that blankets and rain gear are available. Even a mild rain can complicate a normally simple rescue. In high altitudes, extreme cold, or gusting winds, available time is drastically reduced.

(4) High altitudes and gusting winds reduce the ability of fixed-wing or rotary-wing aircraft to assist in operations. Rotary-wing aircraft may be available to remove casualties from cliffs or inaccessible sites, and to quickly transport casualties to a medical treatment facility. Relying on aircraft or specialized equipment is a poor substitute for careful planning.

11-3. MASS CASUALTIES

When there are mass casualties, an orderly rescue may involve further planning.

a. To manage a mass casualty rescue or evacuation, separate stages are taken.

- FIRST STAGE: Remove personnel who are not trapped among debris or who can be easily evacuated.
- SECOND STAGE: Remove personnel who may be trapped by debris, but whose extraction only requires the equipment on hand and little time.

- THIRD STAGE: Remove the remaining personnel who are trapped in extremely difficult or time-consuming situations, such as moving large amounts of debris or cutting through a wall.
- FOURTH STAGE: Remove dead personnel.

b. Evacuation of wounded personnel is based on the victim's condition and is prioritized as follows:

- PRIORITY ONE: Personnel with life-threatening injuries that require immediate emergency care to survive; first aid and stabilization are accomplished before evacuation.
- PRIORITY TWO: Personnel with injuries that require medical care but speed of evacuation is not essential.
- PRIORITY THREE: Injured personnel who can evacuate themselves with minimal assistance.
- PRIORITY FOUR: The logistics removal of dead personnel.

11-4. SPECIAL TRAINING

Before receiving training in basic mountain evacuation, litter teams should receive instruction in military mountaineering and basic first aid. Litter bearers and medics must know the use and care of rope as an item of equipment. The members of litter teams must be proficient in the techniques of belaying and choosing belay points. Proper support and protection must be given to victims and litter bearers when evacuating over steep, difficult terrain.

11-5. PREPARATION FOR EVACUATION

Although the wounded soldier's life may have been saved by applying first aid, it can be lost through carelessness, rough handling, or inadequate protection from the elements. Therefore, before trying to move the wounded soldier, the type and extent of his injury must be evaluated. Dressings over wounds must be reinforced, and fractured bones must be properly immobilized and supported. Based upon the evaluation of the type and extent of the soldier's injury, the best method of manual transportation is selected.

11-6. MANUAL CARRIES

Personnel who are not seriously injured but cannot evacuate themselves may be assisted by fellow soldiers. Personnel who are injured and require prompt evacuation should not be forced to wait for mobile evacuation or special equipment.

a. **One-Man Carries.** The basic carries taught in the Soldier's Manual of Common Tasks (fireman's carry, two-hand, four-hand, saddleback, piggyback, pistol belt, and poncho litter) are viable means of transporting injured personnel; however, the mountainous terrain lends itself to several other techniques. One-man carries include the sling-rope carry and the rope coil carry.

(1) ***Sling-Rope Carry***. The sling-rope carry (Figure 11-1) requires a 4.5-meter sling rope and two men—one as the bearer and the other as an assistant to help secure the casualty to the bearer's back. Conscious or unconscious casualties may be transported this way.

(a) The bearer kneels on all fours.

(b) The assistant places the casualty face down on the bearer's back ensuring the casualty's armpits are even with the bearer's shoulders.

(c) The assistant then finds the middle of the sling rope and places it between the casualty's shoulders.

(d) The assistant runs the ends of the sling rope under the casualty's armpits, crosses the ends, and runs the ends over the bearer's shoulders and back under the bearer's arms.

(e) The assistant runs the ends of the rope between the casualty's legs, around the casualty's thighs, and back around to the front of the bearer. The rope is tied with a square knot with two overhand knots just above the bearer's belt buckle.

(f) The rope must be tight. Padding, when available, should be placed where the rope passes over the bearer's shoulders and under the casualty's thighs.

Figure 11-1. Sling-rope carry.

(2) ***Rope Coil Carry.*** The rope coil carry requires a bearer and a 36 1/2-meter coiled rope. It can be used to transport a conscious or unconscious victim.

(a) Place the casualty on his back.

(b) Separate the loops on one end of the coil, forming two almost equal groups.

(c) Slide one group of loops over the casualty's left leg and the other group over the right leg. The wraps holding the coil should be in the casualty's crotch with the loops on the other end of the coil extending upward toward the armpits.

(d) The bearer lies on his back between the casualty's legs and slides his arms through the loops. He then moves forward until the coil is extended.

(e) Grasping the casualty's arm, the bearer rolls over (toward the casualty's uninjured side), pulling the casualty onto his back.

(f) Holding the casualty's wrists, the bearer carefully stands, using his legs to lift up and keeping his back as straight as possible.

(g) A sling rope around both the casualty and bearer, tied with a joining knot at chest level, aids in keeping an unconscious victim upright. This also prevents the coils from slipping off the carrier's chest.

Note: The length of the coils on the rope coil and the height of the bearer must be considered. If the coils are too long and the bearer is shorter, the rope must be uncoiled and recoiled with smaller coils. If this is not done, the casualty will hang too low on the bearer's back and make it a cumbersome evacuation. A sling-rope harness can be used around the victim's back and bearer's chest, which frees the bearer's hands.

b. **Buddy Rappel.** The carrier can also conduct a seat-hip rappel with a victim secured to his back. In this case, the rappeller faces the cliff and assumes a modified L-shape body position to compensate for the weight of the victim on his back. The victim is top-rope belayed from above, which provides the victim with a point of attachment to a secured rope. The methods for securing a victim to a rappeller's back are described below.

(1) To secure the victim to the carrier's back with a rope, the carrier ties a standard rappel seat (brake hand of choice, depending on the injury) and rests his hands on his knees while the victim straddles his back.

(2) A 4.2-meter sling rope is used. A 45-centimeter tail of the sling is placed on the victim's left hip. (This method describes the procedure for a seat-hip rappel with right-hand brake.)

(3) The remaining long end of the sling rope is routed under the victim's buttocks, and passed over the victim's and carrier's right hip. The rope is run diagonally, from right to left, across the carrier's chest, over his left shoulder, and back under the victim's left armpit.

(4) The rope is then run horizontally, from left to right, across the victim's back. The rope is passed under the victim's right armpit and over the carrier's right shoulder.

(5) The rope is run diagonally, from right to left, across the carrier's chest and back across the carrier's and victim's left hip.

(6) The two rope ends should now meet. The two ends are tied together with a square knot and overhand knots.

(7) The knot is positioned on the victim's left hip. The carrier's shoulders may need to be padded to prevent cutting by the rope.

(8) An alternate method is to use two pistol belts hooked together and draped over the carrier's shoulders. The victim straddles the carrier, and the belay man secures the loose ends of the pistol belts under the victim's buttocks. Slack in the pistol belt sling should be avoided, since the carrier is most comfortable when the victim rests high on his back (see FM 8-35).

(9) A large rucksack can be slit on the sides near the bottom so that the victim can step into it. The victim is belayed from the top with the carrier conducting a standard rappel. The carrier wears the rucksack with the victim inside.

(10) A casualty secured to a carrier, as described above, can be rappelled down a steep cliff using a seat-shoulder or seat-hip rappel. The casualty's and rappeller's shoulders should be padded where the sling rope and rappel lines cross if a seat-shoulder

rappel is used. The buddy team should be belayed from above with a bowline tied around the victim's chest under his armpits. The belay rope must run over the rappeller's guide hand shoulder.

11-7. LITTERS

Many types of litters are available for evacuating casualties in rough mountain terrain. Casualties may be secured to litters in many different ways, depending on the terrain, nature of injuries, and equipment available. **All casualties must be secured.** This should be done under medical supervision after stabilization. It is also important to render psychological support to any victim awaiting evacuation.

If the litter must be carried, belayed, and then carried again, a sling rope should be wound around the litter end and tied off in a 1-meter-long loop. This enables the carriers to hook and unhook the litter from the belay. Slings are available to aid the soldiers with litter carrying. Utility rope or webbing 6 meters long may be used. The rope is folded in half, and the loose ends are tied together with an overhand knot. These slings are attached to the litter rails (two or three to a side, depending on the number of litter bearers) by a girth hitch, and then routed up along the handling arm, over the shoulder, behind the neck, and then down along the other arm. The knot can be adjusted to help the outside arm grip the webbing. These slings help distribute the load more evenly, which is important if a great distance must be traveled.

a. **Manufactured Litters.** The following litters are readily available to mountaineering units.

(1) The poleless, nonrigid litter (NSN 6530-00-783-7510) is best issued for company medics since it is lightweight, easy to carry, and readily available. Casualties should be secured with the chest strap and pelvic straps, which are sewn on one side. This litter may be used when rappelling, on traverse lines, and on hauling lines in the vertical or horizontal position. It can be improvised with poles.

(2) The poleless semi-rigid litter (NSN 6530-00-783-7600) may be used the same as the nonrigid litter. It offers more victim protection and back support because of the wooden slats sewn into it.

(3) The mountain basket-type rigid litter (NSN 6530-00-181-7767) is best suited for areas where several casualties are to be transported. All other litters may be placed inside this litter basket and transported across traverse lines. This litter is rectangular and has no vertical leg divider so that it will accommodate other litters. It is also known as a modified Stokes litter.

(4) The Stokes metal litter (NSN 6530-00-042-8131) is suited for situations as above; however, the casualty must be moved in and out of the litter since no other litter will fit inside it. Some Stokes litter frames have a central weld on the frame end, which is a potential breaking point. Winding the rope around the frame end will distribute the force over a wider area and stabilize the system. (See FM 8-10-6 or USAF TO 00-75-5 for additional information on the Stokes litter.)

(5) The standard collapsible litter (NSN 6530-00-783-7905) (rigid pole folding litter) is most readily available in all units and, although heavy and unsuited to forward deployment, may be rigged for movement over rough or mountainous terrain. The folding aluminum litter (NSN 6530-00-783-7205) is a compact version of the pole litter and is better suited for forward deployment.

(6) The UT 2000 is manufactured in Austria and is specifically designed for mountaineering operations. The litter consists of two parts that join together to form a rigid litter. Each part has shoulder and waist straps that can be used to man-pack the litter making it extremely light and portable. When joined together the shoulder and waist straps are used to secure the casualty to the litter. Strapping is also provided to make a secure hoist point for aircraft extraction and high-angle rescues. Wheel sets are another accessory to the UT 2000 litter (either two wheels or one); they attach to the litter for use during a low-angle rescue.

(7) The patient rescue and recovery system (NSN 6530-01-260-1222) provides excellent patient support and protection (Figure 11-2). However, it is not a spinal immobilization device. A backboard must be used with this system for patients who have injuries to the shoulder area. This system will accommodate long and short backboards, scoop stretchers, and most other immobilization equipment.

Figure 11-2. Rescue and recovery system (NSN 6530-01-260-1222).

b. **Field-Expedient Litters.** A litter can be improvised from many different things. Most flat-surface objects of suitable size can be used as litters. Such objects include boards, doors, window shutters, benches, ladders, cots, and poles. Some may need to be tied together to obtain the required size. If possible, these objects should be padded.

(1) Litters can also be made by securing poles inside blankets, ponchos, shelter halves, tarpaulin, jackets, shirts, sacks, bags, or mattress covers. Poles can be improvised from strong branches, tent supports, skis, and other similar items.

(2) If poles cannot be found, a large item, such as a blanket, can be rolled from both sides toward the center. Then the rolls can be used to obtain a firm grip to carry the victim. If a poncho is used, the hood must be up and under the victim, not dragging on the ground.

(3) A rope litter is prepared using one rope (Figure 11-3). It requires 20 to 30 minutes to prepare and should be used only when other materials are not available. Four to six bearers are required to carry the litter. The rope litter is the most commonly used field-expedient litter.

Note: Above the tree line, little material exists to construct litters.

Figure 11-3. Rope litter.

(a) Make 24 bights about 45 to 61 centimeters long, starting in the middle of the rope so that two people can work on the litter at one time.

(b) With the remainder of the rope, make a clove hitch over each bight. Each clove hitch should be about 15 centimeters from the closed end of the bight when the litter is complete.

(c) Pass the remainder of the rope through the bights outside of the clove hitches. Dress the clove hitches down toward the closed end of the bight to secure the litter and tie off the ends of the rope with clove hitches.

(d) Line the litter with padding such as clothing, sleeping bags, empty boxes.

(e) Make the rope litter more stable by making it about 6 inches wider. After placing the clove hitches over the bights, slide them in (away from the closed end) about 15 centimeters. Take two 3- to 4-meter poles, 8 centimeters in diameter at the butt ends, and slide each pole down through the bights on each side. Dress down the clove hitches against the poles. Take two 1-meter poles, and tie them off across the head and foot of the litter with the remaining tails of the climbing rope.

Note: The above measurements may have to be altered to suit the overall length of rope available.

11-8. RESCUE SYSTEMS

Rescue systems are indispensable when conducting rescue operations. A large number of soldiers will not always be available to help with a rescue. Using a mechanical advantage rescue system allows a minimal amount of rescuers to perform tasks that would take a larger number of people without it.

a. **Belay Assist.** This system is used to bring a climber over a section that he is unable to climb, but will continue climbing once he is past the difficult section.

(1) First, tie off the following climber at the belay with a mule knot.

(2) Tie a Prusik knot with short Prusik cord about 12 inches below the mule knot, and place a carabiner into the loop. Place the tail from the mule knot into the carabiner in the Prusik cord.

(3) Untie the mule knot without letting the following climber descend any more than necessary. Do not remove the belay.

(4) Maintain control of the brake side of the rope and pull all of the slack through the carabiner in the Prusik cord.

(5) Pull up on the rope. The rope will automatically feed through the belay.

(6) If the leader has to pull for a distance he can tie another mule knot at the belay to secure the second climber before sliding the Prusik down to get more pulling distance.

(7) After the climber can continue climbing, the leader secures the belay with a mule knot.

(8) Remove the Prusik cord and carabiner, then untie the mule knot and continue belaying normally.

Notes: With all rescue techniques make sure that you always think everything through, and double check all systems to ensure that you don't accidentally untie the fallen climber or find yourself without back-up safety. Do not compound the problem! When doing any rescue work the rescuers will always be tied in for safety.

b. **Belay Escape.** The belay escape is used when a climber has taken a serious fall and cannot continue. The belayer is anchored and is performing an indirect belay, and must assist the injured climber or go for assistance. To accomplish this he must escape the belay system. The belayer will remain secured to an anchor at all times.

(1) After a climber has been injured, tie off the belay device on your body using a mule knot. To improve this system, clip a nonlocking carabiner through the loop in the overhand knot and clip it over the rope.

(2) Attach a short Prusik cord to the load rope and secure it to the anchor with a releasable knot.

(3) Using a guard knot or Munter mule, attach the climbing rope from the belay device.

(4) Untie the mule knot in the belay device attached to the harness and slowly lower the climber, transferring the weight of the climber onto the Prusik.

(5) Remove the climbing rope from the belay device attached to the harness.

(6) Release the mule knot securing the Prusik, transferring the weight to the anchor.

(7) At this point the climber is secured by the rope to the anchor system and the belayer can now assist the injured climber.

11-9. LOW-ANGLE EVACUATION

Cliffs and ridges, which must be surmounted, are often encountered along the evacuation path. Raising operations place a greater load on all elements of the system than do lowering operations. Since all means of raising a victim (pulley systems, hand winches, and power winches) depend on mechanical advantage, it becomes easy to overstress and break anchors and hand ropes. Using mechanical raising systems tends to reduce the soldier's sensitivity to the size of the load. It becomes important to monitor the system and to understand the forces involved.

a. **Raising Systems, Belays, and Backup Safeties.** Raising systems, belays, and backup safeties are of special importance in any raising operation. The primary raising system used is the Z-pulley system, which theoretically gives three pounds of lift for each pound of force expended. In practice, these numbers decrease due to rope-pulley friction, rope-edge friction, and other variables. A separate belay rope is attached to the litter and belayed from a separate anchor. Backup Prusik safeties should be installed in case any part of the pulley system fails.

(1) ***Raising System.*** When considering a raising system for evacuations, the Z-pulley system is the most adaptable. It can be rigged with the equipment on hand, and can be modified and augmented to handle heavier loads. Although the vertical or horizontal hauling lines can also be used, the Z-pulley system offers a mechanical advantage that requires less exertion by the transport team.

(2) ***Belays.*** Whenever ropes are used for an evacuation operation, the overriding safety concern is damage to the ropes. This is the main reason for two-rope raising systems (raising rope and belay rope).

(3) ***Backup Safeties.*** Because the stresses generated by the Z-pulley system can cause anchors to fail, backup safety Prusiks are used to safeguard the system. These should be attached to alternate anchor points, if possible.

b. **Raising the Litter.** The litter is prepared as already described.

(1) The raising ropes and belay ropes are secured to top anchors and are thrown down to the litter crew.

(2) Padding is placed at the cliff edge and over any protrusions to protect the ropes from abrasion.

(3) The litter attendants secure the ropes to the litter.

(4) The raising crew sets up the Z-pulley system.

(5) One member of the crew secures himself to an anchor and moves to the edge of the cliff to transmit signals and directions. (This is the signalman or relay man.)

Note: If the load is too heavy at this time, another pulley is added to the system to increase the mechanical advantage.

(6) Attendants guide the litter around obstacles while the crew continues to raise the system.

(7) As the litter nears the cliff edge, the signalman assists the attendant in moving the litter over the edge and onto the loading platform, taking care not to jar the casualty.

c. **Descending Slopes.** When descending a moderately steep slope that can be down-climbed, the litter and victim are prepared as described earlier (Figure 11-4, page 11-12).

(1) One man serves as the belay man and another takes his position on the rope in front of the belay man, assisting him in lowering the litter. The litter bearers take their positions and move the litter down with the speed of descent controlled by the belay man.

Figure 11-4. Low-angle evacuation—descending.

(2) The extra man may assist with the litter or precede the team to select a trail, clearing away shrubs and vines. He reconnoiters so that the team need not retrace its steps if a cliff is encountered.

(3) The most direct, practical passage should be taken utilizing available natural anchors as belay positions.

11-10. HIGH-ANGLE EVACUATION

Evacuation down cliffs should be used only when absolutely necessary and only by experienced personnel. The cliffs with the smoothest faces are chosen for the route. Site selection should have the following features: suitable anchor points, good loading and unloading platforms, clearance for the casualty along the route, and anchor points for the A-frame, if used. There are many ways to lower a casualty down a steep slope. As long as safety principals are followed, many different techniques can be used. One of the easiest and safest techniques is as follows (Figure 11-5):

a. Use multiple anchors for the litter and litter tenders.

b. Secure the litter to the lowering rope with a minimum of four tie-in points (one at each corner of the litter). Lengths of sling rope or 7-millimeter cordage work best. Make the attached ropes adjustable with Prussik knots so that each corner of the litter can be raised or lowered to keep the litter stable during descent. Tie the top of the ropes with loops and attach to the lowering rope with a pear shaped locking carabiner.

c. Two litter tenders will descend with the litter to control the descent and to monitor the casualty. They can be attached to separate anchors and either self-belay themselves or be lowered by belayers.

d. Once the steep slope has been negotiated, continue the rescue with a low-angle evacuation.

Figure 11-5. Cliff evacuation descent.

APPENDIX A

LEVELS OF MILITARY MOUNTAINEERING

Military mountaineering training provides tactical mobility in mountainous terrain that would otherwise be inaccessible. Soldiers with specialized training who are skilled in using mountain climbing equipment and techniques can overcome the difficulties of obstructing terrain. Highly motivated soldiers who are in superior physical condition should be selected for move advanced military mountaineering training (Levels 2 and 3) conducted at appropriate facilities. Soldiers who have completed advanced mountaineering training should be used as trainers, guides, and lead climbers during collective training. They may also serve as supervisors of installation teams and evacuation teams. Properly used these soldiers can drastically improve mobility and have a positive impact disproportionate to their numbers. Units anticipating mountain operations should strive to have approximately ten percent of their force achieve advanced mountaineering skills.

A-1. LEVEL 1: BASIC MOUNTAINEER

The basic mountaineer should be a graduate of a basic mountaineering course and have the fundamental travel and climbing skills necessary to move safely and efficiently in mountainous terrain. These soldiers should be comfortable functioning in this environment and, under the supervision of qualified mountain leaders or assault climbers, can assist in the rigging and use of all basic rope installations.

a. On technically difficult terrain, the basic mountaineer should be capable of performing duties as the "follower" or "second" on a roped climbing team, and should be well trained in using all basic rope systems. These soldiers may provide limited assistance to soldiers unskilled in mountaineering techniques.

b. Particularly adept soldiers may be selected as members of special purpose teams led and supervised by advanced mountaineers. Figure A-1 lists the minimum knowledge and skills required of basic mountaineers.

c. In a unit training program, Level 1 qualified soldiers should be identified and prepared to serve as assistant instructors to train unqualified soldiers in basic mountaineering skills. All high-risk training, however, must be conducted under the supervision of qualified Level 2 or 3 personnel.

• Characteristics of the mountain environment (summer and winter). • Mountaineering safety. • Use, care, and packing of individual cold weather clothing and equipment. • Care and use of basic mountaineering equipment. • Mountain bivouac techniques. • Mountain communications. • Mountain travel and walking techniques. • Hazard recognition and route selection. • Mountain navigation. • Basic medical evacuation.	• Rope management and knots. • Natural anchors. • Familiarization with artificial anchors. • Belay and rappel techniques. • Use of fixed ropes (lines). • Rock climbing fundamentals. • Rope bridges and lowering systems. • Individual movement on snow and ice. • Mountain stream crossings (to include water survival techniques). • First aid for mountain illnesses and injuries.

Figure A-1. Level 1, basic mountaineering tasks.

A-2. LEVEL 2: ASSAULT CLIMBER

Assault climbers are responsible for the rigging, inspection, use, and operation of all basic rope systems. They are trained in additional rope management skills, knot tying, and belay and rappel techniques, as well as using specialized mountaineering equipment. Assault climbers are capable of rigging complex, multipoint anchors, and high-angle raising/lowering systems. Level 2 qualification is required to supervise all high-risk training associated with Level 1. At a minimum, assault climbers should possess the additional knowledge and skills shown in Figure A-2.

• Use specialized mountaineering equipment.
• Perform multipitch climbing:
  – Free climbing and aid climbing.
  – Leading on class 4 and 5 terrain.
• Conduct multipitch rappelling.
• Establish and operate hauling systems.
• Establish fixed ropes with intermediate anchors.
	• Move on moderate angle snow and ice.
	• Establish evacuation systems and perform high-angle rescue.
	• Perform avalanche hazard evaluation and rescue techniques.
	• Be familiar with movement on glaciers.

Figure A-2. Level 2, assault climber tasks.

A-3. LEVEL 3: MOUNTAIN LEADER

Mountain leaders possess all the skills of the assault climber and have extensive practical experience in a variety of mountain environments in both winter and summer conditions. Level 3 mountaineers should have well-developed hazard evaluation and safe route finding skills over all types of mountainous terrain. Mountain leaders are best qualified to advise commanders on all aspects of mountain operations, particularly the preparation and leadership required to move units over technically difficult, hazardous, or exposed terrain. The mountain leader is the highest level of qualification and is the principle trainer for conducting mountain operations. Instructor experience at a military mountaineering center or as a member of a special operations forces (SOF) mountain team is critical to acquiring Level 3 qualification. Figure A-3 outlines the additional knowledge and skills expected of mountain leaders. Depending on the specific AO, mountain leaders may need additional skills such as snowshoeing and all-terrain skiing.

• Recognizing and evaluating peculiar terrain, weather, and hazards. • Preparing route, movement, bivouac, and risk management plans for all conditions and elevation levels. • Using roped movement techniques on steep snow and ice. • Performing multipitch climbing on mixed terrain (rock, snow, and ice).	• Performing glacier travel and crevice rescue. • Establishing and operating technical high-angle, multipitch rescue and evacuation systems. • Usiing winter shelters and survival techniques. • Leading units over technically difficult, hazardous, or exposed terrain in both winter and summer conditions.

Figure A-3. Level 3, mountaineer leader tasks.

APPENDIX B

MEASUREMENT CONVERSION FACTORS

MULTIPLY	BY	TO OBTAIN
Millimeters	.03937	Inches
Centimeters	.3937	Inches
Centimeters	.03281	Feet
Meters	39.37	Inches
Meters	3.281	Feet
Meters	1.0936	Yards
Kilometers	.62137	Miles
Knots	1.1516	MPH

APPENDIX C

AVALANCHE SEARCH AND RESCUE TECHNIQUES

The effect of an avalanche can be disastrous. Chances of survival after burial by an avalanche are approximately 90 percent if the victim is located within the first 15 minutes. Probability of survival drops rapidly and, after two hours, chances of survival are remote. Suffocation accounts for 65 percent of avalanche fatalities, collision with obstacles such as rocks and trees accounts for 25 percent, and hypothermia and shock accounts for 10 percent.

The best chance of survival in snow country is to avoid an avalanche; but, if a member of your group **is** *in an avalanche, they are depending on you for rescue!*

C-1. IMMEDIATE ACTION

Survivors at the avalanche site organize into the first rescue team and immediately start rescue operations. If any indication of the location of the victim is found, random probing starts in that vicinity. The tip and edges of the slide are also likely areas to search. A human body is bulky and is apt to be thrown toward the surface or the sides.

C-2. GENERAL PROCEDURES

Establish from witnesses where the victim was located just before the avalanche to determine the point where the victim disappeared—the "last seen" point. Using this and any other information, establish a probable victim trajectory line leading to high priority search areas. Make a quick but systematic check of the slide area and the deposition area, and mark all clues. Look for skis, poles, ice axes, packs, gloves, hats, goggles, boots, or any other article the person may have been carrying—it might still be attached to the victim.

a. Organize initial searchers and probers. If using avalanche beacons, immediately select personnel to begin a beacon search. Ensure all other beacons are turned off or to receive to eliminate erroneous signals. All personnel should have a shovel or other tool for digging or, if enough personnel are available, a digger can be standing by to assist when needed. If the initial search reveals items from the victim, make an initial probe search in that area. This probing should take only a few seconds.

b. Make a coarse probe of all likely areas of burial, and repeat it as long as a live rescue remains possible. Resort to the fine probe only when the possibility of a live rescue is highly improbable. Unless otherwise indicated, start the coarse probe at the deposition area.

C-3. ESTABLISHING THE VICTIM'S MOST PROBABLE LOCATION

In many respects, a moving avalanche resembles a liquid. A human body, with a higher density than the flowing snow, would be expected to sink deeper and deeper into the avalanche; however, several factors influence the body's location. Turbulence, terrain, and the victim's own efforts to extricate himself all interact to determine the final burial position. Study of a large number of case histories leads to the following conclusions.

- The majority of buried victims are carried to the place of greatest deposition, usually the toe of the slide.
- If two points of the victim's trajectory can be established, a high probability exists that the victim will be near the downhill flow line passing through these two points.
- Any terrain features that catch and hold avalanche debris are also apt to catch a victim.
- If an avalanche follows a wandering gully, all debris deposit areas are likely burial spots. The likelihood of a victim being buried in a particular bend is proportional to the amount of debris deposited there.
- Vegetation, rocks, and other obstacles act as snares. The victim tends to be retained above the obstacle. An obstacle may simply delay the victim's motion, leading to final burial down flow from the obstacle.
- Maximum speed of the flowing snow occurs at the avalanche center. Friction reduces flow velocity along the edges. The closer the victim's trajectory is to the center of the slide, the greater will be his burial depth.
- Efforts of the victim to extricate himself by vigorous motion and "swimming" definitely minimize burial depth. Conversely, the limp body of an unconscious victim is likely to be buried deeply.
- An occasional exception to the above is emphasized. The victim may not be buried but may have been hurled away from the avalanche by wind blast. In the case of large and violent avalanches, a search of the surrounding terrain is advisable. Victims have been located in tree tops outside the slide area.

Use of avalanche transceivers is the most efficient method of searching for an avalanche victim, but only if the victim is wearing an active transceiver. Many models of transceivers are available, each with its own manufacturer's instructions for proper use and care. All currently available transceivers are compatible, although they may operate differently in the search mode.

C-4. PROBING FOR AVALANCHE VICTIMS

Probing offers the advantage of requiring simple equipment that can be operated by personnel with no previous training. Although the probers do not need previous training the search leader must be familiar with the technique to ensure proper execution of the probe line.

a. **Probe Poles.** Rigid steel tubing approximately 3/4-inch in diameter and approximately 10 feet long is recommended for the primary probe pole. Longer poles are difficult to manage, especially in a high wind. Although this type of pole performs best, it is difficult to transport to the avalanche site because of its length and weight.

(1) Each person operating in avalanche areas should carry folding sectional poles. These poles are similar to folding tent poles, but are stronger and are connected with cable instead of bungee cord. These poles should be carried on the outside of the pack for immediate access.

(2) If no probing poles are available, initial probing attempts can be started using ski poles in one of two ways: the ski pole can be reversed, probing with the wrist strap down; or the basket can be removed so that the point is down (the preferred method), which allows the ski pole to penetrate the snow more easily.

b. **Probing Lines.** For the probing operation to be effective, probing lines must be orderly and properly spaced. To ensure systematic and orderly probing, the number of personnel per line should be limited. Twenty per line is satisfactory, while thirty is normally the upper limit. The number of probers in the line will be dictated by not only the width of the area to be probed but the number of personnel available. A string may be used to keep the probe lines aligned, but will require added time to maintain.

(1) The probe line maintains a steady advance upslope. Advancing uphill automatically helps set the pace and permits easy probing to the full length of the probe. Probing does not come to a halt when a possible contact is made. The probe is left in contact and the line continues. A shovel crew follows up on the strike by digging down along the pole. Extra probes are carried by the shovel crew to replace those left in contact. Such a plan of operation is especially important when more than one victim is buried.

(2) Striking a body gives a distinct feel to the probe, which is easily recognizable in soft snow but less recognizable in hard compacted snow. A common problem is encountering debris within the snow that can be mistaken for the victim. The only sure check is by digging.

c. **Probing Techniques.** Two distinct probing methods are recognized: coarse probe and fine probe. As evidenced by their names, coarse probing implies a wider spacing of probe pole insertions with emphasis on speed. Fine probing involves close-spaced probing with emphasis on thoroughness. Coarse probing is used during initial phases of the search when live recovery is anticipated. Fine probing is the concluding measure, which almost guarantees finding the body. The coarse probe technique has a 70-percent chance of locating the victim on a given pass, while the fine probe has, essentially, a 100-percent chance of locating the body.

(1) The coarse probe functions as follows:

(a) Probers are spaced along a line 30 inches center to center, with feet about 15 inches apart.

(b) A single probe pole insertion is made at the center of the straddle span.

(c) On command of the probe line commander, the group advances 20 inches and repeats the single probe.

(d) Three commands are used for the complete sequence:

- "DOWN PROBE."
- "UP PROBE."
- "STEP FORWARD."

By using these commands, the leader can maintain closer control of the advancing probe line. It is important that the commands be adjusted to a rhythm that enforces the maximum reasonable pace. A string may also be used along the probe line to keep the probers dressed, although this requires the use of two soldiers to control the string. Strict discipline and firm, clear commands are essential for efficient probing. The probers themselves work silently.

(2) The fine probe functions as follows:

(a) Probers are spaced the same as for the coarse probe. Each man probes in front of his left foot, then in the center of his straddled position, and finally in front of his right foot.

(b) On command, the line advances 1 foot and repeats the probing sequence. Each probe is made 10 inches from the adjacent one.

(c) The commands for the fine probe are:

- "LEFT PROBE."
- "UP PROBE."
- "CENTER PROBE."
- "UP PROBE."
- "RIGHT PROBE."
- "UP PROBE."
- "STEP FORWARD."

(d) Good discipline and coordinated probing is even more important in fine probing than with the coarse probe. Careless or irregular probing can negate the advantages of fine probing. Use of a string to align the probers is especially important with the fine probe. The three insertions are made along the line established by the string, which is then moved ahead 1 foot.

GLOSSARY

AI	alpine ice
ALICE	all-purpose, lightweight, individual carrying equipment
AMS	acute mountain sickness
APS	anchor pulley system
ATC	air traffic controller
BDU	battle dress uniform
BTC	bridge team commander
C	Celsius
CE	community Europe
CEN	combined European norm
cm	centimeter
CMS	chronic mountain sickness
CPR	cardio-pulmonary resuscitation
CTA	common table of allowance
DA	Department of the Army
ECWCS	extreme cold weather clothing system
EN	European norm
F	Fahrenheit
FM	field manual
ft	foot; feet
GPS	Global Positioning System
GTA	graphic training aid
HACE	high-altitude cerebral edema
HAPE	high-altitude pulmonary edema
IAW	in accordance with
LBV	load-bearing vest
LCD	liquid crystal diode
LCE	load-carrying equipment
MEDEVAC	medical evacuation
MPH	miles per hour
MPS	moveable pulley system
MRE	meal ready-to-eat

NSN	national stock number
OP	observation post
PABA	para-amino benzoic acid
PCD	progress capture device
PFD	personal flotation device
RURP	realized ultimate reality piton
SE	southeast
SLCD	spring-loaded camming device
SOF	special operations forces
SOP	standard operating procedure
SOSES	shape, orientation, size, elevation, slope
SPF	sun protection factor
TOE	table of organization and equipment
UIAA	Union des International Alpine Association
U.S.	United States
USAF	United States Air Force
USN	United States Navy
UV	ultraviolet
UVA	ultraviolet A (radiation wavelengths between 320 and 400 nanometers)
UVB	ultraviolet B (radiation wavelengths between 295 and 320 nanometers)
WI	waterfall ice
YDS	Yosemite Decimal System

INDEX

ROPE LOG *(USAGE AND HISTORY)*

For use of this form, seeFM 3-97.61; the proponent agency is TRADOC

ROPE LOG (USAGE AND HISTORY)			UNIT ID MARKING
NSN	DOCUMENT NUMBER	SERIAL NUMBER	MFR LOT NUMBER
DATE OF MFR	ISSUE DATE	DATE IN SERVICE	LENGTH
DIAMETER	FIBER	COLOR	CONSTRUCTION

INSPECT ROPE FOR DAMAGE OR EXCESSIVE WEAR EACH TIME IT IS DEPLOYED AND AGAIN AFTER EACH USE. IMMEDIATELY RETIRE ALL SUSPECT ROPES.

DATE USED	LOCATION	TYPE OF USE	ROPE EXPOSURE	INSPECTOR'S INITIAL/DATE	ROPE CONDITION AND COMMENTS

DA FORM 5752-R, MAY 89

DATE USED	LOCATION	TYPE OF USE	ROPE EXPOSURE	INSPECTOR'S INITIAL/DATE	ROPE CONDITION AND COMMENTS

MANUFACTURER ____________________

____

REVERSE OF DA FORM 5752-R, MAY 89

*TC 21-24

TRAINING CIRCULAR
No. 21-24

HEADQUARTERS
DEPARTMENT OF THE ARMY
WASHINGTON, DC, 10 September 1997

RAPPELLING

CONTENTS

DISTRIBUTION RESTRICTION: Approved for public release; distribution is unlimited.

*This publication supersedes TC 21-24, 24 September 1991.

PREFACE

This circular provides basic rappelling techniques to soldiers and leaders for the conduct of rappelling operations. It serves as the primary reference for both resident and nonresident instruction presented to cadets, officer candidates, and both commissioned and noncommissioned officers. This circular also discusses several advanced techniques dealing with infiltration and exfiltration.

Safety is always the most important consideration when conducting training. This circular provides guidelines for commanders to conduct operations safely. Commanders at all levels must analyze the complete training event to determine the degree of risk involved to men and or equipment (Appendix A). After determining the risks, risk reduction options or controls will be integrated into the training activity. These options or controls may range from a safety briefing or providing additional safety resources, to selecting other means of accomplishing the mission.

Rappelling, FRIES, and SPIES operations are inherently dangerous; therefore, the safety notes and considerations presented in this circular are the minimum acceptable standards.

The proponent of this publication (except for Chapter 5) is the US Army Infantry School. Submit changes for improving this publication on DA Form 2028 and forward it to the Commandant, US Army Infantry School, ATTN: ATSH-RB, Fort Benning, Georgia 31905-5593 (e-mail address: IKNERH@BENNING-EMH2.ARMY.MIL).

The proponent for Chapter 5, Fast-Rope Insertion/Extraction System, is the US Army Special Operations Command. Submit changes or comments for Chapter 5 on DA Form 2028 and forward it to the Commander, US Army Special Operations Command, ATTN: AOOP-TRS, Fort Bragg, North Carolina 28307-5000.

Unless this publication states otherwise, masculine nouns and pronouns do not refer exclusively to men.

CHAPTER 1

TOWER RAPPELLING

The introduction to rappelling is taught on a static tower (Figure 1-1). Using the building block approach to training, soldiers systematically progress to more demanding platforms, including taller static towers, and finally to helicopters. The static tower used may vary in size and height from 34 to 90 feet. The concept of learning the basic rappelling techniques before helicopter operations does not vary. (Units may conduct ground training and wall-side tower rappelling without a rappel master. In this case, the unit commander would be responsible for the safety of the rappelling training.)

Figure 1-1. Rappel tower.

Note: Each static rappel tower should resemble the structure or aircraft being trained to. The Army's standard design for building static rappel towers can be obtained through the Plans and Support Branch of the post Engineers, Building 75, Fort Benning, Georgia 31905 (commercial 706 545-3307 or DSN 835-3307). Strict adherence to the design specifications is required for safety considerations. Rappel towers should be inspected annually by post safety or post engineers.

Section I. PERSONNEL

This section discusses the personnel involved in training rappelling and their duties and responsibilities.

1-1. RAPPEL MASTER

The proponent for accreditation, evaluation, and information for Army rappel master courses is Directorate of Operations and Training, US Army Infantry School, ATTN: ATSH-OTT-F, Fort Benning, Georgia 31905-5593.

a. **Duties and Responsibilities.** The rappel master is responsible for the safety of rappellers. He ensures that all equipment (installation, unit, and personal property) is serviceable. He personally supervises the rappelling operation.

b. **Qualification.** Rappel master qualification is awarded only after the successful completion of a TRADOC accredited rappel master course, which includes the following subjects:

- Duties and responsibilities of a rappel master.
- Safety SOP, regulations, and references.
- Construction of a deployment bag.
- Conduct of an equipment rappel off the rappel tower.
- Conduct of a lock-in.
- Talking a rappeller through completion of a rappel.
- Conduct of ground training.
- Inspecting for proper hook-ups.
- Inspecting and maintaining equipment.
- Inspecting and maintaining snaplinks.
- Inspecting and maintaining rappelling gloves.
- Inspecting and maintaining rappel ropes.
- Identifying the rappel capabilities of aircraft used.
- Controlling rappels from UH-1H or UH-60 aircraft.
- Tying knots (square, bowline, half-hitch, Prusik), safety lines, and rappel seats (Swiss seat, Australian seat).
- Inspecting a rappel seat.
- Aircraft rigging for rappelling operations.
- Aircraft command and control.

c. **Proficiency Maintenance.** To remain current, rappel masters must execute their duties in a tactical or training exercise once every six months. If rappel masters do not execute their duties once every six months, they must take a refresher class taught by a current rappel master. The refresher class includes the subjects listed in the rappel master qualifications paragraph (1-1b).

1-2. RAPPEL SAFETY OFFICER

The RSO is an air assault-qualified or ranger-qualified SFC or above who serves as the OIC for the conduct of rappel operations.

a. The RSO is responsible overall for the safety of all rappellers and ensures that all safety precautions are followed.

b. The RSO briefs VIPs, visitors, and inspecting authorities on training, safety requirements, and layout of training areas.

1-3. RAPPEL LANE NCO

a. **Duties and Responsibilities.** Safety is the rappel lane NCO's number one priority. The rappel lane NCO—

- Ensures proper safety procedures are followed.
- Ensures proper hookup once directed to a rope station.
- Issues commands and maintains eye contact with the rappeller at all times.

b. **Qualifications.** The rappel lane NCO holds the rank of corporal or above, is ranger- or air assault-qualified, and is selected by the commander. Each tower rappel lane must have a qualified rappel lane NCO.

c. **Training.** The rappel lane NCO must also train on the following subjects:

- Responsibilities and safety requirements.
- Inspection and maintenance of equipment.
- Identification of satisfactory anchor points.
- Identification of safe and unsafe hookups.
- Establishment of a rappel point.
- Inspection of a rappel seat.
- Coaching techniques.
- Rappelling procedures.
- Emergency procedures.
- Belay control procedures.

d. **Participation.** The rappel lane NCO must participate in at least seven rappel operations: three as a rappeller; two as an assistant rappel lane NCO; and two performing the duties of a rappel lane NCO under the supervision of a qualified rappel master.

e. **Proficiency Maintenance.** If a rappel lane NCO has not conducted his duties within the last six months, he must complete the training listed in paragraph 1-3c under the supervision of a current rappel master.

1-4. RAPPELLER

Rappel qualification requirements apply to the individual rappeller. Participants in tower rappel training must complete the following requirements under the supervision of a rappel master. The unit commander ensures that personnel successfully complete these requirements before beginning aircraft rappel training.

- Identify all rappelling equipment.
- Demonstrate the construction and attachment of the rappel seat and the rappel rope to the seat.
- Identify unsafe attachments, equipment, rope connections, and seat construction.
- Define terms used in rappelling operations.
- Identify knots used in rappel operations.
- Understand and demonstrate rappel commands.
- Demonstrate rappelling positions.
- Demonstrate belaying procedures.
- Exhibit satisfactory performance from a rappel tower of at least 34 feet in height (two rappels with equipment and weapon, two without equipment and weapon). Two rappels are conducted from the free side of the tower (no wall).
- Demonstrate the ability to lock in.

1-5. BELAYER

Belay requirements are a subtask of basic rappel requirements. Soldiers must know how to belay before conducting rappelling training. The belayer—

- Assumes a position at the base of the lane about one pace away from the tower area.
- Ensures that the rappel ropes are even with the ground during tower rappels.
- Loosely holds the rappel rope with both hands so as not to interfere with the rappeller but still be able to stop the rappeller should he fall.
- Immediately stops the rappeller by pulling downward on the rappel ropes if the rappeller shouts "falling" or loses control of his brake hand during descent.
- Does not wear gloves to ensure a firm grip on the rappelling rope.
- Watches the rappeller at all times, and maintains constant voice or visual contact.
- Wears a helmet to prevent injuries from falling debris.

1-6. BELAY SAFETY

The belay safety must be ranger- or air assault-qualified. He ensures belay personnel are performing their duties properly. Rappel training requires one belay safety for each two rappel stations.

Section II. PREOPERATIONS BRIEFINGS AND SAFETY PROCEDURES

The rappel master ensures participants have a basic understanding of requirements and safety procedures before conducting training.

1-7. SAFETY

The following personnel and equipment must be present during static tower training.

- Two military rappel ropes for each rappel station.
- One safety officer.
- One rappel master for each rappel site.
- One rappel lane NCO per rappel station.
- One medic with medical kit and backboard.
- One safety or medical evacuation vehicle with driver.
- One belayer for each rope station. Rappellers alternate stations.
- One belay safety for each two rappel stations (four ropes).

1-8. SAFETY BRIEFING

As in all training, a safety briefing precedes rappel operations. The rappel master briefs all personnel on safety to include the following instructions.

a. Each rappeller ensures loose clothing and equipment are secured.

b. Rappel seats are tied by the soldier and inspected by the rappel master before climbing the tower. Rappel seats are removed upon completion of every rappel, retied, and reinspected by a qualified rappel master or rappel lane NCO before subsequent rappels.

c. Rappellers climb the tower only when directed by the rappel master or rappel lane NCO.

d. Rappellers stay in the center of the tower until instructed to move to a rappel point.

e. No more than three personnel are behind each rappelling point.

f. If using a troop ladder, only three soldiers are on the ladder at one time. Soldiers do not climb the ladder until told to do so by a rappel master.

g. All rappel masters, rappel lane NCOs, instructors, and anyone else standing near the edge of the top of the tower must wear a restraining strap or safety rope. The strap or rope must be attached to an anchor point.

h. No one should lean or sit on the railing or banisters of the tower.

i. No one is allowed within 3 feet of the edge of the tower without being secured.

j. When attaching the rappel rope to the snaplink, rappellers pull the slack towards the anchor point. The rappel master or rappel lane NCO physically check each hookup.

k. All personnel weighing more than 200 pounds will conduct a standard hookup rappel to determine if they require a friction hookup. A friction hookup is created by placing an additional two ropes in the gate of the snaplink (for a total of six ropes in the snaplink).

l. Combat equipment is positioned on the rappeller so that it does not interfere with the brake hand. The weapon must be slung diagonally across the back with the muzzle pointing down, and on the opposite side of the brake hand.

m. Heavy duty gloves are required for all rappel training.

n. Kevlar helmets with chin straps fastened are worn during tower rappel training.

o. While on the tower, the rappeller maintains eye contact with the rappel master or rappel lane NCO and receives all commands from them.

p. The rappeller ensures that he has a belayer on his rope.

q. The belayer does not wear gloves and keeps both hands on the rope at all times. He also faces the rappeller at all times.

r. All tower rappelling is performed with a double strand of rope.

s. No running is allowed on the tower.

t. No smoking or eating is allowed near the tower.

u. All participants who are unable to rappel, lack confidence, or refuse to rappel are reported to the rappel master or OIC. These participants are immediately removed from the training area.

v. The RSO and rappel master must be aware of overconfidence and carelessness of some rappellers. The rappel master ensures all personnel are tower-qualified before beginning aircraft rappel training.

1-9. TOWER SAFETY AND PREPARATION

The rappel master is in charge of the tower. He conducts a visual and physical inspection of every item of equipment to include the structural lumber and timber, the ladder, the platform floor, and all anchor points.

a. The static tower will not be used during thunderstorms or excessively high winds. If ice is present, or if the platform is slick from rain, rappelling will be delayed until conditions are safe.

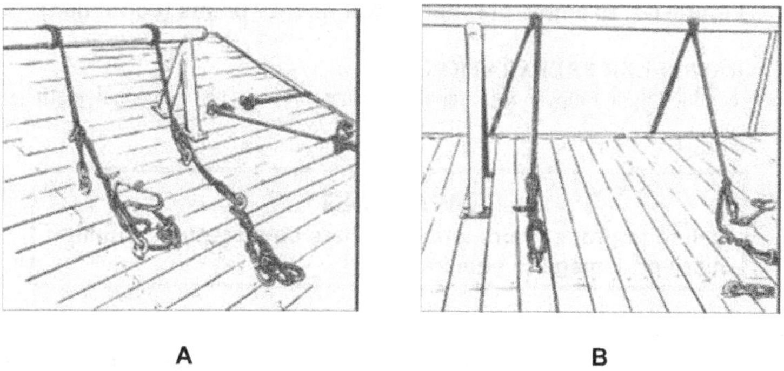

Figure 1-2. Example of rappel tower anchor points.

b. All rope stations are rigged with two anchor points (Figure 1-2A, B, C). The first anchor point is a middle-of-the-rope knot, and the second is an end-of-the-rope anchor knot. The rappel master removes all the slack between the knots to create equal tension on the anchor points. He ensures that no less than 10 feet of rope is on the ground during static rappelling.

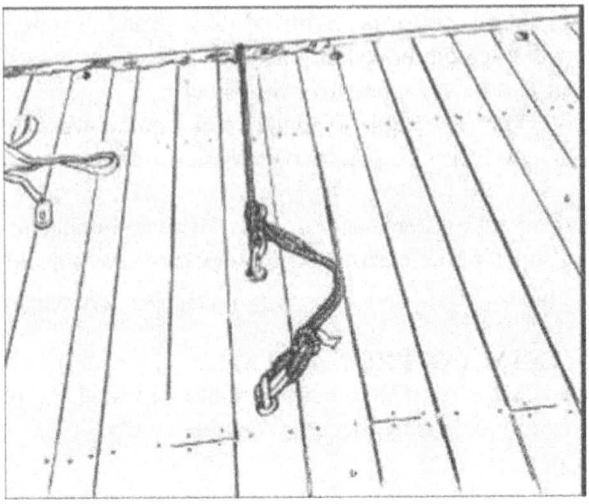

C

Figure 1-2. Example of rappel tower anchor points (continued).

1-10. RAPPELLER PREPARATION

Before conducting a rappel, each rappeller must prepare his individual clothing and equipment.

> **WARNING**
>
> **Failure to properly prepare rappellers could result in bodily injury or damage to equipment.**

a. Secure shirt tails, loose clothing, equipment, straps, and long hair.

b. Wear a helmet during rappelling. Properly fasten all straps and ensure the helmet is in serviceable condition.

c. Wear heavy leather workman's gloves.

d. Wear identification tags.

e. When rappelling with equipment, LBE or LBV should be unfastened in the front or fastened loosely behind the back of the rappeller. The rucksack should be worn high and tight on the back of the rappeller to allow the brake hand to reach the small of the back. Rucksack adjustment straps will be tied across the chest or tucked away.

f. Sling the weapon diagonally across the back with the muzzle down. Ensure the muzzle is on the guide hand side and the stock is towards the brake hand.

Note: Soldiers rappelling with equipment in excess of 50 pounds may want to consider using a friction hookup.

Section III. RAPPELLING PROCEDURES

This section discusses procedures used in tower rappelling.

1-11. SEAT-HIP RAPPEL

When using the seat-hip rappel, friction is created by a snaplink that is inserted in a sling rope seat and fastened to the rappeller. This method provides a faster and more controlled descent than other methods. Wear gloves to prevent rope burns. An alternate technique is to insert a second snaplink into the first snaplink (attached to rappel seat) and run the rope through the second snaplink. This allows easier disengagement from the rappel rope without running the entire rope through the first snaplink. To disengage from the rappel rope using the alternate technique, release the tension from the rope by opening the gate of the first snaplink and removing the second snaplink (with the rope attached).

a. The rappel seat is constructed as follows (Figure 1-3A through 1-3T):

(1) Place the midpoint (center) of the length of the sling rope on the hip opposite the brake hand (the brake hand is the strong hand) (Figure 1-3A, B, C).

(2) Bring the sling rope around the waist above the hip bone. Tie a double overhand knot over the navel (Figure 1-3D, E, F, G, H).

(3) Let the two free ends of the sling rope fall to the ground in front (Figure 1-3I).

(4) Bring the two free ends of the sling rope down between the legs and up over the buttocks. Ensure that the two free ends do not cross (Figure 1-3J).

(5) Pass the ends of the ropes over the rope that is tied around the waist at the two points above the center of the two rear seat pockets (Figure 1-3K).

(6) Grab the free end of the rope that is on the left side of the body with the left hand, and the free end of the rope that is on the right side of the body with the right hand.

(7) Squat down and simultaneously pull on both running ends of the ropes and stand up. This will tighten the seat.

(8) Take the two running ends of the rope down and back over the waist rope from the inside. Bring the running ends back under the ropes that are going across the buttocks (Figure 1-3L).

(9) Tie the two running ends with a square knot and two overhand knots on the hip opposite the brake hand (Figure 1-3M, N).

(10) Place any excess rope in the trouser pocket near the square knot (Figure 1-3O, P).

(11) With the gate down and the hooked end of the snaplink against the navel, place the end of the snaplink through the single rope that is around the waist and the two ropes forming the double-hand knot (Figure 1-3Q).

(12) Rotate the snaplink a half turn so that the gate is facing up and will open away from the body (Figure 1-3R, S, T).

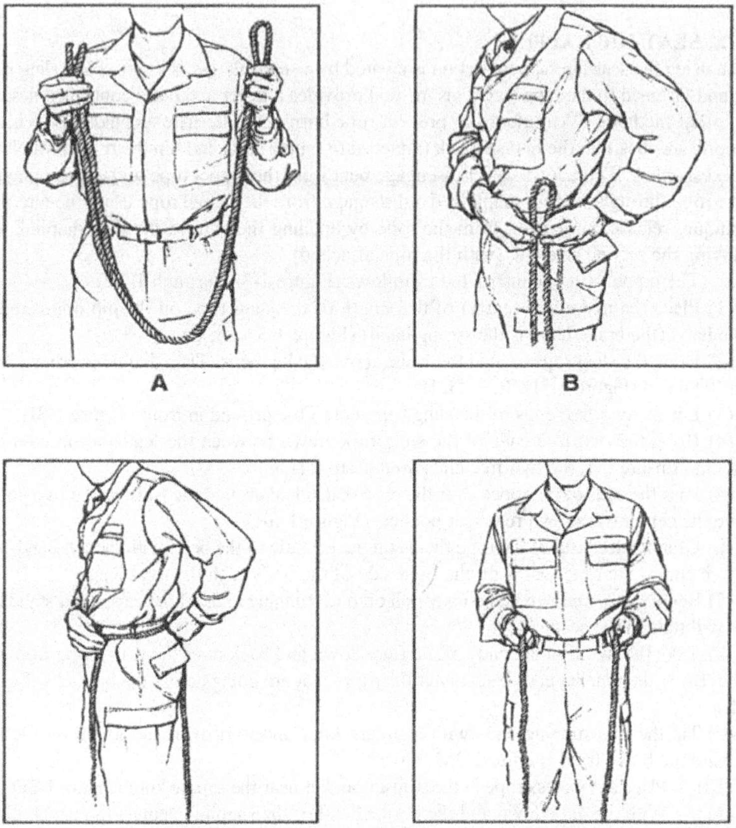

Figure 1-3. Construction of rappel seat.

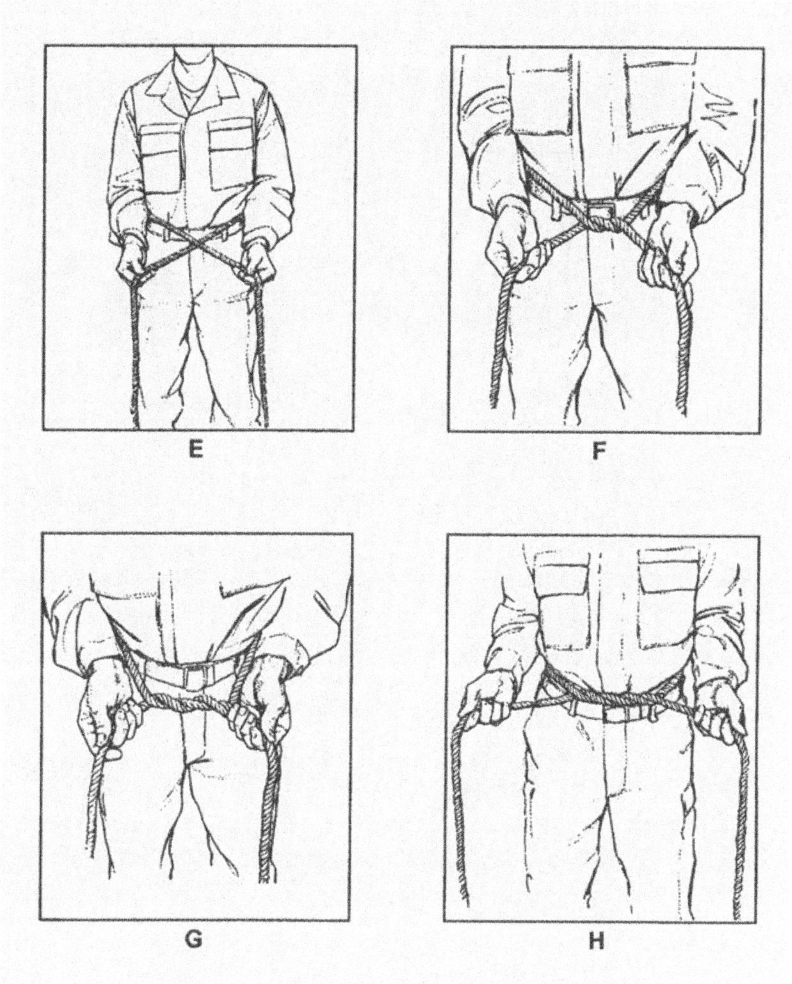

Figure 1-3. Construction of rappel seat.

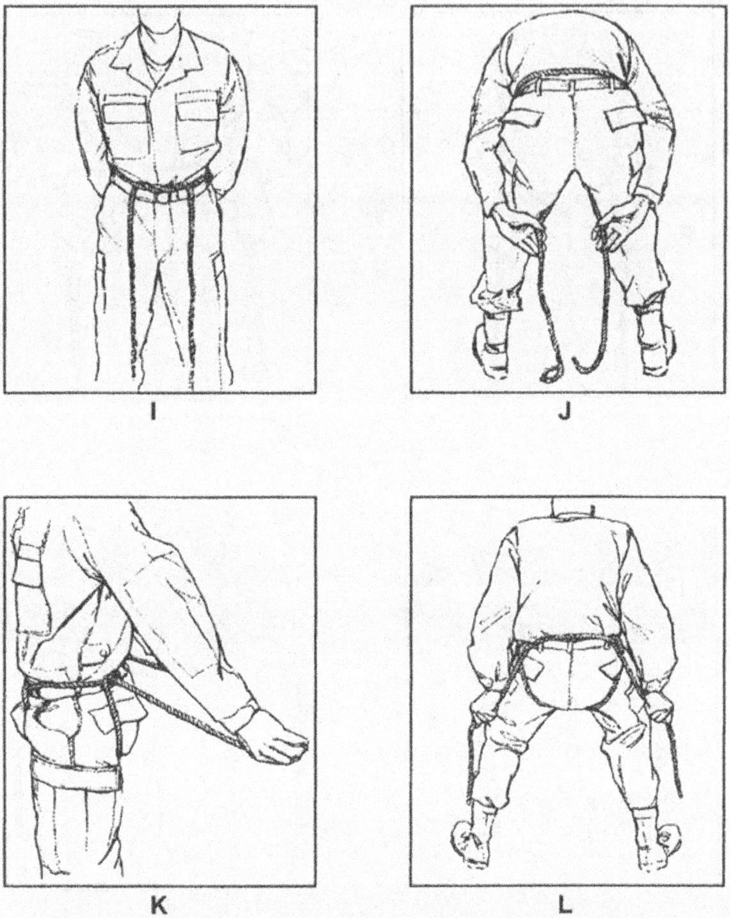

Figure 1-3. Construction of rappel seat.

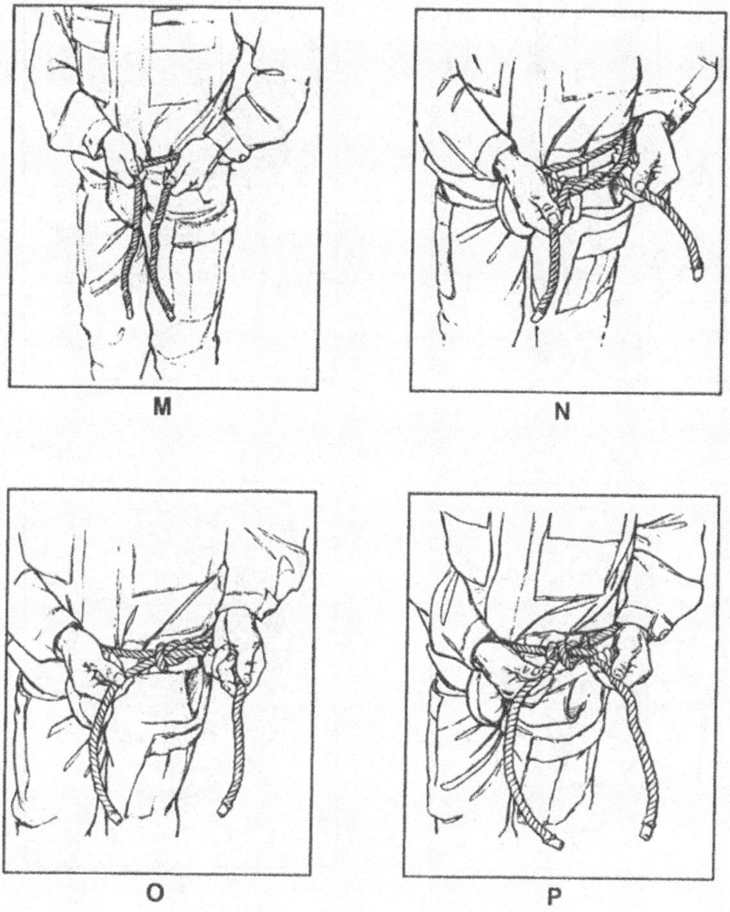

Figure 1-3. Construction of rappel seat.

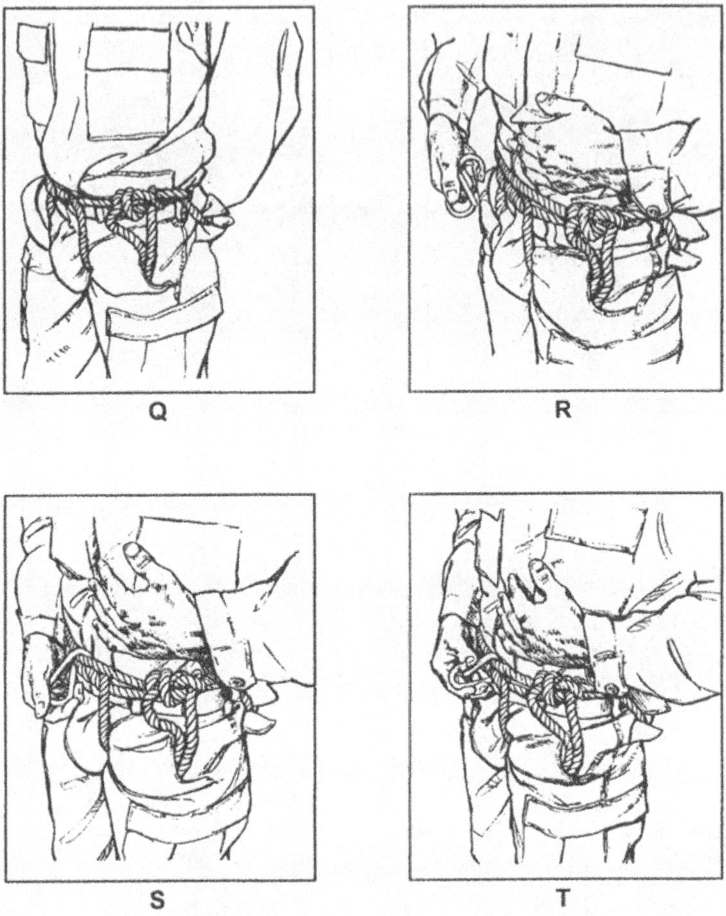

Figure 1-3. Construction of rappel seat.

b. To hook up using the seat-hip method, perform the following:

(1) Place the square knot with two overhand knots towards the anchor point for all seat-hip rappels.

(2) Grasp the two ropes with both hands and drop them through the gate of the snaplink. (At this point, two ropes should be running through the snaplink.)

(3) Using the hand closest to the anchor point, pull the slack towards the anchor point. Rotate the slack under and then over the top of the snaplink.

(4) Drop the two ropes a second time through the gate of the snaplink. (At this point, four ropes should be running through the snaplink.)

(5) Place the guide hand on the rope between the anchor point and the snaplink (palm facing up).

(6) Place the brake hand around the running end of the rope (palm facing down). Place the brake hand with the rope in the small of the back.

1-12. AUSTRALIAN RAPPEL

To hook up for the Australian rappel (Figure 1-4), perform the following:

a. Hook up to the Australian seat.

(1) Grasp the 9-foot sling rope at the midpoint (center) of its length, and double the rope.

(2) Place the doubled sling rope around the back and waist. Ensure that the rope is above the hipbone, but below the ribs.

(3) Tie a *square knot with two overhand knots* over the navel.

(4) Place any excess rope in the trouser pocket nearest the excess.

b. When the Australian seat is donned, face away from the anchor point and to the side of the rappel rope. (Stand to the same side of the rope as the brake hand. Determine left and right of the rope while facing the anchor point.) Place a snaplink onto the seat with the gate facing up, hinge closest to the body. Place it on the hip corresponding to the brake hand. The rappel master grasps the rappel rope and lays the rope into the snaplink. He then places one hand between the snaplink and the anchor point and draws slack toward the anchor point. He rotates the slack down, under, and over the rope and into the snaplink. The rappel master then slides the snaplink directly to the rappellers back.

c. Serve the running end of the rappel rope with the brake hand and prepare to rappel. During descent, brake by drawing the rappel ropes diagonally across the chest with the ropes running from near the waist to the pocket of the opposite shoulder.

Figure 1-4. Australian rappel.

1-13. CLIMBING PROCEDURES

Before climbing the ladder, the rappel master, safety OIC, or rappel lane NCO checks each rappeller's equipment.

a. The rappeller kicks the sand off his boots before climbing.

b. The rappeller grasps the outside of the ladder while climbing, not the rungs (when possible). If stairs are built for the tower, the rappeller grasps the railings as appropriate.

c. Just before climbing up the ladder, the rappeller sounds off, "(name) climbing," and then begins climbing up the ladder. Once at the top and clear of the ladder, the rappeller sounds off, "(name) clear."

d. Once off the ladder, the rappeller waits until the rappel master or lane NCO directs him to proceed to a rope station. At this time, the next rappeller in line may start to climb the ladder.

1-14. TOWER PROCEDURES

After the rappellers climb the tower, the following procedures are adhered to:

a. Once directed to a rope station, the rappel master or lane NCO ensures proper hookup for rappelling.

b. At this time, the rappeller sounds off with "on rappel" and the belayer sounds off with "on belay."

c. While maintaining his brake, the rappeller (on command from the rappel master or lane NCO) steps over the safety rail and faces the anchor point.

d. At this point, the rappel lane NCO sounds off with the following verbal commands and arm-and-hand signals.

(1) *Get Ready*—Rappel master or lane NCO extends both arms to the front with fists clenched and thumbs pointing upward (Figure 1-5). This alerts the rappeller. Each rappeller then looks over his brake hand shoulder to check for the belay man. The rappeller then looks at the rappel master. The rappel master or lane NCO makes his second check of the hookup, rappel seat, snaplink, and equipment.

Figure 1-5. Example of arm-and-hand signal GET READY.

(2) *Position*—The lane NCO extends both arms to the front, elbows locked, forearms pointed downward, and fingers extended. He makes a circular motion with both forearms rotating in opposite directions (Figure 1-6). With the brake hand in the small of the back, the rappeller rotates 180 degrees out onto the wall or skid mock up and assumes an L-shaped position. The feet should be shoulder width apart, balls of feet on the wall or skid, knees locked, and body bent at the waist (Figure 1-7).

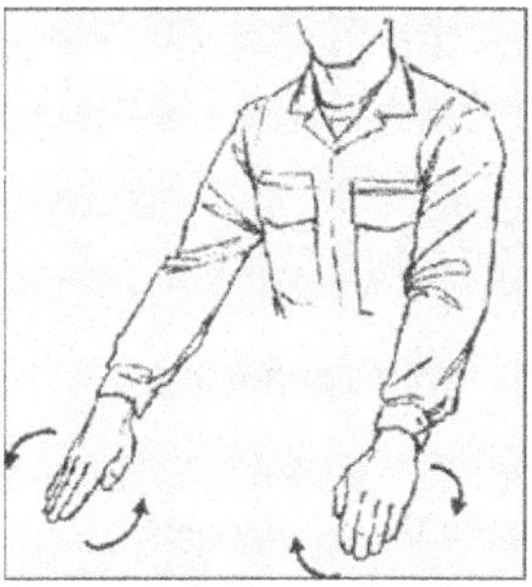

Figure 1-6. Example of arm and hand signal POSITION.

Figure 1-7. L-shape position.

(3) *Go*—The lane NCO extends his right arm with the elbow locked, fingers extended, thumbs around the index fingers, and points directly at the rappeller (Figure 1 8). This initiates the rappel. The rappeller flexes his knees and jumps vigorously backwards. At the same time, the rappeller throws his brake hand out at a 45-degree angle, letting the rope slide through both the brake hand and the guide hand. The rappeller looks over his brake hand shoulder at all times during descent.

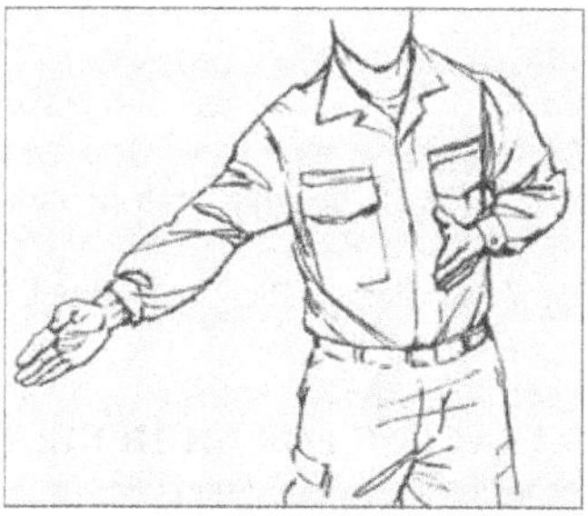

Figure 1-8. Example of arm-and-hand signal GO

e. The rappeller descends in a smooth, controlled manner.

f. The rappeller maintains eye contact with the ground at all times.

g. The rappeller maintains a modified L-shape position during descent with the feet shoulder-width apart, knees flexed, and buttocks parallel to the ground (Figure 1-9).

Figure 1-9. L-shape position while rappelling.

h. When carrying equipment or additional weight, a modified L-shape is used with the legs slightly lower than the buttocks to compensate for the added weight.

i. The rappeller's back is straight. He looks over the brake hand shoulder.

j. The guide hand is extended on the rope with the elbow extended and locked.

k. The rope slides freely through the guide hand, which is used to adjust equipment and to assist in balance during descent.

l. To brake, the rappeller places the brake hand (with rope in hand) in the small of the back and then grasps the rope firmly with the brake hand.

Note: Do not grip the rope firmly with the brake hand while the brake hand and brake arm are extended at the 45-degree angle. If this is done while rappelling, the brake hand and glove may become entangled in the snaplink causing injury to the hand and causing the rappeller to become hung up on the ropes.

m. Releasing tension on the rope and moving the brake hand out to the rear at a 45-degree angle regulates the rate of descent.

n. The rappeller never lets go of the rope with his brake hand until the rappel is completed.

o. After the rope is cleared and the rappeller is off rappel, he acts as the belayer for the next rappeller.

1-15. RAPPEL TOWER TRAINING FOR UH-1H HELICOPTER

Training on the rappel tower for helicopter skid rappelling prepares soldiers to rappel from a UH-1H helicopter (Figure 1-10). The rappeller is hooked up while he sits on the platform just above the helicopter skid. On the rappel master's command GET READY, the rappeller looks over the edge of the tower to ensure the running ends of the ropes are on the ground (Figure 1-10A). On the command SIT IN THE DOOR, the rappeller rotates his feet and legs off the platform and places them on the skid (Figure 1-10B). On the command POSITION, the rappeller turns around and assumes an L-shape position (Figure 1-10C). On the command GO, the rappeller bounds away from the helicopter skid and rappels to the ground. The rappel master is responsible for the proper procedures and safety.

A B C

Figure 1-10. UH-1H helicopter skid rappel training.

1-16. RAPPEL TOWER TRAINING FOR UH-60 BLACKHAWK HELICOPTER

Rappel tower training for the UH-60 Blackhawk is similar to that for the UH-1H with the exception that the UH-60 has no skids on which to stand. Therefore, the edge of the rappel tower is used as a pivot point to assume the L-shape position. All commands are the same except for SIT IN THE DOOR, which does not apply to the UH-60. The stances for each command are also different, as shown in Figure 1-11.

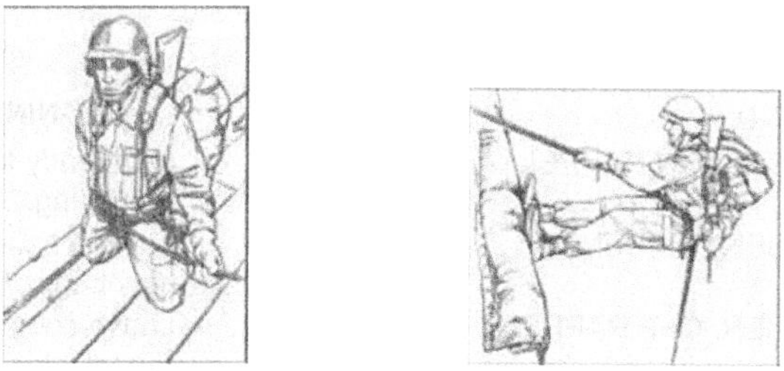

A B

Figure 1-11. UH-60 helicopter rappel training.

1-17. EMERGENCY LOCK-IN PROCEDURES

Mastering the lock-in procedure during tower training is critical before advancing to helicopter rappelling. Using the lock-in procedure allows personnel to hold in position for an extended period of time.

a. If the helicopter gains altitude above the length of the rappel ropes, the rappeller immediately brakes, locks in, and waits for the descent of the aircraft. Procedures for lock-in are as follows:

(1) Place the brake hand in the small of the back and brake to a complete stop.

(2) Release the guide hand from the ropes.

(3) Bring the guide hand around the back and grasp the running end of the two rappel ropes behind the brake hand.

(4) Using the guide hand, bring the two running ends of the rappel ropes around to the front.

(5) Secure these two running ends of the rappel ropes with the two anchor ends of the rappel ropes in the guide hand. This is now the new brake hand.

(6) Take the old brake hand out of the small of the back. Bring it around to the front and grasp the two ropes from the anchor point at a point just above the new brake hand. The old brake hand is now the new guide hand.

(7) Face the rappel master and wait for his command to lower to the ground.

(8) When the command is received from the rappel master to continue the descent, bring the brake hand to a 45-degree angle to the rear. When it is time to brake, bring the new brake hand around to the front diagonally across the chest.

b. If an engine fails or an aircraft emergency occurs during rappelling, the rappellers on the ropes descend as rapidly as possible and move from beneath the aircraft to the sides. Rappellers will maintain control of ropes, if possible.

1-18. COMMUNICATIONS

The rappeller at the top of a rappel point must be able to communicate with those at the bottom. Radios, hand signals, and rope signals are considered during a tactical rappel. For training situations, use the following commands:

COMMAND	GIVEN BY	MEANING
LANE NUMBER, ON RAPPEL	Rappeller	I am ready to begin rappelling.
LANE NUMBER, ON BELAY	Belayer	I am on belay and you may begin your rappel.
LANE NUMBER, OFF RAPPEL	Rappeller	I have completed the rappel, cleared the rappel lane, and am off the rope.
LANE NUMBER, OFF BELAY	Belayer	I am off belay.
LANE NUMBER, FALLING	Rappeller	I am falling. Be alert below—belay man brake.

1-18. DEMONSTRATION

After explaining the procedures to all rappellers, the rappel master should have an assistant demonstrate one complete cycle of rappelling from the static tower. This ensures that the rappellers can hear all the proper commands and see the actions and techniques used on the static tower.

CHAPTER 2

GROUND RAPPELLING

Ground rappelling is a technique that allows soldiers to negotiate mountains and cliffs safely and rapidly. Before rappellers participate in mountain and cliff rappelling they should have completed all the requirements for tower training. To ensure the rappel is safely conducted and meets the mission requirements, rappellers should consider several factors.

2-1. PERSONNEL

The personnel responsibilities for mountain and cliff rappelling are the same as for tower rappelling. As a minimum the following personnel are needed for mountain and cliff rappelling.

a. **Mountaineering Safety Officer (MSO).** The MSO is a SFC or above who is ranger qualified or is a graduate of the Northern Warfare Training Center (NWTC). He has overall responsibility for the safety of all mountaineering participants and ensures that all safety precautions are followed.

b. **Rappel Master**. The rappel master is a qualified officer or NCO in charge of training on ground, tower, and aircraft rappel safety procedures.

c. **Medical Officer or Noncommissioned Officer.** Qualified rappellers and aidmen must know MEDEVAC procedures.

d. **Radiotelephone Operator.** The RATELO monitors the radio and reports to the RSO.

e. **Rappel Lane NCO.** The rappel lane NCO is a qualified officer or NCO in charge of rappelling on a wall or ground rappel lane (Chapter 1, paragraph 1-3).

2-2. SUSTAINMENT TRAINING

Before conducting ground rappel training, the unit or element conducts sustainment training.

a. Review the construction of a rappel seat, seats to be used, and hook-up procedures.

b. Conduct two rappels on the 34-foot tower wall (20-foot tower is acceptable): one with equipment, one without equipment. Conduct two rappels from the free side of a 34-foot or higher rappel tower.

c. Demonstrate the ability to lock in during one of the rappels from the free side of the tower.

2-3. SELECTION OF A RAPPEL POINT

The selection of the rappel point depends on factors such as mission, cover, route, anchor points, and edge composition (loose or jagged rocks). The anchor point should be above the rappeller's departure point. Primary and secondary anchors must be solid (natural anchors are preferred). The rappeller should be sure that the rope reaches the bottom or a place from which he can further rappel or climb. Also, the rappel point should be carefully tested and inspected to ensure the rope will run freely and that the area is clear of obstacles that could be dislodged. If a sling or runner is used for a rappel point, it should be tied twice from two separate loops. Suitable loading and off-loading platforms should be available.

2-4. ESTABLISHMENT OF A RAPPEL POINT

A rappel lane is the area and rope that a rappeller will use during his descent. It should have equal tension between all anchor points by establishing primary and secondary anchor points (Figure 2-1). If one anchor point fails, the rappel rope should not extend. All of the methods discussed below can be performed with a single or double rope. A double rope application should be used for safety.

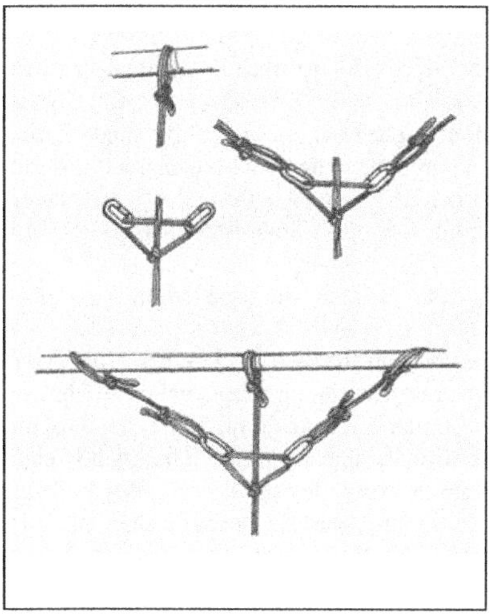

Figure 2-1. Rappel point.

a. If a rappel lane is *less* than half the rope length, the rappeller may apply one of the following techniques:

(1) Double the rope and tie a three-loop bowline around the primary anchor to include the primary anchor inside two loops and enough rope in the third loop to run to the secondary anchor (another three-loop bowline secured with an overhand knot).

(2) Double the rope and tie it around a secure anchor point with a round turn anchor bowline secured with an overhand knot (or any appropriate anchor knot).

(3) Double the rope and establish a self-equalizing anchor system (Figure 2-2) with a bowline-on-a-bight or figure-eight-on-a-bight knot. Tie off on the long standing and with a round turn anchor bowline.

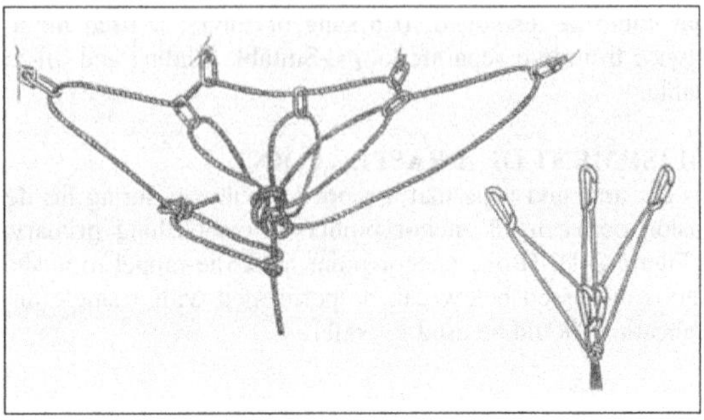

Figure 2-2. Self-equalizing anchor system.

(4) In an emergency, double the rope and place it behind or through a secure anchor point, or tie a runner around an anchor point with a snaplink inserted and place the rope through the snaplink. To preclude a rappeller from sliding off the end of the rappel lane, tie a double figure eight (square knot or double fisherman's knot) at the bottom end of the rope with both ends.

b. If a rappel lane is *greater* than half the rope length, the rappeller may apply one of the following techniques:

(1) Using two ropes, tie a round turn anchor bowline around a primary anchor point. Take the remaining rope (the tail from the primary anchor bowline) and tie another round turn anchor bowline to a secondary anchor point. The secondary anchor point should be in a direct line behind the primary anchor point. The anchor can be either natural or artificial. The ends of the rappel lane ropes should be offset by 15 centimeters (6 inches) so that the rope ends feed freely through the rappeller's snaplink.

(2) Using two ropes, establish a three-piton anchor system using a bowline-on-a-bight knot (or figure-eight-on-a-bight knot) and tie off on the long-standing end with a round turn anchor bowline.

(3) In an emergency, use two ropes and tie the two ends together with a joining knot. Place the joined ropes behind or through an anchor point, or tie a runner around an anchor point with a snaplink inserted and place the joined rope through the snaplink. Offset the joining knot to the left or right of the anchor. Tie off the bottom end of the rope with a joining knot to prevent a rappeller from sliding off the end of the rappel line.

c. Due to the length of the rappel, the rappel rope may not reach the anchor. If the rope is used to tie the knots, it may be too short to accomplish the rappel.

(1) When using a natural anchor—

(a) Tie a sling rope, runner, or another rope around the anchor with a round turn anchor bowline.

(b) Tie a fixed loop (figure eight or butterfly) in one end of the rappel rope, which is attached to the round turn around the anchor through the two snaplinks (opposing gates).

(2) When using an artificial anchor—

(a) Tie off a sling rope, runner, or another rope to form a loop.

(b) Put the loop through the snaplinks that are attached to the artificial anchor point.

(c) Bring the bottom of the loop up and connect it to the snaplinks that are between the artificial anchors (chocks, pitons, or bolts).

(d) Grasp the snaplinks that are between the chocks/pitons and pull them down and together.

(e) Tie a fixed loop (figure eight or butterfly) in the end of the rappel rope and connect this to the snaplinks that have been pulled together.

Notes: 1. Rerouted figure-eight knots can be used instead of bowlines.
2. Runners may be used from one or more anchor points.

d. Backfeed (stack) the rope to ensure it does not snarl when thrown. Take off one wrap at a time and let it fall to the ground, ensuring that no kinks, knots, or twists occur that might hinder the rope from feeding out. When the rope is backfed, anchor off one end of the rope. The two methods used to throw the rope are underhand and overhand. Use the overhand method when trees or shrubs are on or near the rappel point.

e. Set up a retrievable rappel point by applying one of the following techniques:

(1) When the rappel is *less* than half the total length of the rope, double the rope. Place the rope around the primary anchor with the bight formed by the midpoint. Join the tails of the rappel rope and throw the rope over the cliff. Tie a clove hitch knot around a snaplink just below the anchor point ensuring the locking bar inside the snaplink points away from the gate opening end and faces uphill. Snap the opposite standing portion into the snaplink. Upon reaching the bottom of the cliff, pull on the part of the rope to which the snaplink is secured to allow the rope to slide around the anchor point.

(2) When the length of the rappel is *greater* than half the length of the rope used, join two ropes around the anchor point (double fisherman's knot or square knot). Adjust the joining knot so that it is away from the anchor. Tie a clove hitch knot around a snaplink just below the anchor point ensuring the locking bar inside the snaplink points away from the gate opening end and faces uphill. Snap the opposite standing portion into the snaplink. Upon completion of the rappel, pull the rope to which the snaplink is secured to allow the rope to slide around the anchor point.

WARNING

When setting up a retrievable rappel using only a primary point, take care in selecting the point. Ensure soldiers have a safety line when approaching the rappel point, with only the rappeller going near the edge.

2-5. TYPES OF RAPPELS

Types of rappels include body rappel, hasty rappel, seat-hip rappel, seat-shoulder rappel, Australian rappel, and buddy-evacuation. (Seat-hip rappel and Australian rappel are described in Chapter 1, paragraphs 1-10 and 1-11.)

a. **Body Rappel.** Face the anchor point and straddle the rope. Pull the rope from behind and run it around either hip, diagonally across the chest, and back over the opposite shoulder (Figure 2-3). Then, run the rope to the brake hand, which is on the same side of the hip that the rope crosses (for example, the right hip to the left shoulder to the right hand). Lead with the brake hand down and face slightly sideways. The foot corresponding to the brake hand precedes the guide hand at all times. Keep the guide hand on the rope above to guide—not to brake. Lean out at a sharp angle to the rock. Keep the legs spread well apart and relatively straight for lateral stability and the back straight to reduce friction. Turn the BDU collar up to prevent rope burns on the neck. Wear gloves, and use other clothing to pad the shoulders and buttocks. To brake, lean back and face directly toward the rock area so the feet are horizontal to the ground.

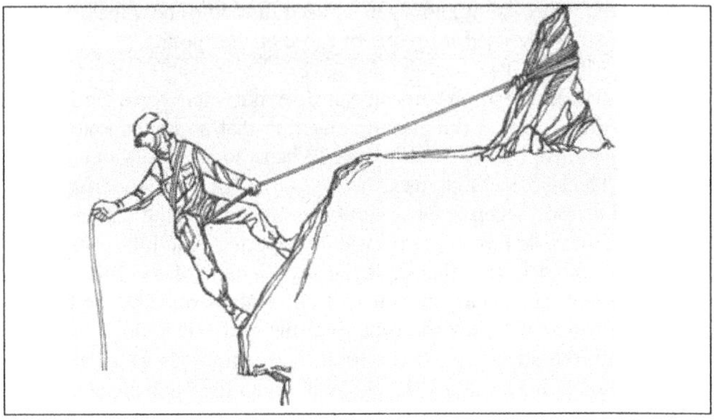

Figure 2-3. Body rappel.

b. **Hasty Rappel.** Facing slightly sideways to the anchor, place the ropes horizontally across the back (Figure 2-4). The hand nearest to the anchor is the guide hand, and the other is the brake hand. Wear gloves to prevent rope burns. To stop, bring the brake hand across in front of the body locking the rope. At the same time, turn to face up toward the anchor point. Use this rappel only on moderate rock pitches. The hasty rappel's main advantage is that it is easier and faster than the other methods, especially when the rope is wet.

Figure 2-4. Hasty rappel.

Figure 2-5. Seat-shoulder rappel.

c. **Seat-Shoulder Rappel.** To hook up for the seat-shoulder method, face the rappel point (Figure 2-5). Snap into the rope that passes up through the snaplink. Bring the rope over one shoulder and back to the opposite hand (left shoulder to right hand). Use the same technique in the descent as in the body rappel. This method is faster than the body rappel, less frictional, and more efficient rappellers with packs and during night operations.

d. **Buddy-Evacuation Rappel.** Use the buddy-evacuation rappel to evacuate an injured soldier from a cliff or steep terrain. Face the cliff and assume a modified L-shape body position to compensate for the weight of the victim on the back. The victim is top-rope belayed from above, which provides the victim with a point of attachment to a secured rope.

Note: To use this rappel, the victim must be conscious.

(1) The method for securing a victim to a rappeller's back is described below:

(a) To secure the victim to the carrier's back with a rope, the carrier ties a standard rappel seat (brake of choice, depending on the injury) and rests his hands on his knees while the victim straddles his back (Figure 2-6.)

(b) A 4.2-meter (14-foot) sling rope is used. A 45-cm (18-inch) tail of the sling is placed on the victim's left hip. (This method describes the procedure for a seat-hip rappel with right-hand brake.)

(c) The long remaining end of the sling rope is routed under the victim's buttocks and passed over the victim's and carrier's right hip. The rope is run diagonally, from right to left, across the carrier's chest, over his left shoulder, and back under the victim's left armpit.

(d) The rope is then run horizontally, from left to right, across the victim's back. The rope is passed under the victim's right armpit and over the carrier's right shoulder.

Figure 2-6. Patient secured to carrier's back.

(e) The rope is run diagonally, from right to left, across the carrier's chest and back across the carrier's and victim's left hip.

(f) The two rope ends should now meet. The two ends are tied together with a square knot and overhand knots.

(g) The knot is positioned on the victim's left hip. The carrier's shoulders should be padded to prevent cutting by the rope.

(2) An alternate method is to use two pistol belts hooked together and draped over the carrier's shoulders. The victim straddles the carrier, and the belay man secures the loose ends of the pistol belts under the victim's buttocks. Slack in the pistol belt's sling should be avoided, since the carrier is most comfortable when the victim rests high on his back. (See FM 8-35.)

(3) A large rucksack can be slit on the sides near the bottom so that the victim can step into it. The victim is belayed from the top with the carrier conducting a standard rappel. The carrier wears the rucksack with the victim inside.

Note: Rucksacks are not designed to support the weight of a soldier and his gear. Therefore, this technique is used only as a last resort.

(4) A casualty secured to a carrier, as described above, can be rappelled down a steep cliff using a seat-shoulder or seat-hip rappel. The casualty's and rappeller's shoulders should be padded where the sling rope and rappel lines cross if a seat-shoulder rappel is used. The buddy team should be belayed from above with a bowline tied around the victim's chest under his armpits. The belay rope must run over the rappeller's guide hand shoulder (Figure 2-7).

Figure 2-7. One-man carry.

2-6. RAPPELLING PROCEDURES

Proper rappelling procedures must be followed to ensure the safety of personnel.

a. The rappel lane NCO is tied off to a separate anchor point for safety, if possible. He—

(1) Ensures that the anchors are sound and the knots are properly tied.

(2) Ensures that loose rock and debris are cleared from the loading platform.

(3) Allows only one man on the loading platform at a time and ensures that the rappel point is orderly run.

(4) Ensures that each man is properly prepared for the particular rapel: gloves on sleeves down, helmet with chin strap fastened, gear prepared properly, and rappe seat and knots correct (if required). He also ensures that the rappeller is correctly hooked up to the rope and is aware of the proper braking position.

(5) Ensures that the proper signals or commands are used.

(6) Dispatches each man down the rope.

(7) Is the last man down the rope.

(8) Ensures that the rope does not run over sharp edges, and pads the rappelling surfaces as necessary to protect the rope.

(9) Ensures that the rope reaches the bottom or is at a place where additional rappels can be made.

b. The first rappeller down—

(1) Selects a smooth route for the rope that is clear of sharp rocks.

(2) Conducts a self-belay with a prusik sling tied from himself to the rappel rope.

(3) Clears the route and places loose rocks, which the rope may dislodge, far enough back on ledges to be out of the way.

(4) Ensures the rope reaches the bottom or is at a place from which additional rappels can be made.

(5) Ensures that the rope runs freely around the rappel point when pulled from below.

(6) Clears the rappel lane by straightening all twists and tangles from the ropes.

(7) Belays subsequent rappellers down the rope.

(8) Takes charge of personnel as they arrive at the bottom (off-loading platform).

Note: A rappeller is always belayed from the bottom except for the first man down. The first man belays himself down the rope by using a safety line attached to his rappel seat that is hooked to the rappel rope with a prusik knot. As the first man rappels down the rope, he "walks" the prusik knot down with him.

2-7. DUTIES OF THE RAPPELLER

To ensure the safe and efficient conduct of rappelling operations, each rappeller must know the duties of his job.

a. Each rappeller down shouts, "Off rappel" (if the tactical situation permits), untangles the ropes, and ensures the ropes run freely around their anchors. When silence is needed, a planned signal of pulling the rope is substituted for the voice signal. After the rope is cleared and the rappeller is off rappel, he acts as the belayer for the next rappeller.

b. All rappellers inspect ropes often when many soldiers are rappelling.

c. The last rappeller to descend constructs a retrievable rappel point and rappels down. Then he gently pulls the rope to prevent the rising rope end from entangling with the other rope. He stands clear of the falling rope and any rocks that it may dislodge.

Note: Soldiers wear gloves for all types of rappels for protection from rope burns. Also, bounding rappels are discouraged since this stresses the anchor and causes undue wear and friction on the rope.

d. Rappellers descend in a smooth, controlled manner. The body forms an L-shape with the feet shoulder-width apart, legs straight, and buttocks parallel to the ground. When carrying equipment or additional weight, a modified L-shape is used with the legs slightly lower than the buttocks to compensate for the added weight. The rappeller's back is straight. He looks over the brake shoulder. The guide hand is extended on the rope with the elbow extended and locked. The rope slides freely through the guide hand, which is used to adjust equipment and to assist balance during the descent. The rappeller grasps the rope firmly with the brake hand and places it in the small of his back. Releasing tension on the rope and moving the brake hand out to his rear at a 45-degree angle regulates the rate of descent. The rappeller never lets go of the rope with his brake hand until the rappel is completed.

e. During training, the lane number must be understood. In combat, a series of tugs on the rope may be substituted for the oral commands to maintain noise discipline. The number of tugs used to indicate each of the commands is IAW the unit SOP.

2-8. BELAYER

The belayer assumes a position at the base of the lane about one pace away from the rock area. He ensures that the rappel ropes are even with the ground during rock and tower rappels. The belayer loosely holds the rappel ropes with both hands so as not to interfere with the rappeller but still able to stop the rappeller should he fall. If the rappeller shouts "Falling" or loses control of his brake hand or descent, the belayer immediately stops the rappeller by pulling downward on the rappel ropes. To ensure a firm grip on the rappelling rope, the belayer does not wear gloves. (Because no friction exists between the belayer's hands and the rappelling rope, gloves are not required for safety. However, whether or not to wear gloves is decided by the OIC/NCO.) The belayer watches the rappeller at all times and maintains constant voice or visual contact with the rappeller. The belayer wears a helmet to prevent injuries from falling debris. All commands are spoken loudly and clearly.

CHAPTER 3
HELICOPTER RAPPELLING

Helicopter rappelling can provide a means of quick insertion with or without an LZ.

Section I. PERSONNEL

The personnel required for helicopter rappelling are the rappel master, the rappel safety officer, the pilot in command, the rappelers, and the belayers.

3-1. RAPPEL MASTER

A qualified rappel master is aboard each aircraft. Safety is the rappel masters number one priority. The rappel master—

- Ensures internal communication between the pilot and rappel master, and external communication between the aircraft and the ground.
- Inspects all equipment and uses only authorized, serviceable equipment.
- Inspects and tests all anchor points and knots before the mission starts.
- Ensures that all rappellers receive a safety briefing and the pilots and aircrew receive an air mission brief.
- Ensures that the rappelers are rappel qualified before conducting helicopter rappelling to include tactical rappelling.
- Maintains communications with the pilot at all times.

3-2. RAPPEL SAFETY OFFICER

The RSO is a SFC or above who is either air assault or ranger qualified. He has overall responsibility for the safety of all rappellers and ensures that all safety precautions are followed. He maintains communications at all times with the pilot and rappel master through FM radio. He alerts the rappel master and pilot of any unsafe acts.

3-3. PILOT IN COMMAND

The pilot in command (PIC) of the aircraft has the following responsibilities:

a. Ensures that the aircrew and all non-aircrew personnel are briefed and understand their responsibilities during rappelling operations, including aircraft safety and action in the event of an emergency.

b. Ensures that the donut ring anchoring device assembly and or aircraft anchor points have been inspected for completeness and functionality, and that they are installed properly.

c. Emphasizes procedural techniques for clearing, recovery, and or jettison of ropes.

d. Keeps the aircraft centered over the target with corrections from the rappel master as required.

3-4. RAPPELLER

In addition to the tower qualification requirements outlined in Chapter 1, the individual rappeller must complete advance training under the supervision of a qualified rappel master in order to participate in tactical helicopter rappelling operations. He must—

- Satisfactorily complete three rappels from a helicopter from a height of 60 feet (two rappels with combat equipment and weapon).
- Demonstrate confidence and proficiency in the techniques, procedures, and equipment used in rappelling from a helicopter.
- Know the rappelling equipment used in helicopter operations and any special equipment required for helicopter rappelling.

3-5. BELAYER

A belayer is assigned to each rope. He is responsible for walking the rope beneath the helicopter during the descent. (Walking the rope is defined as removing the slack from underneath the helicopter by walking backwards with the rope as the helicopter descends to land.) The belayer makes sure the ropes are not caught on the aircraft skids or tires to ensure a safe landing.

Section II. TRAINING

Training for helicopter rappelling includes sustainment training and refresher training.

3-6. SUSTAINMENT TRAINING

Before conducting helicopter rappel training, the unit or element conducts sustainment training.

a. Review the construction of a rappel seat, equipment to be used, and hook-up procedures.

b. Conduct two rappels on the 34-foot or higher tower wall: one without equipment, one with equipment. Conduct two rappels from the open side of a 34-foot or higher rappel tower.

3-7. REFRESHER TRAINING

Refresher training is routinely conducted to maintain acquired skills. Soldiers who have not performed a helicopter rappel during the past six months will undergo refresher training consisting of three satisfactory rappels from a tower (one with weapon and equipment and one executing a lock in) before executing a helicopter rappel.

Section III. PREOPERATIONAL BRIEFINGS AND SAFETY PROCEDURES

This section discusses safety, medical and communications requirements, and the procedures to follow during unusual conditions (adverse weather/terrain conditions, night operations). Personnel must use sound judgement to determine what action to take depending on the nature and severity of the conditions.

3-8. MEDICAL COVERAGE

A qualified and equipped medic will be present to respond to any mishap. Medical transportation must also be available. Absence of a medic, medical equipment, or transportation will terminate training. If the situation warrants and the installation cannot support a MEDEVAC mission, the rappel aircraft may be used as a last resort MEDEVAC vehicle.

3-9. COMMUNICATION REQUIREMENTS

During helicopter rappel training, the RSO will have radio communications with the aircraft. Voice communications are required before starting aircraft rappelling. Additionally, the RSO will inform the PIC to stop operations if an unsafe condition develops.

3-10. ADVERSE WEATHER/TERRAIN CONDITIONS

Rappel operations will not be conducted under the following conditions

- Ambient temperature is 30 degrees Fahrenheit or less.
- Winds in excess of 30 knots.
- Lightning strikes within 1 nautical mile of rappelling operations.
- Wind chill factors caused by the helicopter's rotor wash or extraction cruise air speeds, which could cause cold weather injuries.
- Water or ice on the rope inhibiting the ability of the rappellers to control their descent.
- The rope is exposed to the elements for a sufficient length of time to freeze, thereby reducing its tensile strength.
- Blowing particles produced by rotor wash causes the aircrew or the rappel master to lose visual contact with the ground.

3-11. NIGHT OPERATION REQUIREMENTS

The following requirements are necessary for night rappelling operations.

a. One chemlight will be attached to the end of the rope and one on each rappeller.

b. One chemlight will be secured to the attachment point of the rope.

c. Night vision goggles (NVG) will not be worn by rappellers during the descent. Aircrew members will wear NVG as required during night operations. NVG lighting criteria will be IAW Army regulations, specific aircraft aircrew training manuals, unit SOPs, or the tactical environment.

3-12. SAFETY BRIEFING

The following safety measures are enforced.

a. Loose clothing and equipment are secured.

b. Helmets are worn with chin straps fastened.

c. Rappellers wear identification tags and earplugs, carry identification cards, and role down their sleeves.

d. Weapons are slung diagonally across the back with the muzzle pointing down on the guide hand side.

e. All seats and rappelling equipment must be inspected by a rappel master before rappelling.

f. No running is allowed within 50 feet of the aircraft.

g. Personnel approach and depart the helicopter from the front and forward of the rear of the cargo doors. When approaching or departing the helicopter, personnel bend their bodies forward at the waist to ensure clearance of the rotor blades. At no time will personnel go near the rear of the aircraft.

h. Upon boarding the aircraft, the rappeller sits or kneels down, hooks up, and applies his brake hand to the small of his back.

i. While in the helicopter, the rappeller maintains eye-to-eye contact with the rappel master and receives commands from him.

j. The rappeller ensures that he has a belayer on his rope at all times when conducting training at a hover site.

k. During the descent, the rappeller maintains eye-to-ground contact.

l. If the rappeller sees his rope coming off the ground or sees that his belayer has lost control of his rope, he immediately brakes and executes a lock-in. He then waits for commands from the rappel master.

m. The rappeller brakes once every 30 feet during descent.

n. The belayer does not wear gloves. He keeps both hands on the rope and his eyes on the rappeller at all times.

o. All rappelling will be conducted using a double strand of rope.

Section IV. DEPLOYMENT OF ROPES

Deployment of the ropes from a helicopter is a critical task. It can cause a planned rappelling operation to fail, or it can increase the time required to conduct the operation. This is due to the likelihood of the ropes becoming entangled. To prevent this, ropes must be deployed using a positive control technique. Two of the techniques that may be used are the *deployment bag* technique and the *log coil* technique.

3-13. DEPLOYMENT BAG TECHNIQUE

The rappeller places the deployment bag (D-bag) (standard) on a flat surface with the stow loop facing upward. If the D-bag still has a static line, the rappeller removes it by cutting the static line where it attaches to the bag. A 12-foot section of the static line can be used as a safety line in the helicopter.

a. Use a sandbag (about one E-tool spade full of sand) as a weight in the deployment system. Roll the sandbag into a rectangular shape, tie it, and then place it in a small plastic bag, rolled and taped. Place a retaining band over the middle of the weight, and place the weight on top of the D-bag. The weight should be about the same width as the D-bag.

b. Lay out and inspect the two ropes. The working ends (closest to the D-bag) should be even. Place a round turn with the two working ends of the ropes on the weight's retaining band. Working on top of the deployment bag, start forming figure eights. The stack should consist of 8 to 10 figure eights, one on top of another—do not exceed the width of the D-bag. Then, starting from either side, center a retaining band over the stack. Ensure that it is over all the figure eights in the stack. Repeat the process each time, placing one stack in front of the other. Continue until about 10 feet of rope remains. If one rope is shorter than the other, the end of the shorter rope should be about 10 feet from the last stack.

c. For a primary anchor point, measure down about 4 feet from the end of the shortest rope. Using both ropes, tie a bowline without a half hitch. This knot is the primary anchor point.

d. For a secondary anchor point, tie an end-of-line bowline with a half hitch toward the end of the ropes. The dressed knot should be 18 to 22 inches from the primary anchor point knot when using the UH-1H helicopter and 22 to 30 inches when using the UH-60 helicopter. If excess rope remains, S-fold and tape it between the two knots. Ensure that 3 to 6 feet of rope remain from the last stack of figure eights.

e. To place the ropes in the D-bag, remove the bag from under the stack of rope. Place the weight into the bottom of the D-bag, and place all the stacks of figure eights, in the order they were made, one on top of the other into the bag.

f. Two left, two right, and two rear loops are in the opening of the D-bag. ***Do not use the two rear loops.*** Close the flap on the D-bag and push the loops through the aligning holes on the flap. Using an 8-inch piece of gutted 550-pound cord, tie the two loops together with a square knot and two hitches. Ensure that two 120-foot ropes are coming out of the center of the flap—not to one side. Repeat the process and tie the right loops together. Wrap the excess rope lengthwise around the bag.

Note: The sizes of the loops of the bowline should be no larger than an average size fist.

3-14. LOG COIL TECHNIQUE

The rappeller lays the running end of the double rope along the length of the coiling log (Figure 3-1). He then coils the double rope around both the running end of the rope and the coiling log. The rope must be coiled evenly and tightly.

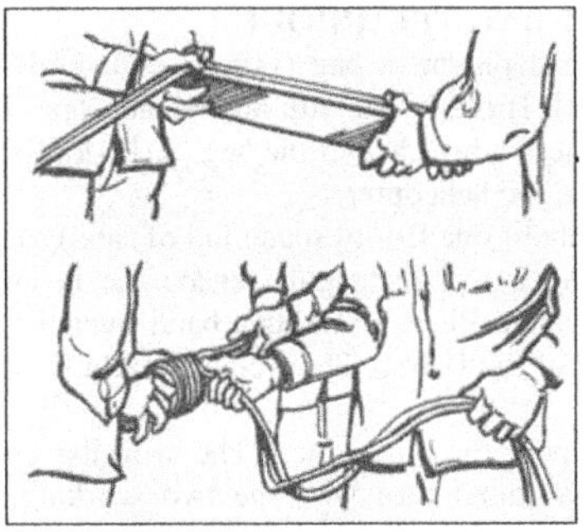

Figure 3-1. Coiling Log

Section V. RAPPELLING OPERATIONS FOR UH-1H IROQUOIS HELICOPTER

The UH-1H helicopter provides a safe, stable aerial platform from which to conduct rappelling operations when landing is not feasible.

3-15. CHARACTERISTICS

The UH-1H is a single-engine, medium-speed, single-main rotor helicopter that can transport eight rappellers, one rappel master, and one three-man crew. It has a lift capacity of 2,300 pounds and is equipped with several floor-mounted tie-down fittings, seven of which are used during rappelling operations.

3-16. RIGGING OF UH-1H HELICOPTER FOR RAPPELLING

The rappel master performs the following to rig the UH-1H helicopter.

a. Removes all seats.

b. Locks the doors in the open position. If no locks are present, removes doors. Also, he removes small cargo doors.

c. Pads and tapes all sharp edges on the floor, door ledge, and all protrusions on the skids. Ensures each door ledge has a scuff pad to protect the rope from contacting the metal door ledge.

d. Secures the donut ring to the center of the floor of the helicopter. The donut ring has six snaphooks numbered clockwise, with 12 o'clock being toward the front of the helicopter. Positions the clamps of the donut ring toward the aft end of the helicopter. Ensures the front two snaplinks and rear two snaplinks are facing outside of the donut ring and hooks them to the tie-down ring with the snaphook facing down. Ensures the center two snaphooks face into the donut ring and are connected to the floor tie-down ring with the gate facing down (Figure 3-2). Attaches the free-floating safety ring to the center floor tie-down ring using a seventh snaphook.

Note: Using this technique has caused equipment failure in many aircraft. The cable clamp nuts must be inspected and certified as airworthy by the pilot or maintenance crew before flying the mission.

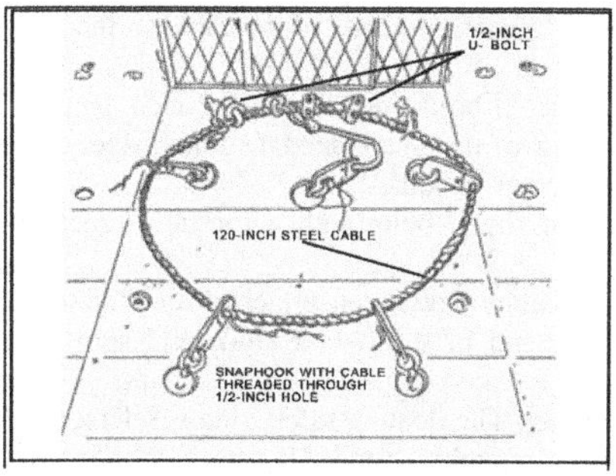

Figure 3-2. Donut ring attached to the floor of the helicopter.

3-17. CONSTRUCTION OF ANCHOR POINTS

In addition to the usual equipment requirements for rappelling operations, an anchor assembly is fabricated. This anchor assembly is commonly known as a *donut ring*. The secondary anchor point is a *floating safety ring*.

a. **Donut Ring.** This primary (or No. 1) anchor point for the rappelling ropes is constructed from a 1/2-inch steel cable with a steel wire core. The cable is 120 inches long, consists of 6 strands (18 wires per strand), and has a tensile strength of 21,000 pounds (Figure 3-2). The completed donut ring has a tensile strength of 3,000 pounds.

(Figure 3-3 shows the donut ring and rappel rope connection in a helicopter.)

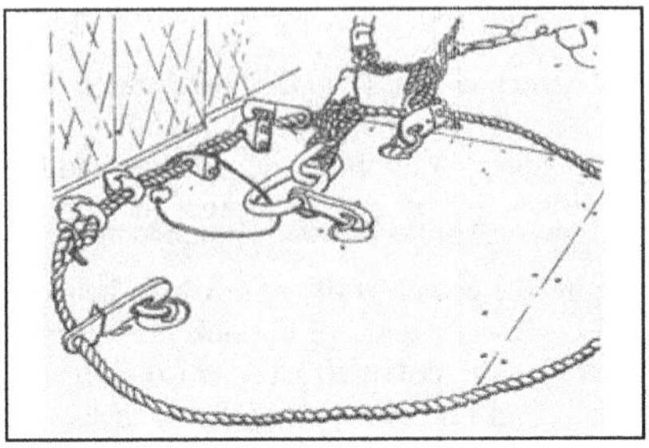

Figure 3-3. Donut ring and rappel rope connection in a helicopter.

(1) Thread six parachute static line snaphooks onto the 120-inch steel cable so that four snaphooks are facing out with the gates down, and the center two snaphooks are facing in with the gates down. Drill the end portion of each snaphook to make a 5/8-inch diameter hole, and thread the cable through the holes in the snaphooks.

(2) Overlap the ends of the 120-inch cable 20 inches to form a circle. Secure the ends with four 1/2-inch U-bolts placed at 2- to 3-inch intervals.

(3) Attach two U-bolts to each dead end of the cable so that the bolts engage the dead end.

(a) Before torquing the U-bolts, position a 12-inch length of chain or 1/8-inch diameter cable on the center of the overlapped 120-inch steel cable so that it remains in position between the two center U-bolts.

(b) Tighten each nut of the U-bolts with a torque wrench (if possible) to 40-foot pounds (480-inch pounds).

(c) After the U-bolt clamps have been attached and tightened, fasten a steel plate (drill to fit) over the open end of the U-bolt studs and spot weld in place to prevent loosening.

b. **Floating Safety Ring**. The floating safety ring is referred to as the secondary (or No. 2) anchor point for the rappelling ropes (Figure 3-4). The snaplink at the end of the rappelling rope is hooked to this connection. Either of the following two types of floating safety rings may be used:

(1) *First method.* Thread an elliptical rappelling ring through the free end of the keeper chain (cable). It is constructed of cold-rolled steel that is 1/4-inch in diameter, with inside dimensions of 2 1/4 inches (minor axis) and 4 inches (major axis). Thread a seventh parachute static line snaphook onto the ring before welding. Weld this ring together so that it withstands a force up to 3,000 pounds.

(2) *Second method.* Attach two snaplinks to the aircraft tie-down ring in the center of the donut ring. Insert the first snaplink through the free end of the keeper chain (cable) and the tie-down ring with the gate down. Insert the second snaplink through the free end of the keeper chain (cable) and the tie-down ring with the gate up. Tape the snaplink

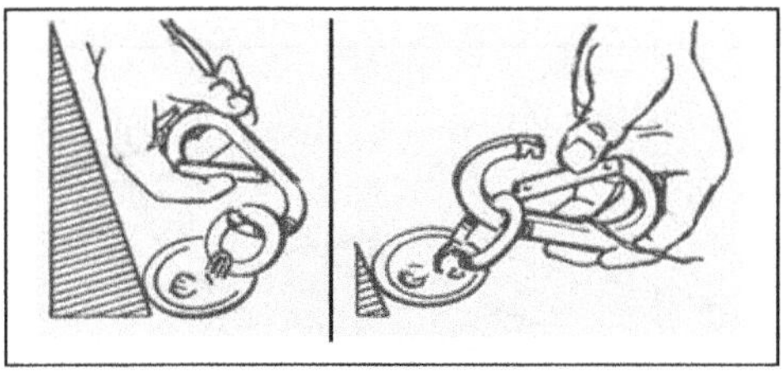

gates closed with masking tape. Then, tape the snaplinks together to ensure that the snaplink gates are on opposite sides of each other.

Figure 3-4. Floating safety ring formed with two snaplinks.

c. **Rappel Rope Anchor Points.** The rappelling rope is connected to the floating safety ring and the donut ring as follows:

(1) *Number 1 anchor point (donut ring).* Using a bight about 5 feet from the end of the standing part of the rope, make one turn through the snaplink forming a round turn. The bowline is the preferred method of attaching rappel ropes to snaplinks or anchor systems. The round turn with two half hitches is a reliable means of attaching ropes to anchor systems. Secure the round turn to the snaplink with two half hitches. Make the connection to the donut ring by attaching the snaplink gate upward. Ensure the gate faces upward with the opening away from the knot.

(2) *Number 2 anchor point (floating ring).* Attach the No. 2 snaplink the same as the first with some exceptions. Using a bight about 2 feet from the end of the standing part of the rope, connect the snaplink to the rope the same as the first connection. Tape the end of the standing part of the rope and the knots with masking tape (or green engineer tape) to secure them in place. Make the connection to the floating safety ring the same as the connection to the donut ring (Figure 3-5). Four ropes can be connected to the floating safety ring using two snaplinks (Figure 3-6).

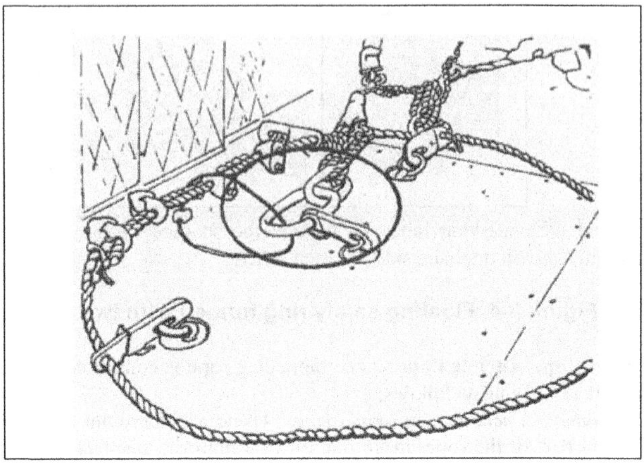

Figure 3-5. Rappel rope connection using two snaplinks for the floating safety ring.

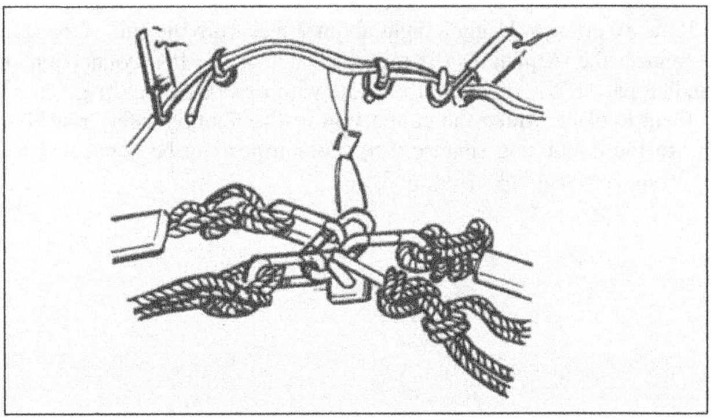

Figure 3-6. Four rappel ropes connected to the floating safety ring (two snaplinks).

3-18. SEATING ARRANGEMENTS AND LOADING TECHNIQUES

Rappelling operations are executed quickly and safely to ensure the success of the operation in a combat or training environment. All rappellers must know the seating arrangements and the procedures for loading (Figure 3-7). They must be rehearsed under the supervision of the rappel master before conducting the operation.

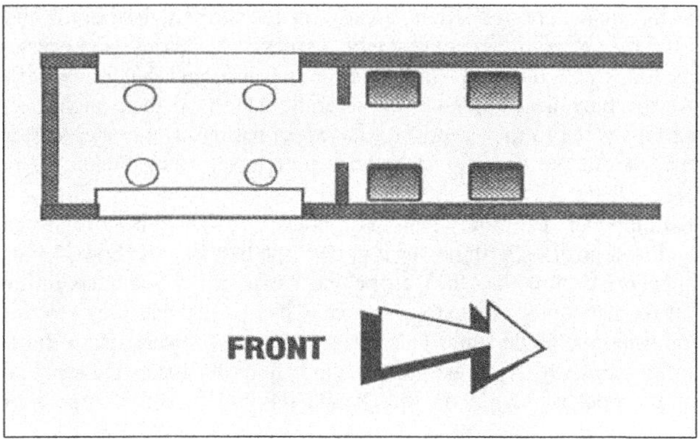

Figure 3-7. Seating arrangement.

a. The first four rappellers are hooked up to the donut ring and use safety belts while in flight. The remaining rappellers use seat belts or safety straps. For safety emphasis during flight, each rappeller's brake hand is in position with no slack in the rope between the brake hand and the donut ring. The rappeller ensures that his rappelling rope is not tangled on any part of the interior of the helicopter or his equipment. He ensures that the coiling log or D-bag is located in such a position that he drops the rope with ease, using his guide hand. The coiling log or D-bag is secured with half of it under the rappeller's inside leg and the other half on top of his outside leg.

b. The rappellers should board the aircraft and sit along the leading edge of each door. They hook up with the rappel seat square knot on the inboard side of the aircraft and the brake hand on the outboard side. The rappel master sits in the center to aid visual inspection of personnel and equipment.

c. For a flight time of less than five minutes and for training and operations where speed is a factor, the rappel master directs the rappellers to be seated in the door facing outward with their feet on the skid. In a tactical situation and before the helicopter comes to a hover, the rappellers assume an upright position facing inside the helicopter.

d. The aircraft is cleared as fast as possible. This reduces the time that the aircraft is in the high-hover attitude to a minimum.

3-19. RAPPELLING PROCEDURES

The rappel master ensures all personnel follow the rappelling procedures for the safe and efficient execution of the operation. These procedures are rehearsed under the supervision of the rappel master.

a. Rappellers (eight maximum) are identified, and rappel masters tell them which door of the aircraft to exit. Rappellers approach the aircraft once it is firmly on the ground, and the crew chief signals when it is safe to approach. The approach to the aircraft should be about 45 degrees to the nose of the aircraft allowing the pilot to see the rappeller.

b. If the rappel master is not already on the aircraft, he boards first and secures himself to the aircraft with a safety harness. He dons a headset to maintain communications with the pilot and positions himself behind the crew chief's chair.

c. As the rappellers approach the aircraft, they hand their anchor attaching points (snaplinks connected to their ropes) to the rappel master. The rappel master connects one set of snaplinks to the floating safety ring and one set of snaplinks (the primary) to the donut ring.

d. All rappellers use safety belts. For safety emphasis during flight, each rappeller's brake hand is in position with no slack in the rope between the brake hand and the donut ring. Rappellers ensure that their rappelling ropes are not tangled on any part of the interior of the helicopter or their equipment. They ensure that they can drop their D-bag or log coil using the guide hand. The D-bag or log coil is placed so it does not fall out of the aircraft. Once over the target area, the rappel master issues the rappel commands and ensures the rappel ropes are on the ground (about 20 feet of rope should be on the ground).

e. On the command GO, the rappeller flexes at the knees and vigorously pushes away from the skid gear, allowing the rope to pass through his brake hand and guide hand. The descent is accomplished smoothly at a rate of about 8 feet per second, avoiding jerky stops. The rappeller initiates his braking action slowly when he is about halfway to the ground. The safety officer or NCO ensures that at least a five-second delay is maintained between each rappeller. Another technique used to deploy rappellers is to issue the command GO to the rappellers diagonally opposite each other. In the early phase of performance, the rappeller's feet are together (particularly if over woods or jungle) and his guide hand is used (not the brake hand) to remove any entanglements.

f. Upon reaching the ground, the rappeller clears the rappel rope through the snaplink (or rappel ring) until the rope is free. If other rappellers are to follow on the same rope, the rappeller on the ground separates or untwists the ropes and becomes the belayer for each subsequent rappeller. If ropes are to be released from the aircraft following the last rappeller, untwisting and separating the ropes is not necessary. A two-second interval is maintained between exit groups (two men exiting at the same time).

g. The rappel master releases the rope from the donut ring after he confirms (by visual inspection) that the rappeller is off rappel. He then drops the rope away from the helicopter.

3-20. RAPPELLING COMMANDS

Helicopter rappelling is conducted in a noise-filled environment. Each rappeller is trained to know the rappel commands and to understand the actions required of him during the execution of each command. Due to the noise created by the helicopter, all rappel commands should be accompanied by arm-and-hand signals. All commands are issued by the rappel master and confirmed orally by the rappellers.

a. The helicopter rappelling commands are as follows:

(1) GET READY. The rappeller checks his combat equipment, looks toward the donut ring, and pulls the rope to check the anchor point connection. He then checks his

snaplink to ensure the rope is properly seated in the snaplink on his rappel seat.

(2) THROW ROPE. The rappeller drops his deployment bag out and away from the helicopter with his guide hand. He ensures that the rope does *not* fall between the side of the helicopter and the skid. The rappel master ensures the rope is touching the ground and is free of tangles and knots. The rappeller then replies, "Rope okay."

(3) SIT IN THE DOOR. The rappeller swings his legs to the outside of the helicopter and takes up a sitting position. (This command applies only to the UH-1H helicopter.)

(4) POSITION. The rappeller pivots 180 degrees on the skid. His feet are shoulder-width apart, the balls of his feet are on the skid, his knees are locked, his body is bent forward in an L-shape position, and his brake hand is on his buttock. The rappel master makes a visual inspection of the snaplink and rappel ring.

(5) GO. The rappel master points at the rappeller to exit. One rappeller from each side rappels at the same time. The rappel master gives the command GO to the rappellers that are diagonal to each other—for example, the front left rappeller and the rear right rappeller. The rappeller flexes his knees and jumps backward, letting the rope run through the brake hand and guide hand.

b. The arm-and-hand signals that are used with rappelling commands are as follows:

(1) GET READY (Figure 3-8). The rappel master extends both arms to the front with fists clenched and thumbs pointing upward.

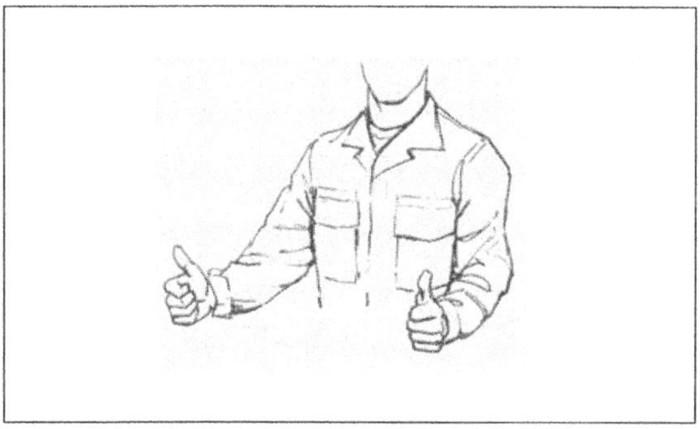

Figure 3-8. Arm-and-hand signal for GET READY.

(2) SIT IN THE DOOR (Figure 3-9). The rappel master extends both arms to the front with elbows locked, fingers extended, thumbs running along the index fingers, and palms facing downward. He bends slightly at the waist so that the arms are below the waist. Then, he moves both arms in a crisscross waving motion, alternating left over right and right over left.

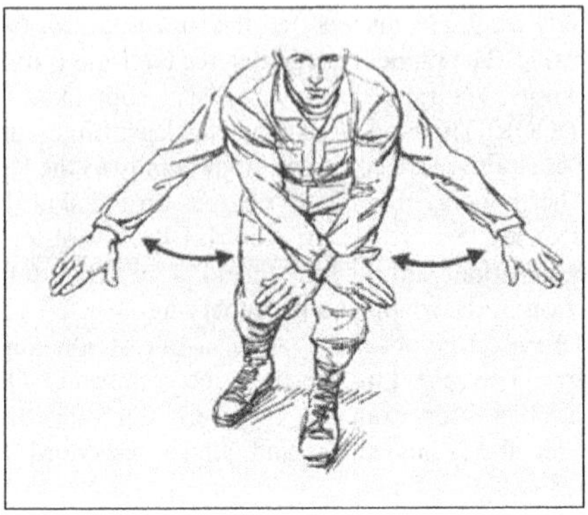

Figure 3-9. Arm-and-hand signal for SIT IN THE DOOR.

(3) THROW ROPES (Figure 3-10). The rappel master extends both arms to the front with elbows locked, fist clenched and index fingers extended. He points at respective rappellers, brings the forearms to an upright position and then down to the elbow-locked position.

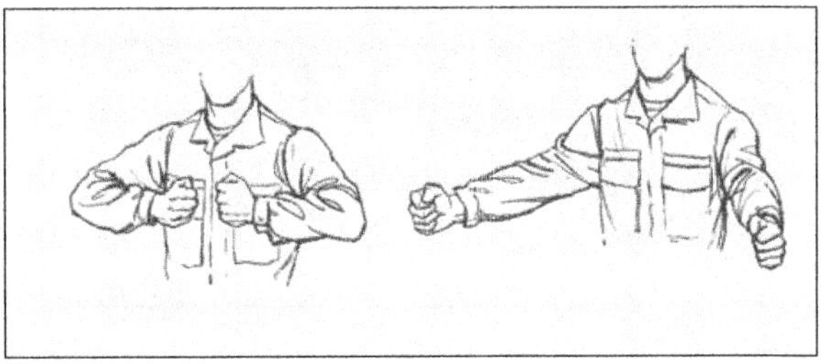

Figure 3-10. Arm-and-hand signal for THROW ROPES.

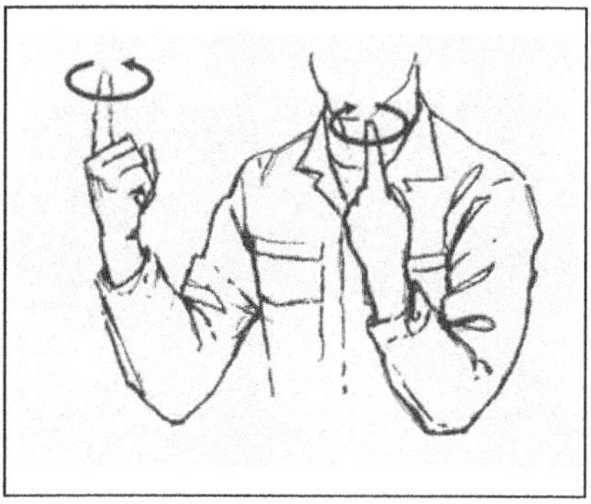

Figure 3-11. Primary arm-and-hand signal for POSITION.

(4) POSITION. The two arm-and-hand signals for POSITION are:

(a) Primary signal (Figure 3-11). The rappel master extends both arms to the front, elbows bent, forearms pointed upward, fists clinched, and index fingers pointing upward. He makes a circular motion with both forearms rotating in opposite directions.

(b) Alternate signal (Figure 3-12). The rappel master extends both arms to the front with elbows locked, fists clenched, and index fingers extended. He bends at the waist so that his arms are below his waist and makes a circular motion with arms rotating in opposite directions.

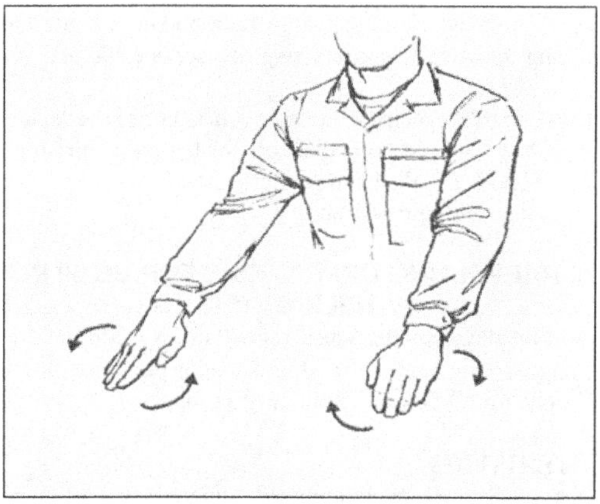

Figure 3-12. Alternate arm-and-hand signal for POSITION.

(5) GO (Figure 3-13). The rappel master extends an arm with elbow locked, fingers extended, and thumb along the index finger, and he points directly at the rappeller.

Figure 3-13. Arm-and-hand signal for GO.

3-21. INSPECTION AND SAFETY CONSIDERATIONS

The rappel master and pilot (or his representative) conduct a joint inspection of the aircraft to ensure the safety of all personnel and serviceability of equipment.

a. Cargo doors are locked in the open position or cleared for closing, depending on the mission.

b. All loose objects in the cargo compartment are removed or secured forward.

c. Sharp edges or protrusions on the cargo floor and door ledges are taped.

d. The donut ring and floating safety ring are serviceable and properly attached to tie-down fittings.

e. A headset and intercom jack for the rappel master are available and operational.

f. Serviceable safety harnesses are available for the rappel master and crew chief.

g. The recovery rope is installed and properly stowed.

h. Unused floor rings are taped down.

Section VI. RAPPELLING OPERATIONS FOR UH-60 BLACKHAWK HELICOPTER

The techniques used by the rappeller when rappelling from different aircraft are similar. However, positioning, seating, and the tie-down anchor point are different. Each rappeller is well trained on each aircraft before conducting a rappel.

3-22. CHARACTERISTICS

The UH-60 is a twin-engine, medium-speed, single-main rotor helicopter that can transport 10 rappellers, 1 rappel master, and 1 four-man crew when the center row of troop seats is removed. It is equipped with four 4,000-pound (load limit) cabin ceiling tie-down fittings or rings that are located in the ceiling of the troop/cargo compartment. These fittings are used as the primary rappelling rope anchoring points. The UH-60 is also equipped with eight 3,500-pound (load limit) cargo restraint net rings. Four of the eight rings are located in the ceiling of the troop/cargo compartment. These rings are used as the secondary rappelling rope anchoring points.

3-23. RIGGING OF UH-60 FOR RAPPELLING

The rappel master rigs the UH-60 helicopter by performing the following:

a. Locks both cargo doors in the open position.

Note: For arctic or other cold-weather operations, or during flights of long duration, the cargo doors are closed and locked until the time specified for opening them.

b. Removes the center row of troop seats.

c. Tapes any sharp edges or protrusions on the cargo floor and door ledges that may come in contact with the rappeller or the rappelling rope.

d. Stows loose equipment forward in the cargo compartment.

e. Extends the rappel master's intercom cord to the rear over the aft utility drain line and tapes the cord to the overhead troop seat support tube.

f. Installs the floor restraint provisions for rappellers No. 1 through 6.

g. Rigs and connects rappelling ropes to the aircraft's primary and secondary anchoring points.

(1) *Primary anchor points*. Ties a bowline with a half hitch about 4 feet from the standing end of the rope. Attaches the two primary snaplinks to the respective cabin tie-down fitting ring with gates facing in the opposite directions (Figures 3-14 and 3-16)

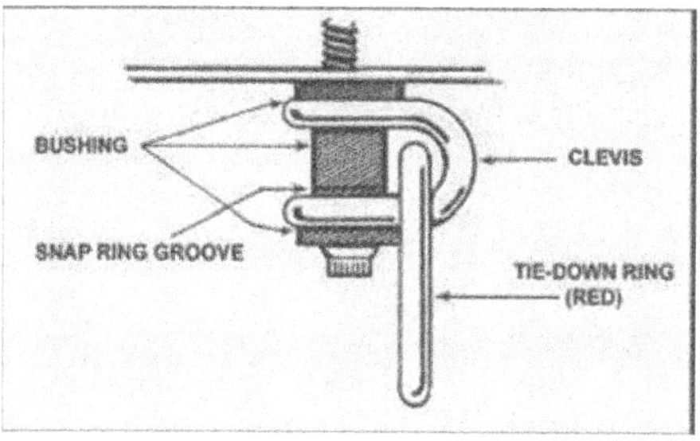

Figure 3-14. Primary snaplink attaching point.

(2) *Secondary anchor points*. Ties a bowline with a half hitch about 1 1/2 feet from the standing end of the rope. Attaches the secondary snaplink to the adjacent overhead cargo restraint net ring (Figures 3-15 and 3-16).

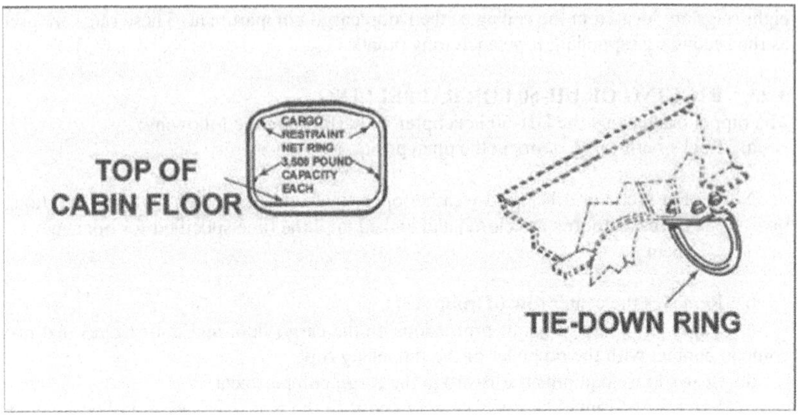

Figure 3-15. Secondary snaplink attaching point.

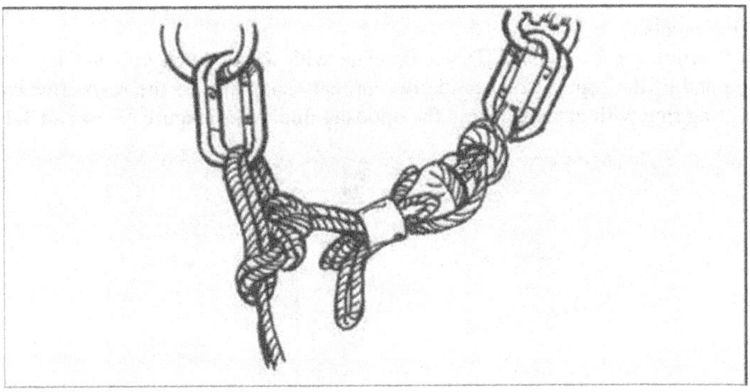

Figure 3-16. Primary and secondary snaplink attaching points.

h. Removes and secures the cargo hook access door and deploys the cargo hook in the DOWN position.

i. Installs the recovery rope for endangered rappellers.

j. Tapes the unused floor rings.

Note: There must be a minimum of 22 inches and a maximum of 30 inches between the anchor knots.

3-24. SEATING ARRANGEMENTS AND LOADING TECHNIQUES

Rappelling operations are executed quickly and safely to ensure the success of the operation in a combat or training environment. All rappellers must know the seating arrangements and the procedures for loading. They must be rehearsed under the supervision of the rappel master before conducting the operation.

a. A maximum of 10 rappellers, with and without combat equipment, and 1 static rappel master is seated and restrained aboard the UH-60 helicopter when the center row of troop seats are removed (Figure 3-17).

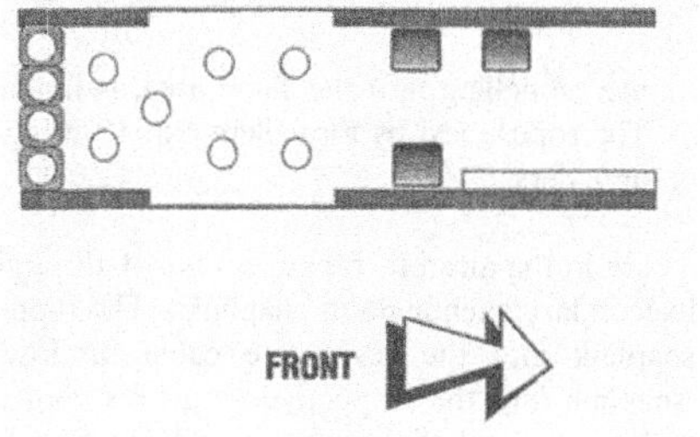

Figure 3-17. Rappelling personnel seating arrangement.

(1) Rappellers No. 1 through 6 and the rappel master are seated on the cargo floor. The rappellers are restrained using 3 1/2 feet of 1-inch tubular-nylon webbing (or equivalent) that is run through the cargo tie-down rings attached to the floor of the aircraft. The webbing is attached to the rappeller by means of a snaplink that is hooked to a Swiss seat. The rappel master is secured by his safety harness.

(2) Rappellers No. 7 through 10 are seated on the troop seats across the aft end of the cargo compartment. They are secured by the seat belts.

b. Rappellers are organized into 10-man sticks. They approach the aircraft from the left or right side in reverse order. They are divided into two groups and approach and enter the aircraft at the same time, depending on the situation. Rappellers No. 9, 7, 5, 3, and 1 enter the aircraft through the left door; they are seated and restrained. Rappellers No. 10, 8, 6, 4, and 2 enter the aircraft through the right door; they are seated and restrained. The static rappel master is restrained in the center of the cargo compartment using a C-3A troop-type safety belt or a safety harness during take-off and landing.

DANGER

BECAUSE THE LOWEST ARC OF THE ROTOR BLADE OCCURS AT THE DIRECT FRONT OF THE AIRCRAFT, APPROACHING THE AIRCRAFT AT THIS POINT COULD RESULT IN PERSONNEL INJURY OR DEATH.

3-25. RAPPELLING PROCEDURES

The rappel master ensures all personnel follow the rappelling procedures for the safe and efficient execution of the operation. These procedures are rehearsed under the supervision of the rappel master.

a. The rappel master is equipped with a headset and maintains communications with the pilot at all times. He wears a safety harness and stations himself in the center of the cargo floor to maintain control of the rappellers, their ropes, and all anchor attaching points.

Note: If all rappellers are rappelling into the same area, individual rappel ropes are not needed. The ropes used by rappellers No. 1 through 4 are used by the succeeding rappellers.

b. After all rappellers are in the aircraft, rappellers No. 1 through 4 hand the rappel master their primary and secondary anchor point snaplinks. The rappel master hooks the primary anchor point snaplink into the respective cabin tie-down fittings and the secondary anchor point snaplink into the respective cargo restraint net rings. After the primary and secondary anchor point snaplinks are secured, the first four rappellers hook onto their respective rappel rope. The rappel master checks the first four rappellers for the correct hookup.

c. After the first four rappellers exit and clear their ropes, the rappel master hands the next four rappellers their rappel ropes to hook up. Left-handed rappellers take extra precaution to ensure the correct hookup. Problems can be avoided by placing left-handed rappellers in the positions where they initially have the rope to their nonbrake side.

WARNING

Extreme care must be taken in hooking up and positioning personnel that require repositioning of their rope to their proper brake-hand side. This could result in bodily injury or damage to Army property.

d. After checking the rappellers for the correct hookup, the rappel master moves the next four rappellers into position on the floor of the aircraft. After the four rappellers exit and clear their ropes, the rappel master hands the last two rappellers their rappel ropes to hook up.

e. After checking the rappellers for correct hookup, the rappel master moves them into position on the floor of the aircraft. After the last rappeller is off the rope, the rappel master releases the ropes.

f. If the rappel master rappels, the crew chief is instructed to release or retrieve the ropes.

3-26. RAPPELLING COMMANDS

The UH-60 rappelling is conducted similar to the UH-1H rappelling with the exception that no skids are on the UH-60 on which to stand. Therefore, the edge of the floor, along the door of the helicopter, is used as a pivot point to assume the L-shaped rappelling position.

a. GET READY. This alerts the rappeller. The rappeller and rappel master perform final checks of the hookup, rappel seat, snaplink, and equipment.

b. THROW ROPES. The rappeller looks down below the aircraft to ensure that no one is under the aircraft. Keeping his brake hand in the small of his back, he tosses the deployment bag with the ropes *out and away* from the helicopter with his guide hand. The rappeller observes that the ropes are touching the ground and are not knotted or entangled.

c. POSITION. From a kneeling position with the brake hand in the small of the back, the rappeller rotates 90 degrees so that he is facing the inside of the aircraft and the rappel master. The rappeller then places his heels on the edge of the floor of the helicopter doorway and leans out into an L-shaped position. His feet are shoulder-width apart, the balls of his feet are on the edge of the helicopter doorway, his knees are locked, and his body is bent at the waist toward the helicopter.

d. GO. The rappel master initiates the rappel with this command. The rappeller flexes his knees and jumps backwards. At the same time, the rappeller throws his brake hand out at a 45-degree angle, letting the running ends of the ropes slide through both the brake hand and guide hand.

3-27. INSPECTION AND SAFETY CONSIDERATIONS

The rappel master and pilot (or his representative) conduct a joint inspection of the aircraft to ensure the safety of all personnel and serviceability of equipment.

a. Cargo doors are locked in the open position or cleared for closing, depending on the mission.

b. All loose objects in the cargo compartment are removed or secured forward.

c. Sharp edges or protrusions on the cargo floor and door ledges that may come in contact with the rappeller or his rappelling rope are taped.

Note: Do not tape the door latches or handles; this can interfere with door operations.

d. Primary and secondary rappelling anchor points are serviceable and securely attached to the aircraft structure.

e. A headset/helmet and intercom jack for the rappel master are available and operational, and the intercom extension cord is secured overhead.

f. Serviceable safety harnesses are available for the rappel master and crew chief.

g. The cabin ceiling tie-downs must have safety wire installed to ensure they do not come undone or unraveled, and the bolt head must be stamped H.

Section VII. RAPPELLING OPERATIONS FOR MH-53 HELICOPTER

The MH-53 helicopter is a highly versatile aircraft. It is used for a variety of missions, usually in conjunction with special operations.

3-28. CHARACTERISTICS

The MH-53 is a two-engine, single-rotor, heavy-lift helicopter. It has a crew of six and can be refueled while in flight. It has a precision navigational and communications package, which makes it excellent for conducting deep infiltration and exfiltration missions. The MH-53 helicopter can fly at night in all weather and terrain conditions. It can follow down to 100 feet and operate from unprepared sites. It is fitted with three gun stations for a mix of 7.62-mm miniguns and caliber .50 machine guns.

3-29. RIGGING OF MH-53 HELICOPTER FOR RAPPELLING

When rappelling from the crew entrance door, rappellers connect the ropes with locking snaplinks to the two 10,000-pound tie-down rings under the left scanner's window. When rappelling from the ramp, rappellers connect the ropes with locking snaplinks to the two 10,000-pound tie-down fittings on the left and right sides of the aircraft at station 522. Any sharp edges that could damage the ropes should be padded or taped. A length of fire hose may also be used over the portion of the rope that comes in contacts with the door or ramp edge.

3-30. SEATING ARRANGEMENTS AND LOADING TECHNIQUES

The MH-53 is a large, versatile helicopter. The seating arrangements and loading techniques are numerous. The using unit develops a seating and loading SOP that ensures the safety of all personnel and permits smooth, efficient execution of the mission.

3-31. RAPPELLING PROCEDURES

The rappel master or his assistant are at the two exit points (ramp and door). The rappel master maintains communications with the helicopter commander and relays all commands and time warnings. The commander issues time warnings at 20-minute, 10-minute, 5-minute, and 1-minute intervals. During limited visibility, the rappel master may use NVDs to observe the safety/belayer. Chemical lights are attached to the top or bottom of the rappel rope. Arm-and-hand signals should be IAW The Air Assault School Handbook and FC 57-2.

Note: Night vision goggles cause limited depth perception and a tunnel vision effect.

a. The rappel master is secured by a harness and ensures the proper hook up of each rappeller. Once hooked up, the rappellers release their safety belts. At the command of the rappel master, the rappellers position themselves to ease immediate deployment.

WARNING

Before conducting deployments off the ramp, the deploying team members must be briefed on the importance of maintaining separation between members. About 24 to 27 inches are desired. This separation helps maintain aircraft center of gravity limits.

b. The rappel master deploys ropes only after the helicopter is in a stable hover over the target area and he has given the command ROPES. He ensures the ropes are on the ground before giving the command GO. After the last rappeller is off the rope, the rappel master retrieves or cuts the ropes away.

3-32. RAPPELLING COMMANDS

The helicopter rappelling commands for the MH-53 are as follows:

a. GET READY. This command is given as the helicopter approaches the rappel site. It alerts the rappellers of the approach to the site and signals them for a final equipment check.

b. POSITION. This command clears the first rappellers into position for deployment.

c. THROW ROPE. This command is given once the helicopter reaches a stable hover.

d. GO. The rappellers exit.

3-33. INSPECTION AND SAFETY CONSIDERATIONS

The rappel master briefs all personnel participating in the operation. He conducts hands-on inspection to ensure the safety of all personnel and serviceability of equipment.

a. Tie-down fittings are serviceable.

b. All sharp or protruding edges that may come in contact with the rappelling ropes are padded or taped.

c. All ropes are retrieved or cut away before forward movement of the helicopter.

d. Only three rappellers are deployed at a time—two from the ramp and one from the personnel door.

CHAPTER 4

SPECIAL PATROL INFILTRATION/EXFILTRATION SYSTEM

The special patrol infiltration/exfiltration system (SPIES) provides an excellent form of exfiltration over short distances. It is not recommended for infiltration because team members are exposed the entire time. Due to the nature of SPIES operations, a thorough briefing is required for all participants before the operation. Personnel being extracted must train extensively in the SPIES extraction before infiltration. For the other personnel involved, a complete preoperations briefing is held before the operation. This is crucial when additional assets are involved other than the extraction helicopter such as gunships, aerial observers, or artillery support.

Section I. SPIES MASTER DUTIES AND QUALIFICATIONS

The SPIES master holds the rank of sergeant or above (may be waived by the commanding officer). The commanding officer ensures that this qualification is entered on the soldier's training record.

4-1. QUALIFICATIONS

To qualify to be a SPIES master, the sergeant must participate in at least three SPIES operations (observe twice and execute SPIES master duties once under observation). For example, he hooks up the helicopter, assists in the preparation of an operation, and conducts a successful operation under the supervision of a qualified SPIES master. A SPIES master is conscientious and thorough, and he knows all aspects of a SPIES operation. He can give an effective pilot's brief, use the aircraft communications equipment, and understand aviation terminology.

4-2. DUTIES

The SPIES master is responsible for the safe conduct of the SPIES operation.

a. **Preflight Duties.**

(1) Inventories and inspects all SPIES equipment.

(2) Briefs the pilot and other concerned personnel about details of the operation, especially the extraction and dismounting procedures.

(3) Ensures that an ICS helmet and gunner's belt are available for the SPIES master's use. Uses a sling rope if no belt is available. Connects and checks the operation of the ICS to be used. (ICS communications is established between the SPIES master and pilots on all SPIES operations.)

(4) Attaches the SPIES rope to the helicopter IAW the guidance in this chapter.

(5) Ensures that nothing is adrift in the aircraft that may fall on a team member.

(6) Checks the location of the emergency axe. Places the axe where readily available, yet secure enough so as not to endanger the men on the SPIES rope. (Inspects the axe to ensure that it is sharp.)

b. **Extraction Duties.**

(1) On arrival at the team's estimated position, assists the pilot in determining the exact location of the team members.

(2) As the aircraft approaches the team's location, aids the pilot (using the clock system) in placing the aircraft directly above the team.

(3) Requests permission from the pilot to drop the SPIES rope when the aircraft is hovering above the team.

(4) Drops the rope, taking care to avoid striking team members on the ground.

(5) Notifies the pilot when the rope is down, and reports all altitude corrections to ensure that team members reach all SPIES attachment points.

(6) Watches for the "thumbs-up" signal from all team members.

(7) On receipt of the "thumbs-up" signal, advises the pilot that the team is ready for extraction and requests a vertical lift-off.

(8) Advises the pilot of the team's position, the location of any potential obstacles, and the avoidance of horizontal movement.

(9) If a team member becomes entangled with an obstacle during the extraction, notifies the pilot and requests that the vertical lift be stopped. If the situation is critical, prepares to cut the SPIES rope (the anchor point or cargo straps) after the team members are secured to the obstacle or on the ground.

(10) When positive that all obstructions are clear, advises the pilot to obtain a safe altitude (about 350 feet AGL for training purposes or as the situation dictates in combat) or to transition into forward flight.

(11) At frequent intervals during the flight, advises the pilot on the safety status of all team members. Maintains a constant visual watch on the team and checks security of the SPIES attachments often.

c. **Dismounting Duties.**

(1) On arrival at the dismounting area, informs the pilot as to the approximate distance of the lower rope end from the ground.

(2) Once the pilot starts the vertical descent, continually informs him as to the approximate distance of the lower rope end from the ground.

(3) Informs the pilot of any horizontal drift that occurs and any obstructions near the SPIES rope. Also informs the pilot of any oscillation that may occur.

(4) Informs the pilot when the rope is about 25 feet above the ground and again when it is 10 feet above the ground. Ensures that the rate of descent is slow enough to enable the team members to land and get out from under team members safely.

(5) Reports when the first man initially touches down, when the last team member starts to safely move away from under the helicopter, and when all team members are disconnected.

(6) On order of the pilot, either retrieves the SPIES rope into the helicopter or disconnects the SPIES rope and drops it to the ground. When using the UH-1H/UH-60 helicopter, the only way to retrieve the SPIES rope while in the air is by having an arranged recovery rope attached with a 16-foot sling rope. In some cases, the SPIES master joins two 12-foot-long sling ropes to haul the SPIES rope aboard and attaches the rope about 5 or 6 feet below the cargo hook or cargo strap hookup point. The type of knot used to connect the sling (or recovery) rope to the SPIES rope is self-tightening in nature (for example, the prusik knot). The SPIES master fastens the standing end of the sling rope to the deck tie-down or uses a snaplink. Although it is important to keep the line out of the way, the primary consideration is its length. The rope must be long enough for any oscillation in the SPIES during flight.

Section II. PREOPERATIONS BRIEFINGS AND PROCEDURES

Before conducting a SPIES training mission, the participants must have a basic understanding of the requirements. The SPIES master conducts briefings to ensure the soldier extracted and the pilot know the procedures. As SOPs are developed and units train together, the SPIES master simply refers to the SOP. He always gives a safety briefing and conducts an equipment inspection.

Note: SPIES has not been approved for Army-wide use. Approval must be obtained from division level or higher commander before the conduct of any training.

4-3. SAFETY

As in all training, a safety briefing must precede operations using the SPIES. The briefing should consist of, but not be restricted to, a review of the following:

- Characteristics of equipment associated with the SPIES.
- Equipment inspection.
- Proper donning of the harness.
- Methods of extraction and insertion used.
- Emergency signals.

4-4. COMMUNICATIONS

Due to the noise level associated in all helicopter operations, radios are used for communication. They are also used before the arrival of the helicopter. Precise arm-and-hand signals are established in case of radio failure or poor communications due to static or noise overriding the audio output of the radio. During the first part of the operations, the SPIES master either observes (daytime) or knows that a definite procedure is taking place (night or jungle) while the teams are hooking up to the SPIES rope for extraction. Headsets and voice suppressors replace the handset for better radio procedures. This allows the RATELO on the ground to use both hands while the helicopter hovers for a faster and safer hookup. If radio communications are hampered, special procedures are used along with hand or light signals.

4-5. EXTRACTION PROCEDURES

After the team has been located, the SPIES master assists the pilot in directing the helicopter to the proper distance over the team.

a. The team leader positions himself so that he can move and approach the rope as it is dropped by the SPIES master. Once the rope is clear of any obstacles, the team leader signals the team to their assigned positions along the 10 hookup points. Using the primary or harness snaplink, each team member hooks to the D-ring on his side of the line. This is the primary hookup. Once the team is hooked up, the team leader hooks into the alternate or second hookup point using the safety line and snaplink. He then faces forward along the line so that he is heading in the direction he is traveling when the aircraft starts its ascent.

b. The SPIES rope is held up and routed over the shoulder closest to the rope. With the other hand, the team members give thumbs-up signals to allow both the team leader and the SPIES master to see that they are ready to go. The team leader physically inspects (if time and situation permits) or hooks himself in on the lowest point along with the RATELO to ensure the running end is clear of all obstacles. He gives the thumbs-up signal to the SPIES master. This thumbs-up signal (at night, an arranged light signal) continues until the team is lifted off the extraction surface. The team leader also signals (arms held straight to the sides) when the team is at safe altitude (approximate 10 feet above the tallest obstacle at the extraction site). The helicopter starts a transition in a horizontal direction on its return flight.

4-6. EMERGENCY PROCEDURES

During the flight, from extraction until the team is safely and quickly detached from the SPIES rope, each team member should be aware of any problems that arise from above or below. The man above checks the man below. At the first sign of danger or if an emergency arises, the team leader or a team member places his free hand on his head. Upon observing anyone on the SPIES rope with his hand on his head, the SPIES master tells the pilot to make an emergency landing in the nearest and safest area.

4-7. DISMOUNTING PROCEDURES

During the familiarization training phase, the SPIES master ensures all members know they should immediately head in the direction of the nose (12 o'clock) of the aircraft. This allows the pilot to see that the team is out from under the aircraft. When an emergency situation arises, the pilot can make a better appraisal of the situation if he can see all the members of the team. When the helicopter is making a scheduled landing, the team ensures that the SPIES rope does not interfere with the aircraft and that the aircraft does not land on the rope.

4-8. INSPECTION OF EQUIPMENT

The SPIES is inspected by a certified SPIES master at six-month intervals and whenever the SPIES master questions serviceability. Outdated, spliced, abraded, or cut rope is removed from service. However, loose or broken stitching in excess of three stitches is not repaired.

CAUTION

Acid contamination, cuts, or fraying of harness or sling webbing constitute nonrepairable damage.

a. The certified SPIES master inspects the harness and suspension sling webbing for signs of contamination from oil, grease, acid, or rust. He also checks for cuts, twists, fading, excessive wear, fusing (indicated by unusual hardening or softening of webbing fibers), fraying, burns, abrasions, and loose or broken stitching. The damaged harness or suspension sling is removed from service and returned to supply for appropriate disposition.

b. The SPIES master inspects all hardware for signs of corrosion, pitting, ease of operation, security of attachment, bends, dents, nicks, burrs, and sharp edges.

Replacement of hardware (except chest strap adapter) that requires unstitching of webbing renders the harness unserviceable.

c. The SPIES master replaces the V-ring by cutting the strap above the stitching. He folds and stitches a new end section of the leg strap. If the leg strap is damaged, the SPIES master returns the harness or suspension sling to supply for appropriate disposition.

d. The SPIES master checks ropes, harnesses, and suspension slings for expiration of service or total life, which is 7 years of service (opening manufacturer's package) or 15 years from date of manufacture, whichever occurs first. The SPIES master also ensures the rope is free of splices. He ensures the eye loop at the end of the SPIES rope is not broken, frayed, or loose.

e. The SPIES master inspects the rope surface for cuts, excessive abrasions, and snags. Cuts on the rope are excessive when there are four or more cut strands in any 5-inch length. The two-to-one braided rope has 12 pairs of strands (24 individual strands) around the circumference. Abrasion is extensive when torn yarns are equivalent to that of four strands of any 5-inch length. Rope that has been subjected to heavy loads may display glazed areas where it has worked against hard surfaces. This condition may be caused by paint or fused fibers. After long use, the rope may become fuzzy on the surface (although this should be minimized with the surface coating). However, the effect on the rope's strength is negligible.

f. The SPIES master inspects the rope for signs of contamination by acid, alkaline compounds, salt water, fire extinguishing solutions, and petroleum-based solvents. Although the ropes gradually change color, such changes do not indicate decrease in strength unless the change is due to contact with strong chemicals. Changes in color caused by chemicals are usually spotty; changes caused by use are uniform throughout the length of the rope.

4-9. RIGGING OF UH-1H HELICOPTER FOR SPIES OPERATION

Personnel who are not familiar with SPIES are encouraged to observe or take part in the rigging of the helicopter. This builds confidence in the equipment and assists in a more comprehensive training of new SPIES masters.

a. **Equipment.**

- One 120-foot SPIES rope with deployment bag.
- Two 11-foot, 3- or 4-loop cargo slings or two 9-foot, 3- or 4-loop cargo slings.
- Two Type IV connector links (four if aircraft does not have cargo hooks).
- Heavy duty tape (100-mph tape).
- One 12-foot sling rope.
- Five oval snaplinks (nine if the aircraft does not have cargo hooks).

NOTE: If the aircraft does not have cargo hooks, use four 11-foot or four 9-foot, 3- or 4-loop cargo slings.

b. **Installation.**

(1) The primary attachment point for the SPIES rope is the cargo hook. The end of the SPIES rope having a polyurethane encapsulated eye is attached to the cargo hook.

The two 9- or 11-foot long cargo suspension slings are joined to form one continuous sling, using a Type IV link. The sling is stretched out on the helicopter deck. One end is taken under the helicopter and through the eye of the SPIES rope. It is then connected on the other end of the sling using a Type IV link assembly. The sling must pass between the helicopter skids and the fuselage.

(2) Once the the SPIES rope and cargo straps are in place, the straps running across the deck of the helicopter are secured in place by at least four and as many as eight snaplinks. The snaplinks are evenly spaced across the deck and alternated from one side of the strap to the other and top and bottom, so that the first snaplink is to the rear of the strap and going around the bottom two straps. The next snaplink is in the front of the cargo strap and around the top two sections of the strap. This process continues until at least four points are established.

(3) If eight snaplinks are available, then each tie-down has two reversed. If no hook is available or is not working properly, the SPIES can be used safely by doubling the cargo slings and Type IV links. Thus, two cargo straps are side by side for a total of four slings and four Type IV links. Padding is used around the edge of the fuselage to protect the sling from damage.

(4) Care must be taken when using the UH-1H helicopter because of the ways in which it can be outfitted. Some helicopters have a step attached at the doorway. This is an added obstruction not only during installation but also during the operation. Others may have rocket pods or machine guns mounted. Not all of the UH-1 helicopters are hooked up the same way each time.

4-10. RIGGING OF UH-60 HELICOPTER FOR SPIES OPERATION

The SPIES master inspects all materials required to rig the helicopter.

a. **Equipment.**

- One 120-foot SPIES rope with deployment bag.
- Two 11-foot, 3- or 4-loop cargo slings or two 9-foot, 3- or 4-loop cargo slings.
- Two Type IV connector links (four if aircraft does not have cargo hooks).
- Heavy duty tape (100-mph tape).
- One 12-foot sling rope.
- Five oval snaplinks (nine if the aircraft does not have cargo hooks).

Note: If the aircraft does not have cargo hooks, use four 11-foot or four 9-foot, 3- or 4-loop cargo slings.

b. **Installation.**

(1) The primary attachment point for the SPIES rope is the cargo hook. The end of the SPIES rope has a polyurethane encapsulated eye that is attached to the cargo hook. The two 9- or 11-foot cargo suspension slings are joined to form one continuous sling using a Type IV link. The sling is stretched out on the helicopter deck. One end is taken under the helicopter and through the eye of the SPIES rope. It is then connected on the other end of the sling using a Type IV link assembly. The sling passes between the helicopter skids and the fuselage. Padding is used around the edge of the cargo hatch to protect the sling from damage.

(2) Once the SPIES rope and cargo straps are in place, the straps running across the deck of the helicopter are secured in place by at least four and as many as eight snaplinks. The snaplinks are spaced evenly across the deck and alternated from one side of the strap to the other and top and bottom. Thus, the first snaplink can be to the rear of the strap, wrapping around the bottom two straps. The next snaplink is in the front of the cargo strap and around the top two sections of the strap. This process continues until at least four points are established.

(3) If eight snaplinks are available, each tie-down has two snaplinks connecting the same spot, and the swing gates are reversed. If no hook is available or is not working properly, the SPIES can be used safely by doubling the cargo slings and Type IV links. Two cargo straps are side by side with a total of four slings and four Type IV links.

4-11. RIGGING OF CH-46/CH-47 HELICOPTER FOR SPIES OPERATION

The SPIES master inspects all the materials required to rig the helicopter.

a. **Equipment.**

- Two 11-foot, 3-loop slings.
- Two 9-foot, 3-loop cargo slings.
- Four Type IV connectors.
- One 13-foot sling rope.

b. **Installation.**

(1) The SPIES rope is attached using two 9- or 11-foot cargo suspension slings and four Type IV links. The cargo slings are passed through the encapsulated eye of the SPIES rope and attached to the outboard cargo tie-down rings on the aircraft floor. Two tie-down rings are used for each sling. Padding is used around the edge of the cargo hatch to protect slings from damage.

(2) Not all of the tie-down rings will be in the exact same position on the helicopters. This is one of the main considerations in deviating from the installation procedures. When possible, the cargo straps are placed to form two U shapes. One strap is placed forward of the cargo hole in the center of the aircraft floor and one aft or toward the rear of the helicopter. Installed properly, the cargo straps hold the SPIES rope comfortably in the center of and slightly below the opening of the cargo hatch.

(3) The use of snaplinks attached close to all four faulty tie-down points not only ensures a backup in case of a faulty tie-down ring but also reduces the amount of movement in the cargo suspension straps. A total of eight snaplinks are used for added security at each point with the swing gates reversed.

Section III. LAND AND WATER EXTRACTION

If the situation, mission, or terrain suggests the possibility of a SPIES extraction, the leader includes the SPIES harness in each individual's equipment list. If the mission or insertion precludes the wearing of the harness, it will be carried on the top and just inside the flap of the pack being used. Once the extraction helicopter has been requested, the harness is retrieved and donned.

4-12. LAND EXTRACTION PROCEDURES

The SPIES is used only when the team requires immediate extraction or is unable to move to a clear (open) position suitable for helicopter landing.

a. The extraction helicopter proceeds to the area and establishes radio or visual contact with the team. The backup helicopter equipped with the SPIES remains aloft and away from the area. The helicopter maintains visual contact with the LZ and monitors radio communications.

b. After the team has indicated readiness for pickup and the tactical situation has stabilized, the extraction helicopter moves to the LZ by the safest route. When the helicopter is above the team's location, the SPIES master drops the SPIES rope on order of the pilot after the aircraft has obtained a stable hover at slightly above treetop height.

c. The team hooks up the same as in the familiarization procedures. Team members sling individual weapons over the shoulders, securing weapons and equipment to withstand the wind. A safety line is attached to prevent lost weapons during SPIES operations. When the team is ready, the team leader gives a thumbs-up signal.

d. During extraction, the team RATELO maintains communications with the extraction helicopter. He gives a verbal backup to the thumbs-up signal and relays all information during the flight. His location is near or at the bottom hookup point so he can assist in giving accurate information about the extraction, clearing of obstacles, and descent.

e. Lift-off of the extraction helicopter is vertical until the SPIES rope has cleared all obstacles. With the barrel directed downward at a 45-degree angle and outward, the team members fire their individual weapons using the hip position.

f. Once the SPIES rope is clear of all obstacles, the extraction helicopter changes to horizontal flight and departs the area by the safest route. Airspeed is limited to 70 knots in moderate climates and 50 knots in cold climates while team members are attached to the SPIES rope.

g. When the extraction helicopter has reached a safe dismount area, a transaction is made into hover flight at an altitude of 250 feet AGL. A vertical descent is started with the SPIES master continuously providing information to the pilot on the distance from the ground to the lower end of the SPIES rope. The vertical descent rate of the aircraft (at touchdown) is less than 5 feet per second.

h. When the team reaches the ground, they immediately move out toward the front of the helicopter. All team members rapidly unhook themselves and teammates that need assistance. Once this is completed, they either move away from the area and set up security or assist in clearing the rope if the helicopter is going to land.

4-13. WATER EXTRACTION PROCEDURES

The SPIES is suitable for extracting teams from the water. For this procedure, three inflatable life vests or any type of floatation device is tied to the SPIES rope to provide buoyancy for the rope while in the water. One floatation device is tied at each end of the attachment points, and one floatation device is tied in the middle of the attachment point area just above the middle two sets of D-rings. The team members to be extracted wear their SPIES harnesses under their individual life vest. They can also wear swimming fins, masks, and snorkels (amphibious operations) to ease hooking up to the SPIES rope within the spray area beneath the hovering helicopter.

a. After the extraction helicopter has attained a stable hover above the team member's location, the SPIES master drops the SPIES rope (with floatation attached) on order from the pilot. When the team members complete the hookup to the SPIES rope, the team leader signals the SPIES master to start lift-off. Lift-off of the extraction helicopter is vertical until all team members and the bottom end of the rope have cleared the water. During the initial lift-off, team members know that they are to be dragged through the water. The team members prepare to roll on their backs until clear of the water.

b. Flight speed and altitude are the same as over land. The dismounting procedures remain the same, except when landing on a ship. Once on board, the team members take their orders from personnel in charge of the deck.

Section IV. AFTER-OPERATIONS PROCEDURES

After-operations procedures include repairing, cleaning, and proper storage of the rappelling equipment.

4-14. REPAIRING AND CLEANING OF EQUIPMENT

The contaminated ropes are washed with a mild detergent, such as liquid dish soap, and cold water followed by a rinse in clean, fresh water. Ropes are dried at a temperature not to exceed 140 degrees Fahrenheit. Stubborn oil, grease, hydraulic fluid, and other petroleum stains can be removed with the cleaning agent xylene (Grade A or B, TT-X 916).

4-15. STORAGE OF EQUIPMENT

To avoid ultraviolet deterioration, the nylon materials should be protected from direct sunlight. The SPIES rope is stowed in an aviator's kit bag for protection when not in use. Bins or similar facilities are used for storage of SPIES equipment. Shelves used for storage should be at least 4 inches from the walls and 12 inches from the floor. Areas used for storage should be well ventilated and free of oil, acid, cleaning compounds, and other contaminants. Equipment must *not* be stowed above or near hot water pipes or heating apparatus.

CHAPTER 5

FAST-ROPE INSERTION AND EXTRACTION SYSTEM

The fast-rope insertion and extraction system (FRIES) is employed by small units to make rapid infiltration and exfiltration using rotary-wing aircraft in confined areas. The use of FRIES is normally restricted to situations where aircraft cannot land, and when mission requirements preclude the need for large amounts of equipment, and ammunition or heavy crew-served weapons.

Note: Headquarters, Department of the Army (HQDA) policy specifies that FRIES is not approved for Army-wide use and names the Commanding General, United States Army Special Operations Command (CG USASOC) as the executive agent for FRIES doctrine. The use of FRIES is restricted to special operations forces, pathfinders, long-range surveillance units, and HQDA-approved schools with a USASOC-approved FRIES program of instruction. Requests for other units to conduct FRIES operations should be submitted to Commander, USASOC, ATTN: AOOP-TRS, Fort Bragg, NC 28307-5000.

Section I. GENERAL

FRIES information is subject to change. Units that recognize the need to update this chapter should submit recommendations to Commander, USASOC, ATTN: AOOP-TRS, Fort Bragg, NC 28307-5000.

5-1. OBJECTIVES

The objectives of this chapter are to—

- Prescribe the requirements for FRIES qualification and training, including proficiency sustainment, safety, and administrative procedures.
- Establish the scope of the duties and responsibilities of key personnel who conduct FRIES operations.

5-2. GUIDANCE FOR COMMANDERS

Units that have HQDA approval to perform FRIES operations are authorized to conduct initial FRIES qualification and FRIES master (FRM) qualification training. Training requirements are established in this publication, USASOC Regulation 350-6, and applicable SOAR policies. Aircraft material requirements are specified by the US Army Technology Applications Program Office (TAPO) .

a. FRIES training and operations are inherently risky; therefore, safety is paramount. All training and operations require in-depth attention to detail from planning through preparation and execution.

b. The following basic guidance applies to FRIES training and operations:

(1) Unit commanders must personally approve FRIES training sites.

(2) A risk analysis and or assessment must be completed before FRIES operations.

(3) Night FRIES operations are medium risk or higher.

(4) Units authorized to conduct FRIES operations may incorporate FRIES activities in field training exercises only after personnel have become FRIES qualified.

(5) Field exercises using FRIES activities are subject to the limitations imposed by applicable directives, including mandatory support requirements.

(6) Requests for waiver of procedures or restrictions prescribed herein must be submitted in writing with endorsement of the first general officer in the requesting unit=s chain of command to HQ USASOC.

5-3. TRAINING PREREQUISITES

> **CAUTION**
>
> **Trainers must ensure all personnel preparing to participate in FRIES training are tested *with equipment* to ensure they have sufficient upper body strength to safely perform the full scope of roping duties.**

Before allowing soldiers to participate in FRIES training and operations, the unit commander must make sure the soldiers are physically fit. Minimum screening standards that apply to each training participant include—

- Having passed the Army physical fitness test within the past six months.
- Having passed a medical examination within the past two years.
- Being free of any injury or physical condition that could cause a potential safety hazard during FRIES training.
- Demonstration of ability to perform a controlled descent from a height of 15 feet while carrying a 40-pound load (not including the basic duty uniform and combat boots).
- Demonstration of ability to hold a static position on a FRIES rope for 20 seconds, using hands and feet to lock in while carrying a 40-pound load.
- Being qualified and current to fly FRIES activities IAW HQ USASOC and the Special Operations Aviation Regiment (SOAR) Aircrew Training Program and applicable policies (for aircrews).

Note: The peacetime maximum soldier load will not exceed 50 pounds. This includes helmet, weapon, vest, web gear, and rucksack. Rucksack weight will not exceed 35 pounds.

5-4. PERSONAL EQUIPMENT REQUIRED

The minimum personal equipment required for FRIES training or operations includes—

- Heavy leather gloves.

- A helmet with a chin strap.
- Protective goggles.
- A long-sleeve shirt or jacket, long pants, and boots.
- Hearing protection and identification tags for helicopter operations.

Section II. FRIES QUALIFICATION TRAINING

Each major unit (commanded by an O5 or higher) is responsible for conducting its own FRIES qualification and sustainment program. FRIES trainers will be FRIES master (FRM) qualified and current. Unit training should be tailored to unit and situation needs, however proponent requisites will not be reduced without written approval of the CG USASOC.

5-5. INITIAL FRIES QUALIFICATION TRAINING

Before participating in fast-rope operations, personnel are briefed on the FRIES and its purpose, capabilities, and limitations. The briefing also covers the duties and responsibilities of the pilot in command (PIC), safety, FRIES master (FRM), the assistant FRIES master (AFRM), and any ground assistants. Once the FRIES briefing is conducted, the remainder of the initial training is hands-on practice of the proper FRIES operational techniques.

a. Individuals are shown the proper techniques for boarding the aircraft, moving to the door, grasping and descending, locking-in, and clearing the rope. Individuals are also shown the proper wear, hook-up, and use of the STABO extraction harness.

b. After the demonstration, all ropers participate in a practice exercise in which they properly perform the following tasks using a tower and subsequently an aircraft. All ropers must demonstrate a minimum of six properly executed FRIES descents (three without equipment and three with equipment) and at least one successful lock-in.

(1) *Perform FRIES descent and lock-in from a tower at the 34-foot level.* Training should be progressive, starting from a tower at the 34-foot level without equipment and then with equipment. All ropers must complete a successful lock-in at this level. Ropers will not progress above the 34-foot level until after demonstrating the ability to stop descent, lock-in, and hold a stationary position for 20 seconds with equipment.

(2) *Perform FRIES descent and lock-in from a tower at the 60-foot level.* The recommended succession will be no less than three successful FRIES descents on the 60-foot tower, again starting without equipment, completing a lock-in, and finishing with equipment.

(3) *Perform FRIES descent from a helicopter at 60 feet or higher.* The recommended completion should be at least two helicopter descents from 60 feet or higher (one without equipment and one with equipment).

WARNING

During initial qualification, students will not perform rapid exits or have more than three soldiers on a rope at any time. During advanced training, the limit is no more than five personnel with equipment on any rope at any time. In all cases (insertion or extraction), the maximum load allowed on any fast-rope system will not exceed 1,500 pounds.

CAUTION

All ropers should be limited to no more than a total of 10 roping events in a 24-hour period; no more than 6 events should be with equipment.

5-6. FRIES PROFICIENCY SUSTAINMENT TRAINING

Commanders must make sure soldiers participating in FRIES operations receive sustainment training on equipment and procedures within 24 hours before the FRIES operation. Ropers who do not attend FRIES sustainment training will not be allowed to participate in FRIES operations. As a minimum, training will include a review of the following:

- Arm-and-hand signals.
- Individual equipment riggings.
- Aircraft familiarization.
- Safety procedures.
- Any rehearsals the FRM or commander deems necessary.

Section III. FRIES MASTER SELECTION AND QUALIFICATION TRAINING

Selection and qualification of FRIES masters (FRMs) is a unit prerogative. Units will maintain records of qualification and proficiency. These records will accompany SM to new units when a permanent change of station occurs.

5-7. FRIES MASTER SELECTION

Units may train and certify soldiers as FRIES-qualified FRMs. Soldiers selected by units to qualify as FRMs should be the rank of E5 or higher and should have demonstrated leadership abilities with emphasis on maturity and decisiveness.

5-8. FRIES MASTER PREREQUISITES

Soldiers performing FRM, AFRM, and safety duties must have completed initial FRIES qualification training, be experienced in FRIES operations, have completed FRM training, and have executed at least 10 FRIES descents and 2 extractions. An FRM candidate's experience must include no less than five descents with combat load and at least one night descent with equipment from 60 feet or higher.

5-9. FRIES MASTER TRAINING AND CERTIFICATION

During FRM training and certification, FRM candidates must participate in three FRIES helicopter operations (twice observing another FRM and once executing FRM duties under observation of a current FRM) and be certified by either the parent unit or an authorized school. FRMs must be proficient in FRIES operations and must demonstrate proficiency at the following:

- Inspecting, preparing, and rigging all FRIES gear.
- Inspecting and preparing aircraft for FRIES operations.
- Controlling and coordinating actions of AFRMs, safeties, and ropers.
- Preparing and conducting both insertion and extraction operations.
- Conducting troop and pilot briefings.
- Giving arm-and-hand signals.
- Giving time warnings and associated commands.
- Deploying and retrieving FRIES ropes.
- Rigging and lowering equipment.

5-10. FRIES MASTER REFRESHER TRAINING

Units will conduct refresher training to maintain acquired skills. FRMs who have not participated in FRIES operations during the past six months will receive refresher training by a current FRM and serve as an AFRM before performing FRM duties. Refresher training for ropers consists of an FRM briefing and participation in FRIES training.

Section IV. KEY PERSONNEL DUTIES AND RESPONSIBILITIES

The following personnel duties and responsibilities provide baseline requirements for the safe conduct of FRIES operations. Unit standing operating procedures (SOPs) may increase (but will not reduce) training safety requirements.

5-11. AIR MISSION COMMANDER

The air mission commander (AMC) provides overall command and control when conducting large, complex missions. The AMC is selected or appointed by the headquarters responsible for the overall mission. He ensures that all operators, aircrew members, and support elements synchronize their actions during the conduct of the mission and the FRIES operation. An AMC is not usually needed during small-unit qualification or proficiency training events when only one helicopter is used.

5-12. FRIES TRAINING OFFICER

Units that conduct FRIES training appoint a training officer in charge (OIC) or a noncommissioned officer in charge (NCOIC) to manage the training. The OIC is responsible for planning and coordinating various aspects of training to include—

- Coordinating all support activities, such as procuring FRIES equipment, aircraft, training areas, medical support, and communication.

- Assigning qualified personnel to perform FRM, AFRM, ground safety, and medic duties.
- Adhering to procedures for planning, preparing, and executing the operation IAW this TC, USASOC Regulation 350-6, and the participating unit=s SOPs.
- Ensuring the FRMs, AFRMs, PICs, aircrews, safeties, and medics are briefed on the operation.
- Ensuring FRIES training events are not conducted on or over unsafe ground. (The ground must be free of loose material that could become airborne and cause an unsafe condition or obscure vision for aviators or ropers.)
- Ensuring a medic, an aid bag, a backboard, and a dedicated vehicle and driver are on site during all training.
- Ensuring positive communication with aircraft is maintained throughout training.

5-13. PILOT IN COMMAND

The aviation unit providing helicopter support for the FRIES training and operation appoints the PIC. The PIC is responsible for all aspects of the flight and ensures—

- Aircrew members are current and qualified to conduct FRIES operations
- Aircrew members are informed of and understand their responsibilities in fast-rope operations.
- Procedures for planning, preparation and execution are adhered to IAW this TC, USASOC Regulation 350-6, and applicable SOPs and policies.
- Aircraft is properly configured to perform the mission.
- Aircraft is at the proper altitude, airspeed, and location as briefed.
- The command to deploy ropes is not given until the aircraft is at a stabilized hover.
- The aircraft position is maintained to keep ropes in contact with the surface until all descending ropers are on the ground (or extracting ropers are securely attached for extraction).
- Ropes are fully recovered inside the aircraft (or jettisoned) before the aircraft departs the stabilized hover position at the infiltration site.
- All personnel are briefed on in-flight emergency and safety procedures.
- Ropes are never deployed with anything other than night illumination attached to the free end.
- All personnel are briefed on fire support to be provided by the aircraft including:
  - — The nature of fire support the aircraft can provide.
  - — The time the fire support starts, shifts, and stops.
  - — The primary and alternate commands and signals that start, stop, lift, or shift fires.

5-14. FRIES MASTER

Units conducting FRIES operations will designate one overall FRM to organize, coordinate, and supervise the activities of the day, and AFRMs as needed. AFRMs are additional FRM-qualified soldiers who assist the primary FRM. An AFRM will be designated for each roping point. AFRMs are responsible for all preparation, inspection, and command and control of all roping activities on their points. FRMs also must—

- Ensure all FRIES roping personnel, including FRMs, AFRMs, and safeties, are qualified and current and understand the proper roping procedures for FRIES operations.
- Perform safety and serviceability checks on all FRIES equipment and rigging.
- Ensure the attachment bar or points are serviceable and free of any defects or contamination and that the quick-release mechanisms and safety pins are present and serviceable and operate correctly.
- Inspect the rope to make sure it has no contamination, damage, or defects that could make it unsafe.
- Check the rope to make sure it is the correct type and length for the operation (smooth rope for infiltration or looped for exfiltration).
- Ensure the rope is properly attached with safety pin in place and back coiled.
- Ensure the extraction harnesses are serviceable and properly worn.
- Ensure the rope chemlights are correctly rigged and illuminated when needed (two at the mount, two at the end, and two 15 feet from the end).
- Confirm the proper seating arrangement, including seat belts or a safety strap for ropers.
- Coordinate with the PIC and brief the aircrew, safeties, and ropers on the following:
  - — The aircraft flight, approach, flare, and position over the target site.
  - — The height of the aircraft above the target.
  - — A description of the FRIES site (rooftop, small clearing) and size, as well as the actions of the ropers upon arrival at the site.
  - — The signal to deploy the rope, who will deploy it, and how it will be done.
  - — Time warnings and actions at each time warning.
  - — Arm-and-hand signals.
  - — The use of radios and intercoms.
- Brief the PIC and crew chief on verbal commands and final adjustments over the target.
- Confirm the correct target and inform the pilot before the deployment of ropes or ropers.

WARNING

The FRM must ensure that nothing, other than rope illumination, is ever attached to the free end of the FRIES rope during deployment and that equipment is never attached and dropped.

- Deploy the ropes when the PIC gives the command DEPLOY ROPES. (This action may be delegated to the aircrew or safety.)
- Ensure safeties and AFRMS understand the time and activity sequence and all arm-and-hand signals.
- Ensure safeties know the proper procedures to—
- Keep the aircraft on target.
  - — Clear the ropes.
  - — Recover the ropes.
  - — Jettison the ropes.
- Observe the target and pass to the PIC, through the crew chief, final aircraft position adjustments over the target to make sure conditions are safe.
- Deploy the ropes when the PIC gives the command DEPLOY ROPES. (This action may be delegated to the safety.)
- Ensure all ropers know how to hook up for extraction and know the correct arm-and-hand signals.
- Direct the aircrew or safety to inform the pilot when the aircraft is on target (the PIC replies with the command DEPLOY ROPES).

CAUTION

When equipment is to be lowered, it should be lowered before the FRIES is deployed or personnel begin roping. When simultaneous, equipment and ropers will exit from opposite doors of the aircraft.

- — The safety repeats, " DEPLOY ROPES." The FRM signals the AFRMs to deploy ropes and makes a final check, ensuring all ropes from his aircraft reach the surface and are safe for descent (at least 15 feet of rope must be on the surface).
- — When conditions are safe, the FRM commands "GO" to the ropers and points to the rope or leads the stick out.

Note: If the FRM leads the stick out, the AFRM automatically assumes control. He completes marshaling the remaining ropers and is the last man to exit.

- Coordinate actions with the crew chief for the positioning or storage of seat belts, safety straps, rucksacks, door bundles, and communication cords.

Note: During night operations, upon completing descent, the AFRM observes the the ropes of his aircraft. When he sees that all ropes are clear, he issues the prearranged ALL CLEAR hand signal to the aircrew safety. This hand signal tells the aircrew that they are free to jettison or recover the ropes, and fly away.

5-15. FRIES AIRCREW MEMBER AND SAFETY

FRIES safeties are aircrew members who are FRM-qualified and perform the additional duties of the safety during helicopter roping operations. Safeties assist FRMs and pilots as needed. Aircrewmen performing safety duties differ from FRMs and AFRMs in that safeties are members of the aircrew (crew chief or gunner) whose orientation is the flying portion of the operation. Conversely, AFRMs are FRM-qualified members of the ground unit who are performing FRM duties as assistants to the primary FRM and who exit with the ground unit. Safeties remain in the aircraft and operate the doors, extend and retract the rope mount bars, and wear night vision goggles (NVGs) when appropriate. Safeties must have completed FRM training IAW this TC and the aircrew training program (ATP).

a. During roping operations, the safety's four major areas of responsibility are:

(1) Observe all activities in the aircraft and assist the pilot and FRM or AFRM as needed.

(2) Serve as the communication link between the PIC and the FRM. The safety monitors aircraft communication and relays information between the PIC and the FRM, constantly keeping both informed.

(3) Assist the FRMs and AFRMs in lowering cargo and clearing the aircraft of ropers.

(4) Clear and jettison the ropes, when appropriate.

b. The safety performs the following duties:

(1) Relays time warnings, aircraft positioning commands, and other information to assist the FRM.

(2) Conducts aircraft equipment rigging and serviceability inspections, making sure mounts, doors, and ropes are ready and safe for operations both pre-flight and en route.

(3) Verifies that the aircraft is at a stabilized hover with ropes in contact with the surface and remains so (or stops stick until corrections are made).

(4) Checks the rope to make sure 15 feet of rope are on the target surface.

(5) Tells the PIC, "ROPES DEPLOYED", when appropriate.

(6) Tells the PIC, "ROPERS OUT", when the first roper has exited.

(7) Helps the AFRM in controlling and spacing each roper's exit.

(8) Monitors the rope and ropers to make sure they do not get hung up, dragged, or dropped off site.

(9) Observes the exit of the last roper until the roper reaches the surface, moves clear of the rope, and gives the ALL CLEAR signal.

(10) Tells the PIC, "ROPERS DOWN", (station name: aft, forward, left, right) when all ropers are on the surface and conditions are safe to jettison the ropes.

(11) Passes the signals as required to position the aircraft, including HOLD, MOVE LEFT, RIGHT, FORWARD, BACK, UP, or DOWN.

(12) Acts as belay man during cargo lowering when needed.

(13) Jettisons ropes upon receiving the command JETTISON ROPES from the PIC. Tells the PIC, "Ropes are jettisoned," when the ropes have been recovered or jettisoned.

(14) Tells the PIC when ALL CLEAR.

CAUTION
This command tells the pilot the aircraft is cleared for flight.

(15) Deploys the rope for extraction upon receiving the PIC command DEPLOY ROPES.

(16) Observes the extracting rope and personnel and informs or guides the PIC through aircraft positioning and throughout the recovery flight.

5-16. FRIES ROPER

Ropers are responsible for notifying the FRM, AFRM, safety, or pilot if they observe any unsafe act or condition. Ropers may halt or call for a halt of roping for safety at any time. During FRIES insertion training, the number of ropers on the fast rope at one time is limited to three. During extraction training, the total weight per extraction bar or rope will not exceed a total of 1,500 pounds. When rucksacks are worn by ropers, the rucksacks must not exceed a weight of 35 pounds (total roper equipment weight will not exceed 50 pounds). Ropers will—

- Keep hands at head level.
- Maintain visual contact with lower ropers during their descent.
- Maintain a minimum of a 1-second interval on exit to avoid collisions.
- Keep at least two points of contact on the rope (both hands) at all times.
- Use their feet for additional breaking any time needed.
- Execute descents at a safe speed.
- Slow the rate of descent halfway down the rope to avoid landing on each other.
- Move quickly away from the rope upon arrival on the surface.
- Know correct wear of extraction (STABO) harness and extraction procedures.

WARNING
The FRM, AFRM, and safety must make sure all personnel are on the surface and have moved clear of the ropes before aircraft fly away or ropes are jettisoned.

Section V. FRIES RIGGING OF AIRCRAFT

The aviation unit is responsible for preattachment inspection and installation of FRIES anchor point mount bars and associated equipment. All aircraft used to conduct FRIES operations with US Army personnel must be rigged and equipped IAW current parent service directives--for Army aircraft, USASOC, SOAR, and TAPO directives apply. Rigging of the aircraft includes installation and removal of the seats and seat belts (or personnel restraints), and attaching and detaching the ropes.

Figure 5-1. UH-60 helicopter

> **CAUTION**
> **Army aircraft that have the FRIES mounts installed must have an airworthiness release for FRIES operations from the Aviation Troop Support Command (ATCOM) TAPO.**

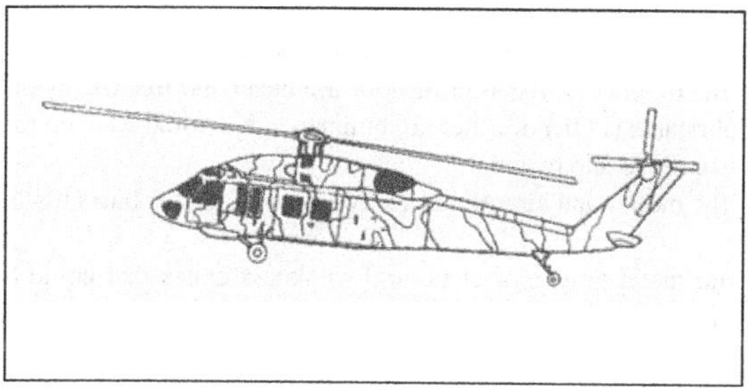

5-17. RIGGING OF FRIES IN UH/MH-60

The UH-60 is a twin-turbine, medium-speed, single-main-rotor helicopter (Figure 5-1).

The aviation aircrew or safety is responsible for—

- Inspecting the FRIES mount bar assembly, attachment hardware, and associated equipment to make sure they are serviceable and ready to use.
- Ensuring the bar assembly is correctly mounted to the aircraft (Figure 5-2).

Figure 5-2. Bar assembly correctly mounted to aircraft.

- Ensuring the mount is complete, functional, and free of any defect.
- Ensuring the mount and the aircraft floor are clean and free of any petroleum, oils, or lubricants (POL) or other contaminants that could weaken the rope or cause personnel to slip or fall.
- Ensuring the mount and aircraft are free of sharp edges or burrs that could cut the rope.
- Ensuring no metal fatigue or structural weakness exists that could cause the system to fail.
- Extending locking and retracting mount bars as needed.
- Lowering and belaying equipment as needed.
- Opening, locking, and closing cargo doors as needed.

Note: For cold-weather operations or during flights of long duration, the cargo doors are closed and latched until the time specified for opening. The aircrew will operate doors or directly supervises those who do. In good weather, doors should remain locked open.

- Removing or reconfiguring troop seats.
- Providing seat belts or floor-mounted personnel restraints for ropers' use during flight.
- Attaching and detaching ropes as needed.

Note: Clusters of rucksacks or small door bundles (weight limited to 250 pounds) may be lowered (belayed) on rappelling ropes. Cargo lowered this way must never be free dropped (the belay must start at the time the bundle is pushed [tipped] out the door).

5-18. RIGGING OF FRIES IN CH/MH-47

The CH-47 is a tandem-rotor, medium-transport helicopter (Figure 5-3).

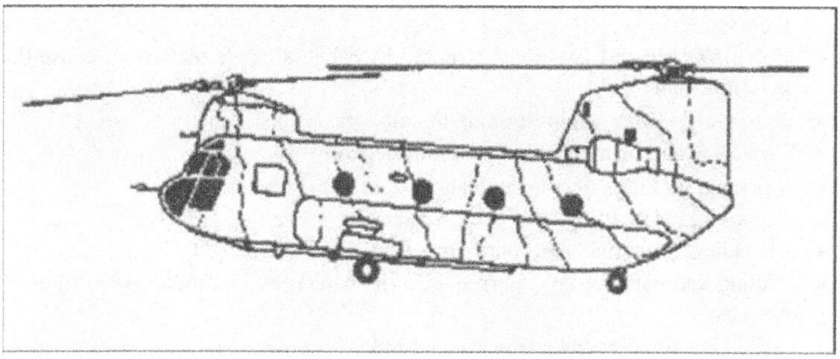

Figure 5-3. CH-47 helicopter.

The aviation aircrew or safety is responsible for:

- Inspecting the FRIES mount bar assembly, attachment hardware, and associated equipment to make sure they are serviceable and ready to use.
- Ensuring the bar mounts are correctly mounted to the aircraft (Figure 5-4, rear; Figure 5-5, winch).

Figure 5-4. Bar mounts correctly mounted, rear.

Figure 5-5. Bar mounts correctly mounted, winch.

- Ensuring the mount is complete, functional, and free of any defect.

- Ensuring the mount and the aircraft floor are clean and free of any POL or other contaminants that could weaken the rope or cause personnel to slip and fall.
- Ensuring the mount and aircraft are free of sharp edges or burrs that could cut the rope.
- Ensuring no metal fatigue or structural weakness exists that could cause the system to fail.
- Extending locking and retracting mount bars as needed.
- Lowering and belaying equipment as needed.
- Opening, locking, and closing cabin doors as needed.
- Removing or reconfiguring troop seats.
- Providing personnel restraints for ropers' use during flight.
- Taping and padding any sharp edges that can come in contact with ropes or personnel
- Attaching and detaching ropes as needed.

5-19. RIGGING OF FRIES IN MH-53

The MH-53 is a USASOC twin-engine, single-rotor, medium-transport helicopter used by the Army (Figure 5-6).

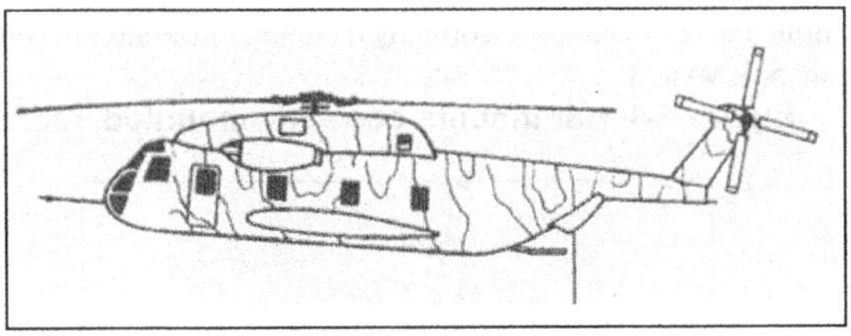

Figure 5-6. MH-53 helicopter.

The MH-53 primary FRIES mount bar assembly is located overhead, just aft of the rear ramp. The ramp mount bar assembly spans the cabin width and has a two-pin release mechanism on each end. The ramp position mount assembly provides two points; each will accommodate one FRIES rope or one cargo lowering rope. The ropes are carried back-coiled on the ramp.

a. The alternate position for fast-rope descent from the MH-53 is the crew entrance in the right side of the fuselage immediately behind the pilot's position. In this position, the fast-rope mount must be connected to the external hoist point. The FRIES master is responsible for inspecting the MH-53 mount system for its presence, security, and operability.

b. The aviation unit is responsible for—

- Installation and maintenance of the fast-rope mount assemblies at each position to be used.

- Configuring the aircraft to accommodate the number of soldiers and type of mission planned.
- Ensuring seat belts or restraint straps are present to secure ropers during flight.

Note: The MH-53 aircrew can improvise handrails to assist roper exits. When used, improvised handrails will be rigged IAW USAF special operations SOPs.

- Configuring the MH-53 for fast-roping when equipped with external 650-gallon fuel tanks.
  - Remove the ramp door to install the cross bar.
  - Place the bar in the brackets with the two quick-release paddles facing forward.
- Configuring aircraft armament (50-caliber machine gun and the GAU-2B ramp minigun) for FRIES operations.
  - When using the aft right roping position, point the 50-caliber machine gun to the 6 o'clock position and secure the gun with bungee cords.
  - When using the left aft rope, disconnect the minigun ammunition chute and stow it in the ammunition can, or disconnect it at the ammunition can and securely wrap it around the minigun. Use bungee cords to secure the ramp minigun in the firing position facing the 6 o'clock position.

Note: Ideally, only the right aft rope is used when the GAU-2B ramp minigun is installed. Although not recommended, the left aft rope may be used, if required.

Section VI. EQUIPMENT MAINTENANCE AND INSPECTION

Before conducting a FRIES operation, the FRM must inspect all FRIES equipment (ropes, STABO harnesses, and mount bars) for serviceability and readiness for use.

WARNING

FRIES ropes and harnesses are critical life support equipment and must be stored, cared for, and maintained as life support items.

5-20. ROPES AND HARNESSES

FRIES equipment should be maintained in the same manner as a parachute. The unit rigger section can provide detailed guidance on appropriate inspection, care, and maintenance of FRIES equipment.

a. FRIES ropes and harnesses must always be stored in a clean, cool, dry space out of direct sunlight and free of chemicals or chemical vapors. Equipment that becomes wet with fresh water should be hung up to dry (indoors) on hardwood pegs. Equipment that is exposed to salt water or becomes imbedded with dirt or mud should be washed and rinsed in fresh water (within 72 hours) and hung up to dry (indoors) on hardwood pegs (out of direct sunlight).

b. Before conducting a fast-rope operation, the FRM—

- Inspects the fast rope thoroughly and carefully. Checks the rope length to ensure it is the correct rope for the operation planned.
- Checks the woven loop on the mount end for excessive wear or chemical contamination. Checks the rope along its entire length for fraying, cuts, and chemical contamination.
  - — Do not use a rope that is severely frayed. (Light fraying on the rope from normal use does not weaken the rope.)
  - — Do not use a rope when any single strand is cut halfway through or has two or more cuts that penetrate one-third or more through any strand's thickness within 1 foot of the running length of the FRIES.
- Inspects the rope for contamination of acid, alkaline compounds, saltwater, fire extinguisher solutions, or petroleum-based solvents. Changes in color caused by chemicals are usually blotchy and have an unusual odor. Although used ropes gradually change color, such changes do not indicate a decrease in strength unless the change is due to contact with strong chemicals. Changes occurring because of use are usually uniform throughout the length of the rope.

WARNING

Any rope known or suspected to be chemically contaminated, cut beyond allowable limits, excessively frayed, or sun damaged must be taken out of service, marked for destruction, and turned in to the rigger section.

- Inspects the extraction loops to the same standard as the main rope. Ensures the woven attachment loops are secure.
- Inspects the harness to ensure—
  - — Ropers are wearing the harness under all load-carrying equipment.
  - — Ropers have properly fastened all connectors.
  - — Harness material and stitching are not cut, torn, or contaminated, and all hardware is free of corrosion and is in operable condition.

Note: Serviceability inspection and maintenance of the extraction harness are the responsibility of the rigger-qualified personnel.

5-21. FRIES MOUNT BARS

The aviation unit is responsible for the installation, removal, storage, and maintenance of FRIES mount bars. The US Army Special Operations Aviation Regiment, Ft Campbell, KY, prepares and provides checklists for these procedures. Aviation units participating in FRIES operations must have and comply with current SOAR guidance.

Aviation units:

- Install and maintain the FRIES mount bar; check mount bolt torque and witness marks.
- Check the FRIES mount bar for security and operability.
- Check the condition of the mount bar; inspect for bends, dents, distortion, cracks, corrosion, and contaminants.
- Ensure the fast-rope quick-release pins are installed and operable (two each side).
- Check the security and condition of the rope-release system, as follows:
  - The mounting bolts are installed and properly torqued.
  - The safety pins are installed, and lanyards are present.
  - The system locks and unlocks as designed.
- Ensure the release cables are attached at both ends and are not broken or kinked.
- Check the release handles to ensure they are securely mounted and can move freely.
- Check the latch as follows:
  - Latch or close the release mechanism and apply a downward pressure on the release arm. Attempt to pull the release handles the release should not open.
  - Relax pressure on the release mechanism and pull the handles the release should open.
  - With no pressure on the release mechanism and with the safety pin installed, attempt to pull the release handles--the release should not open.

Section VII. OPERATIONAL REQUIREMENTS AND LIMITATIONS

This section establishes medical coverage and communication support requirements for FRIES operations, including during adverse weather conditions and night operations.

5-22. MEDICAL SUPPORT

Dedicated medical support is required for FRIES training and demonstrations. A qualified medic with assistant, aid bags, two backboards, two stretchers, and a dedicated vehicle with a driver must be present to respond to any mishaps. The nearest medical evacuation (MEDEVAC) unit will be notified of where and when FRIES training is scheduled. MEDEVAC response will be coordinated as on call. Training or demonstrations will not be conducted without medical support present. The ground force commander may direct the FRIES aircraft to transport life or death (urgent) patients when no other MEDEVAC unit is available, providing the medical team is trained and properly equipped with the appropriate equipment for aerial transport of patients. This must be included in the air mission briefing prior to training and coordinated with the medical team providing coverage. FRIES operations that are evaluated as high risk or that occur at night must have a certified combat lifesaver on location and medics (equipped as stated above) in the immediate vicinity.

5-23. COMMUNICATIONS REQUIREMENTS

During peacetime training and operations, FRIES lift aircraft and ground elements will maintain continuous radio communications. Aircrew personnel clearing ropes at each FRIES station must be on the intercom system, monitoring aircraft internal communications. They must also be proficient in the use of arm-and-hand signals as precoordinated between FRIES users and aircraft crews. The designated aircrew member will relay time warnings between the PIC and FRM. Routine time warnings are 10-, 6- and 1-minute warnings to the FRM. Ground-to-air communications must be continuously maintained throughout FRIES training.

5-24. ADVERSE WEATHER OR TERRAIN CONDITIONS

FRIES training and operations will ***not*** be conducted under the following conditions:

- Anytime temperatures are below 32 degrees Fahrenheit or wind-chill factors that could cause cold-weather injuries exist including those caused by helicopter rotor wash or extraction cruise speed.
- Anytime water or ice on the rope prevents personnel from controlling their descent.
- Anytime the rope is exposed to freezing conditions for a sufficient length of time to freeze, thereby reducing its tensile strength.
- Anytime dust out or white out conditions restrict the aviators or the ropers vision.
- Anytime trees or vegetation on the LZ is too dense for safe training.
- Anytime the LZ is too steep or too slippery for safe training.
- Anytime winds exceed 30 knots or are too gusty for safe operations.

5-25. NIGHT OPERATION REQUIREMENTS

Night operations require the following procedures.

a. Attach two chemlights at the free end of the rope and two above the extraction attachment loops (15 feet from the end of the rope) to aid in determining when adequate rope is on the surface.

b. Attach two chemlights (one primary and one backup) to the loop of the rope, or to the rope mount, to guide personnel exiting the aircraft to prevent glare from obscuring vision.

c. Attach one chemlight or luminous tape to the top of each roper's helmet. If a chemlight is used, select a color different from the colors used on the ropes.

CAUTION

Ropers will not wear NVGs during descent. NVGs limit depth

perception and cause a tunnel-vision effect. Static FRM and Safeties may wear NVGs to help confirm that fast ropes are fully deployed. The aircrew will wear NVGs as required by flight mission needs.

Section VIII. FRIES PROCEDURES

The risks of personal injury and equipment damage during FRIES operations can be reduced by detailed planning, thorough training, and strict adherence to safe procedures. Conducting FRIES operations with heavy loads requires personnel to be proficient in these operations. FRIES users must limit ropers to no more than 6 descents with combat load and no more than 10 total descents in any 24-hour period. The total weight of equipment soldiers may carry during training is limited to never more than 50 pounds. This section provides standardized procedures for FRMs, PICs, and ropers. These procedures are common to all units and aircraft, except as noted.

5-26. FRIES MASTER

The FRM performs the following series of actions.

- Briefs members of his team and aircrew.
- Inspects team members and equipment.
- Installs the FRIES rope in the aircraft and conducts safety checks.
- Relays time warnings to team members. (Time warnings are a tool to help keep aircrew and ropers' actions synchronized and can be modified according to user needs; however the 1-minute warning should always be used.)
- Breaks chemlights, if required. Two chemlights or luminous tape will be attached at the anchor point to identify the handhold point.

WARNING

Equipment must never be attached to the FRIES and dropped. As little as 10 pounds of weight can exceed the load limit of the FRIES mounts when the load stops at the end of a 60-foot drop. The only items to be attached to the fast rope during deployment are the position markers.

- Ensures the rope is properly configured for deployment (back coiled to prevent tangles).
- Ensures the team members are in order of exit no later than the 1-minute warning.
- Confirms target on final approach.
- Deploys the rope and ensures it reaches the ground. During night

operations, the rope is marked with six chemlights: two at the mount, two on the end, and two 15 feet from the end.

- Deploys personnel.
- Accounts for personnel, and signals aircrew.

> **WARNING**
> **Because the FRIES mount in the CH-47 helicopter is narrow, special care must be taken by ropers to avoid fouling weapons or equipment on the frame during exit.**

5-27. PILOT IN COMMAND

The PIC briefs the FRM, safety, and roping personnel before FRIES operations.

a. The PIC briefing includes:

- Approaching, loading, unloading, and departing the aircraft.
- Actions in the aircraft.
- Flight route and checkpoints en route.
- Altitude or height for roping.
- Time warnings to FRM as needed (recommend 10 minutes, 6 minutes, and 1 minute).

b. When the PIC is over the intended objective he calls, "DEPLOY ROPES" Fast-rope warnings and expected aircrew coordination calls are as follows:

(1) The aircrew or FRM provides any final position adjustments of the aircraft with the commands of MOVE (left, right, forward, back, up, down, or hold) as needed to place the ropes ON TARGET. (When the pilot is to hold the position, an explanation will be provided as soon as possible.)

(2) When the FRM deploys the ropes, the aircrew member calls, "ROPES DEPLOYED."

(3) When the first roper exits the aircraft, the aircrew member calls, "ROPERS OUT."

(4) The safety watches the last roper until he is on the ground and gives the prearranged signal. The safety then tells the PIC, "ALL ROPERS AWAY," and identifies his station. This alerts the PIC that the safety is going to either jettison or recover the ropes.

(5) The safety will then either recover or jettison the rope as planned, clear the aircraft all around for obstructions, and call, "ROPES CLEAR (station)" This informs the PIC that the aircraft is clear for flight.

c. The PIC also briefs on emergencies that could occur in the following situations:

- En route (when ropers are inside the aircraft).
- Before deployment of ropes.

- When fast-rope personnel are on the ropes.
- After the ropers are on the surface.

5-28. ROPERS

Ropers perform the following actions:

a. At the command STAND BY (given at 1-minute warning), ropers make final check of themselves and take the prepared-to-exit position.

b. At the command GO, the first roper exits the aircraft, rotates his body 90 to 180 degrees to make sure his equipment clears the aircraft, places the fast rope between the arches of his feet, and begins descent. The roper uses his hands and feet to slow his descent approximately two-thirds of the way down.

c. Subsequent ropers exit at 1-second intervals using the same procedure, but begin slow descent approximately half-way down to avoid landing on other ropers.

d. During descent, ropers keep a lookout and break as necessary to avoid landing on obstructions or on fellow ropers.

e. Ropers prepare to land just before reaching the ground by spreading their legs about shoulder-width apart with their knees slightly bent.

f. At landing, ropers quickly move clear of the ropes to avoid collisions with descending ropers.

CAUTION

According to AR 95-3, anytime operations occur over water, all personnel will wear an approved personal flotation device as identified in FM 57-220 or USASOC Regulation 350-2.

5-29. EQUIPMENT-LOWERING PROCEDURES

Anytime a roper's load exceeds a total of 50 pounds or his rucksack weight exceeds 35 pounds, equipment will be lowered (as clusters or bundles) on separate ropes. Lowering equipment from heights great enough to cause equipment damage necessitates belaying the equipment during descent to avert damage. Belaying devices used in mountaineering allow ropers or safeties to control equipment descent.

Figure 5-7. Rucksacks upright on cargo deck.

CAUTION

When equipment is to be lowered, it should be lowered before the FRIES is deployed or personnel begin roping. When done

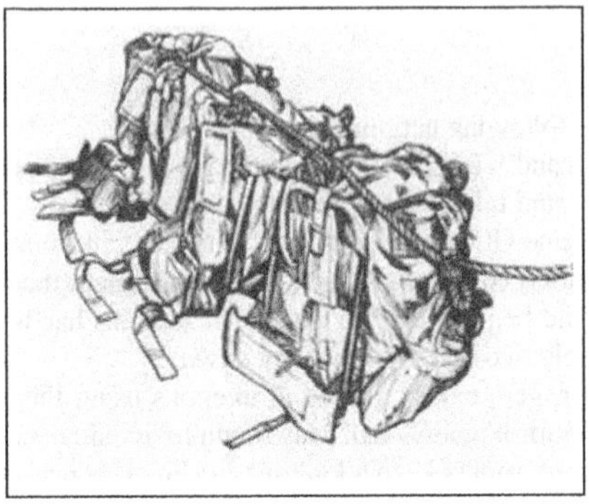

simultaneously, equipment and ropers will exit from opposite doors of the aircraft.

a. Rigging equipment to be lowered requires rigging rucksacks in clusters. Clusters of up to five rucksacks may be lowered (smaller clusters work better). FRMs will ensure the total weight of any cluster does not exceed 250 pounds.

(1) Place the rucksacks upright on the cargo deck of the helicopter (Figure 5-7) .

(2) Lay a sling rope across the frames between the shoulder strap attachment points.

(3) Tie a bowline knot at each end of the sling and a figure eight knot at each rucksack.

(4) Attach a snaplink to the frame of each rucksack, to the loops in the sling rope, and to the end of the lowering rope.

(5) For Lowe rucks, install the issue H-harness on the rucks and join them into clusters with a sling rope tied to the quick-release attachment buckles.

b. Rigging of the lowering rope to belay the rucksacks requires the following actions.

(1) Backcoil a rappel rope on the helicopter floor at the gunner=s position.

(2) Attach a snaplink and belay device (either a figure eight [Figure 5-8] or an SMC ladder [Figure 5-9]) to the floor tiedown (with a snaplink) inboard of the ceiling rappelling anchor point.

(3) Attach a snaplink to the ceiling rappelling ring.

(4) Route the belay rope from the backcoil, through the belay device, then up through the ceiling snaplink, and down to the cluster snaplink.

(5) Tie the end of the belay rope into a bowline with halfhitch and snap it into the cluster snaplink.

(6) The FRM identifies two ropers to push the cluster or bundle out of the aircraft. He also identifies an AFRM or a safety to belay the cluster or bundle. The FRM briefs these personnel on the time to lower the bundle, the signals to be used, and the procedure to clear the belay system before deploying the FRIES. In performing his

belaying duties, the AFRM or safety must The aviation aircrew or safety is responsible for:

- Provide sufficient slack to allow the pushers to eject the cluster from the aircraft.
- Control the descent until the cluster reaches the surface.

Figure 5-8. Belay Device figure eight SMC **Figure 5-9. Belay device ladder**

- When the cluster is on the surface, unsnap both the upper and lower snaplinks, drop the remainder of the rope from the aircraft, and immediately look out to make sure conditions are safe to deploy the FRIES.
- Inform the FRM and pilot when conditions are safe.

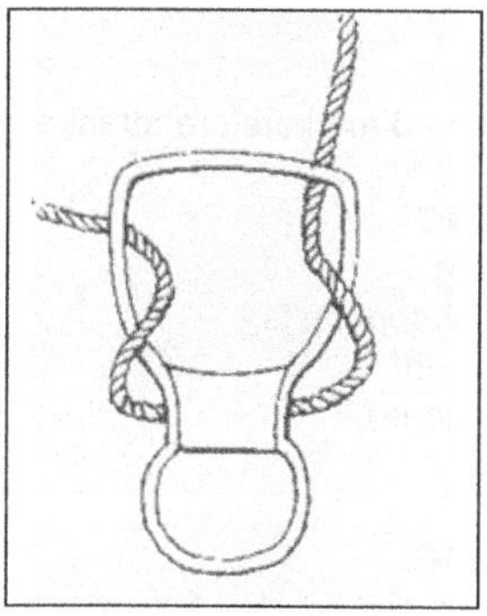

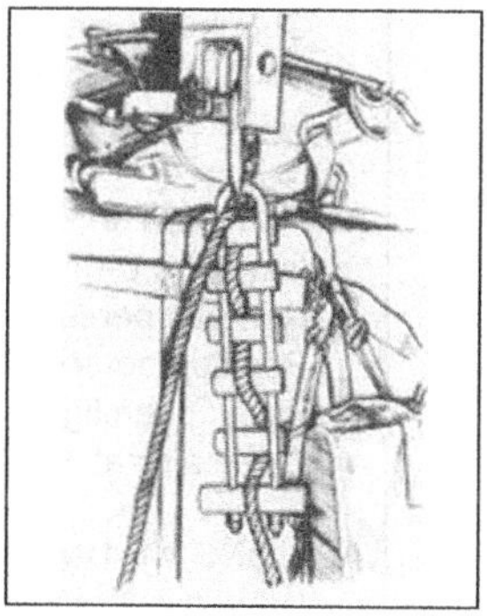

- Deploy FRIES on command

Note: Expect to reposition aircraft before ropers exit to avoid ropers landing on equipment.

5-30. HELICOPTER OPERATIONS FRM CHECKLIST

Table 5-1 shows a checklist of duties of key personnel in fast-rope operations.

PRE-ROPING ACTIONS
Receive a briefing from the OIC or the AMC. Coordinate and brief all participants. Rig aircraft and conduct a joint inspection. Brief ropers, AFRMs, and safeties. Rig and inspect ropers.

Conduct a static rehearsal.

AIRCRAFT LOADING

Position equipment and personnel.
Make sure all personnel have straps or seat belts.

ACTIONS IN FLIGHT

Monitor the command net.
Monitor the aircrew net.
Monitor the flight route.

ACTIONS AT 10-MINUTE WARNING (applies to long infiltrations)

Issue the 10-minute time warning and GET READY.
Check equipment and belay system hookup.
Check fast-rope hookup.
Secure fast-rope bar in position.
Make sure rope is back coiled and markers are attached.

ACTIONS AT 6-MINUTE WARNING

Issue the 6-minute time warning.
Remove personnel restraints or seat belts.
Position personnel and equipment.
Break chemlights (night operations).
Open aircraft doors, if required.

ACTIONS AT 1-MINUTE WARNING

Issue the 1-minute time warning and STAND BY.
Position ropers in stick formation.

ACTIONS AT FLARE

Identify the target area.
Deploy bundles/equipment (Safeties) and clear ropes.
Deploy FRIES ropes (FRM and AFRM).
Check ropes to make sure 15 feet of rope are on the surface (FRM).

Table 5-1. FRM helicopter fast-rope operations checklist

ACTIONS FOR DESCENT

~ FRM, AFRM, or safety positions the number 1 man at the rope. (The FRM may exit first or last.)
~ FRM issues the command GO. AFRM echoes GO command, and the ropers exit the aircraft.
Safety informs PIC, "ROPERS OUT."

AFRM or safety controls the rate of exit of the ropers. AFRM exits last. Aircrew or safeties observe the last roper. Safety tells the PIC, ALL ROPERS AWAY, after the last roper is on the surface and signals. PIC issues command JETTISON or RECOVER ROPES. Aircrew or safety jettisons or recovers ropes, and issues ROPES CLEAR report to the pilot.

Table 5-1. FRM helicopter fast-rope operations checklist (continued).

5-31. FRIES COMMANDS AND SIGNALS

The following standardized arm-and-hand signals are provided for FRIES operations. Ground-to-aircraft arm-and-hand signals are military standard as found in the Pathfinder field manual 57-38.

a. Time warnings (10-, 6-, and 1-minute) are given by holding up the appropriate number of fingers and verbally sounding off.

b. The signal for GET READY is the same as the standard Airborne signal—arms extended horizontally, fingers extended vertically and joined.

c. The signal for STAND-BY is arms extended downward at a 45-degree angle from the body with hands closed except for index fingers, which are extended, pointing to the floor

d. To execute the signal for DEPLOY ROPES, the FRM or safety forms both hands into fists at mid-thigh level and sweeps arms horizontally outward. At full extension, he opens his hands as if dropping the ropes out of the helicopter and sounds off with DEPLOY ROPES.

e. The signal for GO is arms raised to horizontal, hands closed with index fingers extended pointing to the ropes.

f. The signal for STOP or ABORT is hands closed into fists with arms raised across the forehead .

g. The signal for SIT-DOWN and DON SEAT BELTS is pointing to the floor, followed by moving both fists around waist, as if donning a belt.

h. The signal for OK is hands raised to eye level, index finger and thumb forming a circle with other fingers extended.

Section IX. EMERGENCY ACTIONS

Procedures for correcting an emergency condition that could reasonably be encountered are described in this section. Multiple emergencies, adverse weather, or other unusual conditions may require modification of these procedures. The nature and severity of the emergency dictate the response necessary; therefore, personnel must use sound judgment in determining the correct action to be taken.

5-32. EMERGENCIES BEFORE ROPING STARTS

All personnel sit down, don their seat belts, and take further instructions from the pilot or crew chief.

5-33. EMERGENCIES AFTER ROPING STARTS

Emergency procedures to be followed after roping starts are discussed in this paragraph.

a. **Unsafe Drift or Premature Liftoff.** Anyone who detects the aircraft having drifted off the site must immediately stop training and inform the FRM, AFRM , safety, or PIC.

(1) The FRM, AFRM, safety, or roper stops stick.

(2) The ropers stop descent and lock in.

(3) The FRM or crewman informs the PIC, and guides him in moving the aircraft back on target.

(4) The unit continues operations.

b. **Rope Hung or Snagged (On Ground, Tree, Building).** Anyone who detects the rope being snagged or hung must immediately stop training and inform the FRM, AFRM, safety, crew chief, or PIC.

(1) The safety ensures ropers are off the rope and are clear.

(2) The aircraft descends or lands as needed.

(3) The safety frees the rope and informs the pilot.

(4) The unit resumes the operation.

c. **Lost Communications.** All training and operations must be planned to include the use of the intercom between the PIC or crewmen and the FRM. If the intercom fails, the following arm-and-hand signals can be used until the rope can be cleared and the intercom is restored.

- Wrists crossed diagonally above the forehead with clenched fists to signal stop stick.
- Finger pointed toward the exit and a verbal GO to signal ropers to exit.
- Open palm moved and faced in the required direction to signal aircraft movement.
- Clenched fist to stop aircraft movement.

Note: Training will not be conducted without radio communications between the pilots or aircrew and the FRM and the ground.

Section X. SAFETY

Safety is everyone's responsibility. All personnel involved in FRIES operations are responsible for identifying hazardous situations and preventing injuries to personnel. Anyone who observes an unsafe condition or act is authorized to halt the operation and inform the FRM or the PIC.

5-34. PREFLIGHT AND INFLIGHT

Anyone who detects an inflight emergency will immediately notify the PIC, crew chief, or safety and follow aircrew instructions.

a. Before a FRIES operation, the FRM will coordinate with the PIC both routine and emergency procedures as well as any unique conditions or potential problems. He also discusses the procedures to be used during flight, during roping, and before the aircraft's departure from the roping site. The FRM then discusses the information with all the AFRMs, safeties, and crew chiefs as well. He reviews the time warnings (including arm-and-hand signals) and appropriate actions such as who will deploy ropes, verify rope status, and perform rope-clearing procedures.

b. The mission prebrief covers any aircraft equipment that, if disturbed, could cause an aircraft or fast-rope problem. The problem areas should be made known to all personnel concerned and preventive measures taken to neutralize or eliminate them where possible.

c. The FRM gives the crew chief the final position adjustment over the target, using hand signals.

d. All safety straps and communications cords that are not in use are securely stowed. All safety straps and communication cords in use should be positioned so they do not present a hazard to personnel movement in the aircraft.

e. After the PIC gives the command DEPLOY ROPES, the FRM is responsible for deploying the rope and making sure it is in contact with the surface before ropers exit.

Note: The safety is responsible for observing rope and ropers and for making sure the pilot maintains aircraft position until all ropers are on the surface.

CAUTION

According to AR 95-3, anytime operations occur over water, all personnel will wear an approved personal flotation device as identified in FM 57-220 or USASOC Regulation 350-2.

5-35. DURING ROPING

The following safety procedures apply during roping.

a. During the actual descent, fast-ropers will not use NVGs because they limit depth perception and create a tunnel vision effect.

b. During descent, ropers must watch for obstructions and lower ropers, and must break as necessary.

c. Individual ropers will lock in during emergencies by wrapping one leg around the rope once or twice and standing on the fast rope with the other foot.

d. Ropers must maintain sufficient intervals on the rope to prevent contact with lower ropers. Individual ropers exit when the head of the man in front of them disappears below the ramp.

e. Ropers execute descents at speeds commensurate with their experience and proficiency in FRIES infiltration operations. (For example, less-experienced ropers should descend slowly until they become more proficient.)

f. The number of ropers on each rope at one time must not exceed the safe limit of the rope mount point in use, as follows:

(1) Do not exceed 1,500 pounds per rope on the MH/UH-60.
(2) Do not exceed 1,500 pounds per rope on the MH/UH-47.
(3) Do not exceed 1,500 pounds per rope on the MH/UH-53.
(4) Do not exceed 1,300 pounds per rope on the MH-60G.

5-36. SAFETY BRIEFING

All units must present a safety briefing before conducting FRIES training or operations. The safety briefing must cover the following (as a minimum):

- The planned event (day or night, tower or helicopter, high or low hover, insertion or extraction).
- Actions on the aircraft (or tower).
- Operation (or training) location and any hazards within 500 meters.
- Identification of key personnel, their location, and their duties.
- FRIES equipment and its characteristics.
- FRIES equipment rigging or inspection.
- Roping personnel preparation (shirttails tucked in, chin strap fastened, loose straps taped, gloves and goggles on).
- Arm-and-hand signals.
- Emergencies and emergency procedures.
- Medical procedures (medic with aid bags, backboards, litters, and dedicated vehicle with driver on site).
- Communications requirements (frequencies, call signs, required reports).
- Any other information essential to safety.

5-37. SAFETY REMINDERS

The following safety reminders will be reviewed by all key personnel before conducting operations.

a. Ropers must wear long-sleeve shirts or jackets, full-length pants, laced boots, helmets, protective goggles, and heavy leather gloves.

b. MILES harnesses must not be worn during fast-roping.

c. Ropers will not wear NVGs.

d. Commanders must personally approve the rope site selected for night operations. The FRIES site must be large enough to permit all ropers, upon reaching the surface, to move clear of the rope and for units to consolidate or re-form.

e. All night rope operations must be assessed as medium risk or higher.

f. FRIES descent training is not authorized with a fast rope longer than 90 feet.

g. Ropers must not carry more than 50 pounds of equipment during FRIES descents. The maximum weight of rucksacks is limited to 35 pounds. When rucksacks exceed 35 pounds (or total load exceeds 50 pounds), gear must be lowered (belayed) on a separate system, preferably from the opposite door.

h. Ropers will not carry bulky equipment that could interfere with FRIES operations. Equipment that cannot be carried by the soldier during roping (due to its weight or bulk) can be lowered (belayed) during insertion. During extractions, rucksacks must be attached to the extraction loops of the FRIES rope using snaplinks.

i. Ropers must not make more than 6 combat equipment descents or a total of 10 descents in a 24-hour period.

j. FRIES training must not be conducted if—

- The aircraft is unstable.
- The fast rope is not fully deployed (15 feet of the rope must be on the surface).
- Any obstacle is present that could interfere with the descent of the ropers.
- The rope is frozen or slippery
- Anyone identifies any unsafe condition.

k. Safety personnel must—

- Position themselves so they can observe active FRIES points with out obstructing ropers.
- Conduct final rigging checks of the mount, equipment, and ropes at the 10-minute warning.
- Ensure the aircraft is over the target area and the ropes reach the ground.
- Constantly keep both the PIC and FRM informed—
  - When the aircraft is in position for deployment.
  - When anything is being deployed from the helicopter (ropes, bundles or personnel).
  - When 15 feet of rope are on the surface.
  - When the first roper exits on the fast rope.
  - When all ropers on the surface are clear.
  - When the ropes are recovered or jettisoned.
  - When the aircraft is cleared for flight.
  - When 15 feet of the extraction rope is on the surface.
  - When extraction personnel are attached to the rope and ready for lift off.
  - When extracted personnel are safely lowered to the ground.
  - When extracted personnel are out from under the aircraft.
  - Any other conditions that could impact safe operations.

Note: If the FRM and the safety exit, the crew chief performs rope clearing and or jettisoning duties.

WARNING

Using the word CLEAR tells pilots the aircraft is ready to fly. Before any use of the word CLEAR, FRMs and safeties must ensure that all personnel are well away from the ropes and the ropes have been fully recovered or released (during insertion), or all personnel are securely attached to the ropes (during extraction).

l. FRIES training must not be conducted in densely wooded areas or areas prone to blowing dust, sand, or snow, which can produce a hazard for ropers or the aircraft.

m. FRIES training and operations must not be conducted during severe weather such as rain, sleet, ice, or extreme cold.

n. The possibility of personnel injury during FRIES training and operations is always present. Every soldier involved in FRIES activities must maintain constant vigilance to prevent personnel injury and equipment damage.

CHAPTER 6
EQUIPMENT

This chapter discusses the basic equipment used for rappelling operations (Appendix B). It covers the types of ropes, the procedures for rope selection, and the care for each rope type. It also discusses the two methods of coiling and the different snaplinks used to join the equipment and rope.

Section I. ROPES

Ropes are the most important items of equipment for the rappeller. They provide access down the obstacle or aerial platform, while ensuring individual safety.

6-1. TYPES OF ROPES

Two types of ropes are available to the rappeller—*static* and *dynamic*.

a. **Static Rope.** This rope allows for minimal stretch of the rope. The static rope stretches about 5 to 15 percent at the point of failure and about 2 percent under a working load. The minimum tensile strength for an 11-mm (7/16-inch) static rope for military use is 4,500 pounds. *Kernmantle* static ropes range from 3 millimeters to 11 millimeters in diameter and are constructed similar to the 550-parachute cord. (*Kern* means core and *mantle* means sheath.) The internal core of the rope is constructed of a continuous multifilament-nylon yarn, which is spun into continuous parallel strands. This internal core is then covered with a nylon-braided outer sheath. Due to the internal parallel strand construction of the static rope, there is less spinning and kinking than with a dynamic rope. Static ropes can be used for slings, harnesses, etriers (stirrups), rappelling, rope installations, hauling lines, and where stretch of the rope is not desired.

b. **Dynamic Rope.** This rope allows for stretch within the fibers of the rope, which can be a disadvantage in rappelling, prusik climbing, and other applications. Dynamic ropes are more susceptible to abrasion and wear. They have about a 5 to 10 percent working stretch. The minimum tensile strength for an 11-mm (7/16-inch) rope for military use is 4,500 pounds. Two types of dynamic ropes are nylon-laid and kernmantle:

(1) *Nylon-laid ropes.* Synthetic fibers are the best climbing and rappelling material. Nylon has become the standard material for climbing ropes and has replaced manila, flax, hemp, and sisal. The military mountaineering critical application rope is constructed in a *laid* fashion with continuous multifilament-nylon fiber yarn twisted into strands. Three strands are twisted into a climbing rope of a specific diameter (Figure 6-1). Nylon-laid ropes are easy to inspect for serviceability by twisting the fibers. Laid ropes tend to untwist slightly when under a load, causing kinking and spinning. They are also susceptible to abrasion. The following specifications pertain to the standard military nylon-laid climbing/rappelling rope:

- 36 1/2 meters (120 feet) long.
- 11 millimeters (7/16 inch) wide.
- At least a 4,500-pound tensile strength.
- Right-hand lay.
- One-third stretch factor (at point of failure).

- May lose as much as 15 percent of its strength when wet.
- Weighs 6 pounds when dry.

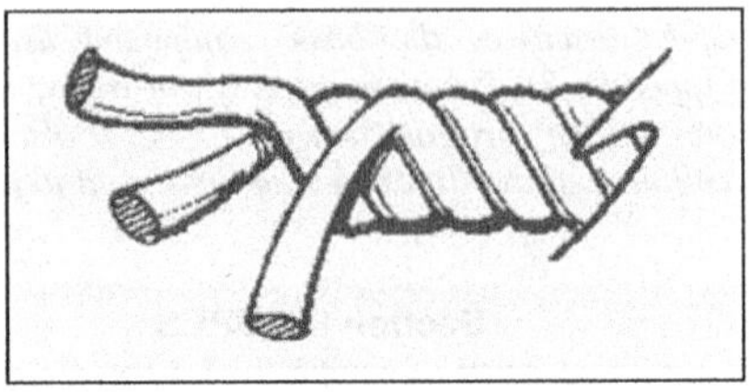

Figure 6-1. Nylon-laid rope.

(2) *Kernmantle ropes.* These ropes are similar to their static counterparts in that they consist of an inner core and an outer sheath. The core is constructed of continuous twisted nylon filaments, which are laid or braided together and enclosed in a tightly braided outer sheath. These ropes are well suited for climbing on rock, snow, or ice where a brief elastic stretch of the rope occurs during a fall. They have a stretch of about 40 percent at the point of failure. The breaking strength is high, and there are no exposed strands for rock crystals to work between to damage the rope (Figure 6-2). There is less sliding friction through the snaplink and over other surfaces, since the outer sheath is smooth. Kernmantle ropes are available in many sizes, lengths, stretch factors, tensile strengths, and fall ratings.

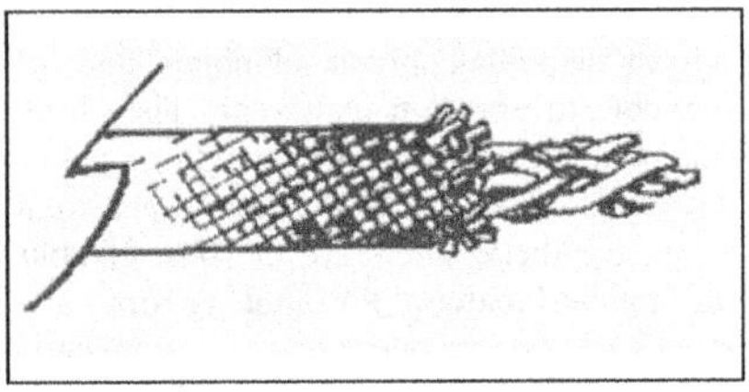

Figure 6-2. Kernmantle (core and sheath) rope.

6-2. ROPE SELECTION

Rope selection for military operations requires ropes of various types, lengths, and diameters; and depends on its intended use (static or dynamic), environment (water-repellant), climate, mission (lightweight ropes), and other factors. Static ropes should have minimal stretch; be smooth, flexible, and strong; resist abrasion and cutting; and have a low-impact force.

a. *Laid rope* is easier to inspect (by twisting the strands apart to observe the interior surfaces), has greater friction in holding knots, and is less expensive than kernmantle ropes. Its disadvantages include: much lower abrasion resistance, extremely high stretch factor for both static and dynamic loads, greater friction on rock, readiness to kink, tendency to spin a free-hanging rappeller, and greater absorption of water and dirt.

b. *Kernmantle* rope is easy to work with, has less friction, does not kink easily, and has less susceptibility to abrasion. The outer sheath protects the internal fibers from dirt, abrasion, and ultraviolet radiation. Its disadvantages are that it is hard to determine damage to the core fibers, and it is more expensive than laid rope. Kernmantle rope is not readily identifiable as dynamic or static and must be marked accordingly.

c. Rope diameter and length vary with intended use. A standard diameter for many uses is 11 millimeters (7/16 inch). For stirrups, utility ropes, and hauling lines, 7 millimeters (5/16 inch), which can also be used in double strands, is adequate. Standards to remember are that the smaller the diameter, the less the tensile strength; but the larger the diameter, the heavier the rope. The standard lengths are 36 1/2 meters (120 feet), 40 meters (135 feet), 45 meters (150 feet), and 50 meters (165 feet). Normally, standard lengths should not be cut. If a nonstandard length is needed, the rope should be precut before the mission and clearly marked.

d. A rope should be selected for a specific use. There are different types of ropes for various uses. A dynamic rope is designed for climbing, while a static rope is designed for rappelling, rescue operations, load hauling, and rope installations.

WARNING

Using a rope for other than its intended use could result in personal injury or damage to equipment.

6-3. CARE OF ROPES

Because the rope is the rappeller's lifeline, it requires a great amount of care and maintenance.

a. Before and after use, the rope should be carefully inspected for excessive wear, cuts, mildew, or rotten spots. If any of these defects are found, the rope is unserviceable. An unserviceable rope is destroyed by cutting it into usable parts. If no sections are usable, the rope is cut into smaller parts and discarded.

b. Ropes should never be spliced together, stepped on, or dragged. This greatly reduces the tensile strength of the rope.

c. The rope must not come in contact with sharp edges. If this is likely, the areas must be padded or taped before rappelling.

d. The rope must be kept as dry as possible. If it becomes wet, it must be dried out to prevent rotting. (Avoid excessive heat/heating during the drying process.)

6-4. COILING

Two methods of coiling are the *mountaineer's coil* and the *farmer's coil* (Figure 6-3).

a. **Mountaineer's Coil.**

(1) To use the mountaineer's coil, grasp the rope about 3 feet from the end with the left hand. The right hand is then placed next to the left hand and runs along the rope until both arms are outstretched. Grasping the rope firmly, bring the hands together to form a loop, which is laid in the left hand. Repeat, forming uniform loops that run in a clockwise direction until the rope is completely coiled. If the rope tends to twist or form figure eights, give it a slight twist with the right hand when each loop is formed. Coil the rope in a clockwise direction to conform with the lay.

(2) To complete the coil, form a bight about 30 centimeters (12 inches) long with the starting end of the rope and lay it along the top of the coil. Uncoil the last loop and use it to wrap around the coil and the bight. Wrap toward the closed end of the bight, binding the final wrap across itself to lock it in place. Make six to eight wraps to secure the coil. Push the end of the rope down through the closed end of the bight. Carry the running end of the coil either in the pack (by forming a figure eight, doubling it, and placing it under the flap) or by placing it over the shoulder and under the opposite arm. If the rope to be coiled is anchored as in coiling a belay or rappel lane, start the coil near the end closest to the anchor so that the kinks work themselves out of the free end.

b. **Farmer's Coil.**

(1) The farmer's coil is easier and faster to coil than the mountaineer's coil. Establish the center of the rope either by locating its center mark or by grasping both ends of the rope and feeding them out until a bight comes up that is the center of the rope. Begin coiling with the doubled rope the same as the mountaineer's coil method. Leave about 4.6 meters (15 feet) of rope uncoiled. Then, squeeze the coils together and make four to six wraps around the middle of the coil, ensuring that the first wrap is held in place by the other wraps.

(2) Form a bight with the two running ends and place it through the bight formed by the top of the coil. Then, run the two running ends over the top of the coil and through the bight that is formed. Dress them in the center of the back of the carrier. Run the two ends over the shoulders to form shoulder straps. Then, bring them under the arms, cross them in the back over the coil, bring them around the body of the carrier, and tie them off with a square knot at the stomach.

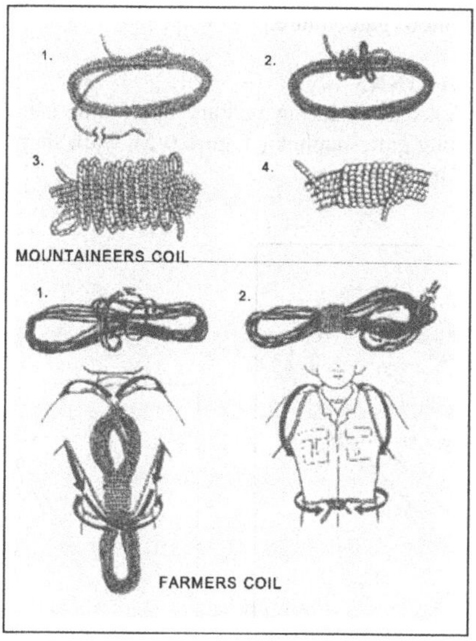

Figure 6-3. Two methods of tying a coil.

Section II. SNAPLINKS

Snaplinks are used to join equipment, rope, and people into a functioning system. When used properly, they are strong, versatile items. Snaplinks are metal (steel, aluminum, or alloys) loops with a hinged, spring-loaded gate on one side. They come in many sizes, shapes, and strengths (with and without locking gates) and in many types of metal. Steel is the strongest snaplink, but it is heavier.

6-5. DESCRIPTION

Snaplinks are available in many sizes and shapes to suit different needs. A locking snaplink is best for rappelling. Nonlocking snaplinks are easy to operate, but should be used only where they cannot be accidentally opened. Hollow snaplinks should be avoided since their use is limited. The following information applies to all snaplinks.

a. The weakest part of a snaplink is the gate. The gate must be closed before applying a load. When the gate is open, snaplinks should have little or no lateral movement of the gate.

b. Locking pins should be checked to ensure that they are not loose, worn, or corroded.

c. The metal should be checked for any cracks, grooves, burrs, flaws, or rust.

d. The spring-loaded gate should automatically close securely from an open to a closed position with no gap between the locking pin and notch.

e. If an engraver is used to mark snaplinks, it should be applied only to the gate, never to the load-bearing side.

f. A snaplink should never be side-loaded (across the gate) since this reduces the overall strength to the point of gate failure.

6-6. TYPES OF SNAPLINKS

The two types of snaplinks used during military operations are the standard snaplink (Figure 6-4) and the locking gate snaplink (Figure 6-5). Both snaplinks are available in a D-shaped variation or an oval variation.

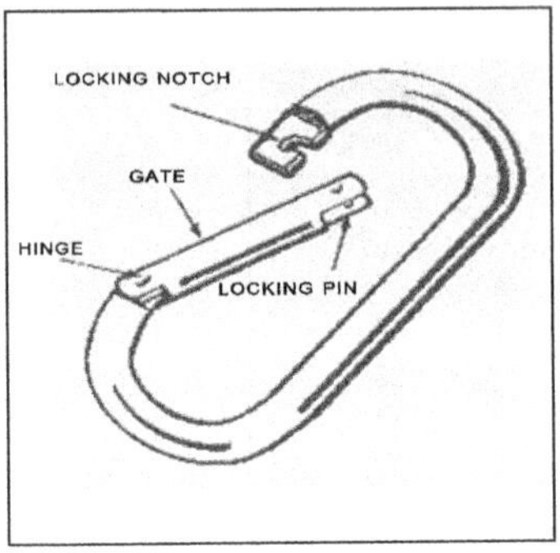

Figure 6-4. Standard snaplink (oval).

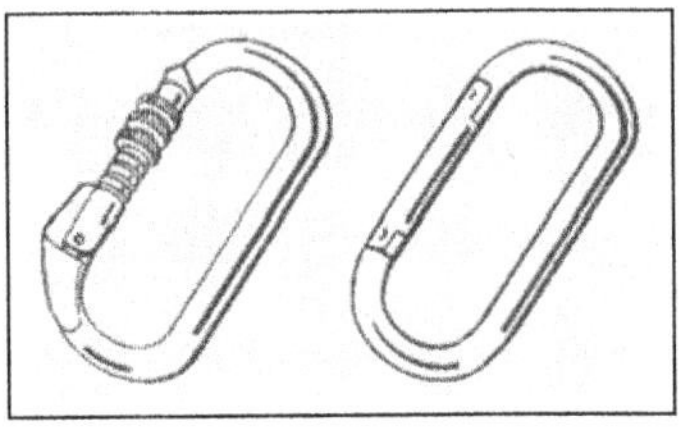

Figure 6-5. D-shaped snaplinks.

a. **D-Shaped Snaplinks.** D-shaped snaplinks are stronger than the oval type because the shape directs the largest part of the load to be applied to the longer, stronger side opposite the gate (Figure 6-5). D-shaped snaplinks are made of steel or aluminum alloys, and are available in many sizes and thicknesses, all of which can be with or without locking gates. Their strength and durability vary, which must be considered before use.

b. **Oval Snaplinks.** Oval snaplinks are versatile and have different applications. They are made of steel or aluminum alloys and are available in many sizes and thicknesses. Both sides of an oval snaplink bear the strain equally under load weight. Many modified ovals are available and most can be with or without the locking gate. All must meet military standards or UIAA specifications.

c. **Locking Snaplinks.** Locking snaplinks have a locking mechanism with a threaded sleeve on the gate. The sleeve screws tightly over the gate opening end or hinge end to hold the gate closed (Figure 6-6). A reverse locking gate is needed to prevent a moving rope from unscrewing the sleeve. The locking sleeve and threads should be kept free of dirt and grit. If the sleeve is forced to close, it may strip the threads. The locking mechanism ages and weakens after repeated use and should be routinely inspected for wear. The characteristics of locking snaplinks are as follows:

- Material: steel.
- Approximate strength: 2,000 pounds with the gate closed.
- Gate: spring-loaded without locking sleeve.
- Weight: 4.3 ounces.

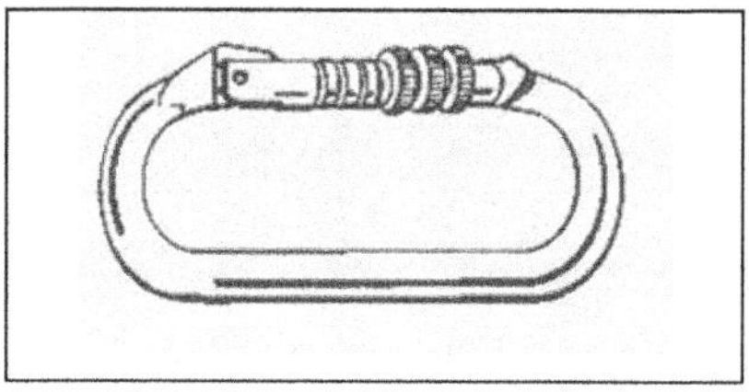

Figure 6-6. Locking snaplink.

6-7. INSPECTION

All rappelling material and individual rappelling equipment must be inspected before conducting a rappelling operation. The rappel master is responsible for ensuring the serviceability of rappelling material and individual rappelling equipment.

a. **Snaplinks.** Snaplinks should be inspected daily before, during, and after use.

(1) The metal should be checked for cracks, grooves, burrs, rust, and flaws. The gate should open and close freely without binding. There should be no lateral movement when the gate is open. The gate spring action should snap shut when released. The locking notch should have a slant or slot so that the gate remains shut under the impact of a rappeller's fall. The gate pins should not work their way out of their holes and should not be shorter than their holes. If there is a locking mechanism, it should be inspected to ensure that threads are not stripped and that the sleeve tightly locks the gate.

(2) If burrs, grooves, or rough areas are identified, the snaplink should not be used. Rust should be removed with steel wool and the spot rubbed with oil or solvent. The spring should be lubricated as needed. The snaplink must be boiled in water for 20 to 30 seconds to remove the cleaning agents since solvents and oils cause dirt to cling to the snaplink and to rub off on the ropes. It is better to use a dry graphite-based lubricant on snaplinks since such lubricants do not attract dirt.

b. **Ropes.** The ropes are inspected for serviceability. A rappel rope is unserviceable if it is saturated with petroleum products, is mildewed, shows excessive wear, has soft spots (internal damage), or if one strand is cut more than one-half of its diameter. Ropes should be inspected daily—before, during, and after use. An applicable annotation should be made in a corresponding rope log (DA Form 5752-R) (Figure 6-7).

ROPE LOG (USAGE AND HISTORY) For use of this form, see TC 90-6-1; the proponent agency is TRADOC			UNIT ID MARKING D-11-86-5
NSN 8455-7356036 6	DOCUMENT NUMBER W330SV 61890007	SERIAL NUMBER 6-6-6933	MFR LOT NUMBER G620135115103
DATE OF MFR 08 Feb 87	ISSUE DATE 01 Oct 87	DATE IN SERVICE 17 Nov 87	LENGTH 45M (150 FT)
DIAMETER 11mm (7/16 inch)	FIBER TYPE 6.6 NYLON	COLOR OD Green	CONSTRUCTION DYNAMIC KERMANTLE

INSPECT ROBE FOR DAMAGE OR EXCESSIVE WEAR EACH TIME IT IS DEPLOYED AND AGAIN AFTER EACH USE. IMMEDIATELY RETIRE ALL SUSPECT ROPES.

DATE USED	LOCATION	TYPE OF USE	ROPE EXPOSURE	INSPECTOR'S INITIAL/DATE	ROPE CONDITION AND COMMENTS
20 Nov 87	Yonah Mtn GA	Direct Aid Climb 5.7 A1	65°F Clear & dry Smoothe Granite	DLE 20 NOV 87	
21 NOV 87	Yonah Mtn GA	Top Rope Belayed Balance Climb	68°F Clear & dry Smoothe Granite	ACG 21 NOV 87	
15-18 Jan 88	Gulcana Glacier ALASKA	ROPED Glacier Travel	2°F snow storm	JEB 15-18 Jan 88	
27 Jan 88	Smugglers Notch VT	Front Point Ice Climb	35°F Cloudy [illegible]	ATS 27 Jan 88	SUSTAINED one 15 FT Leader FALL
28 Jan 88	Smugglers Notch VT	Front Point Ice Climb	20°F Cloudy & Snowy	JRL 28 Jan 88	
8 Feb 88	Mt Shasta CAL	FIAT Foot ICE Climb	8°F Lightly snowing	DCG 8 Feb 88	Approx one inch of sheath slippage on end of rope
22-24 Feb 88	DeCos Mtn NM	Two man Party Rock Climb 5.7+	45°F Clear & DRY GRANITE	[illegible] 22 Feb 88	
10-22 Jun 88	Mt McKinley ALASKA	ROPED PARTY Climb CLASS VI	20-53°F one DAY of RAIN	[illegible] 10-22 Jun 88	Abrasion to rope at points of tie-in (ends & middle)

DA FORM 5752-R, MAY 89

Figure 6-7. Example of completed DA Form 5752-R.

c. **Gloves.** The gloves are inspected for serviceability. They are unserviceable if holes are found in the friction-bearing surfaces or along the seams.

d. **Helmet.** The helmet is inspected for correct assembly and serviceability. It is unserviceable if it cannot be properly assembled or if any straps are cut or torn one-quarter of its width.

(1) Ensure that the chin strap attaching tabs are secured to the helmet by the locking nuts.

(2) Ensure that the pull-the-dot tab has four plies of nylon and that the chin strap is routed from the inside of the buckle around the horizontal bar and back toward the inside of the buckle.

Section III. ALTERNATE METHODS OF DESCENT

Snaplinks or carabiners that are used for rappelling are placed through the sewn loops of the seat harness or the sewn loops of the seat-chest combination.

6-8. FIGURE-EIGHT DESCENDER

The rappeller attaches a locking carabiner to the harness (Figure 6-8). He routes the rappel rope up through the large hole and places the bight over the collar. He inserts the figure eight with the rappel rope attached into the locking carabiner, and locks it down. The rappeller brakes to the rear and descends as in a seat hip rappel. If heavy loads are to be descended, he uses a double wrap around the collar.

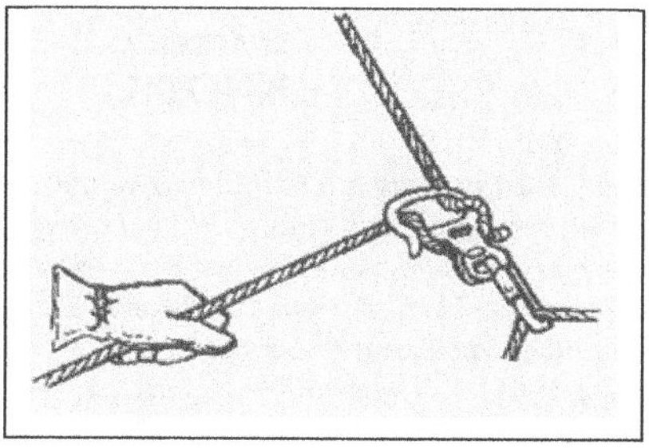

Figure 6-8. Figure-eight descender.

6-9. MUNTER HITCH

The rappeller attaches a large radius snaplink to the harness or rappel seat. He ties a Munter hitch and clip into the snaplink (Figure 6-9, page 6-9). He brakes to the front and descends with an L-shape body position.

Notes: 1. The guide hand should remain on the standing end of the rope to keep the rappeller's head and face away from the hardware.

2. The Munter hitch creates significant rope-to-rope friction and may cause premature wear on nylon ropes.

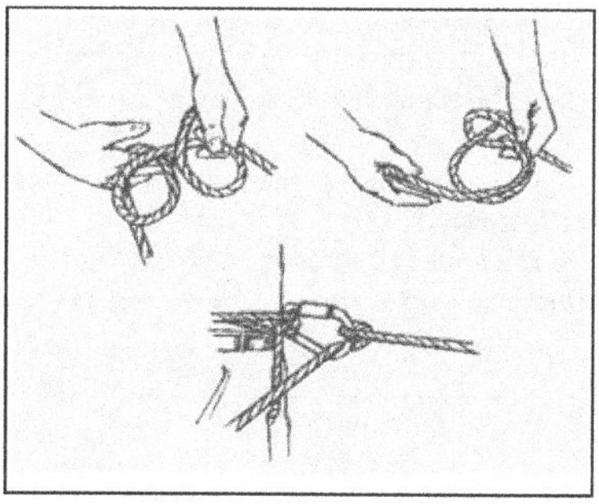

Figure 6-9. Munter hitch.

CHAPTER 7
KNOTS

A knot is a fastening made by intertwining or tying pieces of string, cord, rope, or webbing. All knots used by a rappeller are divided into four classes: Class I—joining knots, Class II—anchor knots, Class III—middle rope knots, and Class IV—special knots. These classes of knots are intended only as a general guide since some of the knots discussed herein may be appropriate in more than one class. The skill of knot tying may be lost if it is not used and practiced. With experience and practice, knot tying becomes instinctive and helps the soldier in many situations.

7-1. MILITARY MOUNTAINEERING TERMINOLOGY

The terms explained below are the most commonly used in military mountaineering (Figure 7-1).

a. A *bight of rope* is a simple bend of rope in which the rope does not cross itself.

b. A *loop* is a bend of rope in which the rope does cross itself.

c. A *half hitch* is a loop that runs around an object and locks itself.

d. The *running end* (working end) of the rope is the free end of the rope that can be used.

e. The *standing part* of the rope is the part that is static (anchored, coiled); the remaining part of the rope not being used (also called static end).

f. The *lay* of the rope is the same as the twist of the rope.

g. A *round turn* is a single complete wrap of the rope around an object providing 360-degree contact. The running end leaves the completed circle in the same direction as the standing part. In a round turn, the rope is wrapped around an object 1 1/2 times.

h. The *pigtail* is the short length of rope remaining at the end after tying a knot or coiling a rope.

i. *Back feeding* (or stacking) is taking off one wrap at a time from a coil, and letting it fall naturally to the ground.

j. *Dress down the knot* is tightening down a knot to its functioning form.

k. An *overhand* is a loop with the running end pulled through the loop.

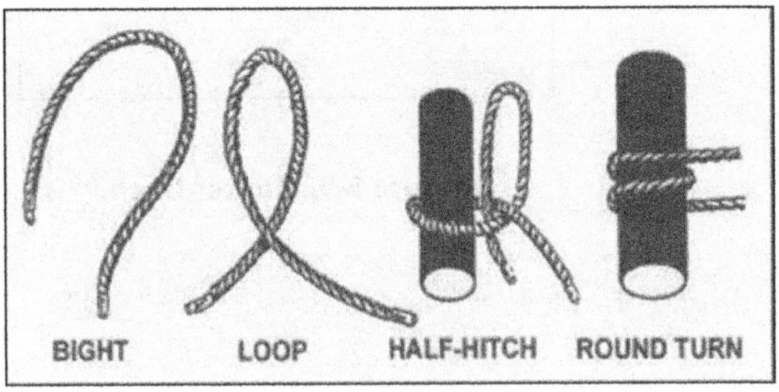

Figure 7-1. Examples of rope terminology.

Section I. JOINING KNOTS

7-2. SQUARE KNOT

The square knot is used to tie the ends of two ropes of equal diameter (Figure 7-2).

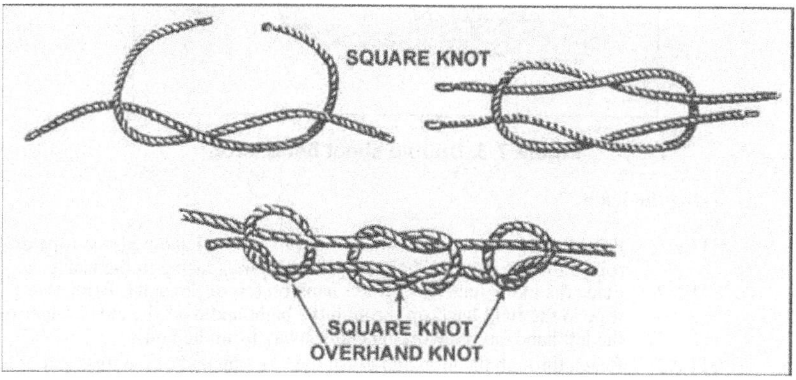

Figure 7-2. Square knot.

a. **Tying the Knot.**

STEP 1. Holding one working end in each hand, place the working end in the right hand over the one in the left hand.

STEP 2. Pull it under and back over the top of the rope in the left hand.

STEP 3. Place the working end in the left hand over the one in the right hand and repeat STEP 2.

STEP 4. Dress the knot down and secure it with an overhand knot on each side of the square knot.

b. **Checkpoints.**

(1) There are two interlocking bights.

(2) The standing parts are on the same side and properly secured with overhand knots.

7-3. DOUBLE SHEET BEND KNOT

The double sheet bend knot is used to tie the ends of two or more ropes of equal or unequal diameter (Figure 7-3). When a single rope is tied to a multiple of ropes, the bight is formed with the multiple of ropes.

Figure 7-3. Double sheet bend knot.

a. **Tying the Knot.**

STEP 1. Form a 30-centimeter (12-inch) bight in the left hand with the rope or ropes. Ensure that the short end of the bight is facing to the inside.

STEP 2. Place the index finger of the left hand on top of the bight. Bring the rope in the right hand up through the bight and over the index finger of the left hand, so the working end is away from the body.

STEP 3. Reach through the loop and grasp the working end of the rope and bring it back towards the body. Place it between the index finger and the bight, forming a round turn. Repeat this one more time, wrapping toward the tip of the index finger.

STEP 4. Remove the finger and dress the knot down.

b. **Checkpoints.**

(1) The two wraps around the bight are held in place by a locking bar.

(2) The two standing parts of the ropes exit parallel and from opposite ends of the knot. The two working ends form an L and exit the knot at a 90-degree angle and on opposite sides of the knot.

7-4. FISHERMAN'S KNOT

The fisherman's knot is used to tie two ropes of similar or dissimilar materials (Figure 7-4).

Figure 7-4. Fisherman's knot.

a. **Tying the Knot.**

STEP 1. Tie an overhand knot in one end of the rope.
STEP 2. Pass the working end of the other rope through the first overhand knot. Tie an overhand knot around the standing part of the first rope with the working end of the second rope.
STEP 3. Tightly dress down each overhand knot and tightly draw the knots together.

b. **Checkpoints.**

(1) The two separate overhand knots are tied tightly around the long standing part of the opposing rope.

(2) The two overhand knots are drawn snug.

7-5. DOUBLE FISHERMAN'S KNOT

The double fisherman's knot (also called double English or grapevine) is used to tie two ropes of similar or dissimilar materials (Figure 7-5).

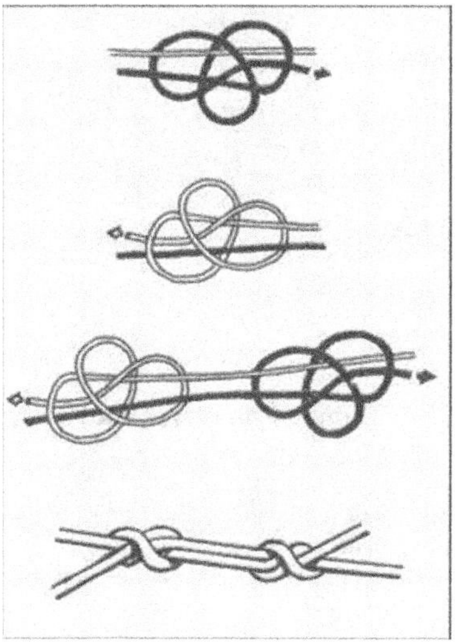

Figure 7-5. Double fisherman's knot.

a. **Tying the Knot.**

STEP 1. With the working end of one rope, tie two wraps around the standing part of another rope.
STEP 2. Insert the working end (STEP 1) back through the two wraps and draw it tight.

STEP 3. With the working end of the other rope, which contains the standing part (STEPS 1 and 2), tie two wraps around the standing part of the other rope (the working end in STEP 1). Insert the working end back through the two wraps and draw tight.

STEP 4. Pull on the opposing ends to bring the two knots together.

b. **Checkpoints.**

(1) Two wraps with the working end running through and drawn tight.

(2) Both wraps are drawn together with all the slack out of the knot.

7-6. WATER KNOT

The water knot is used to attach two webbing ends (Figure 7-6). It is also called a ring bend, overhand retrace, or tape knot. It is used in runners and harnesses.

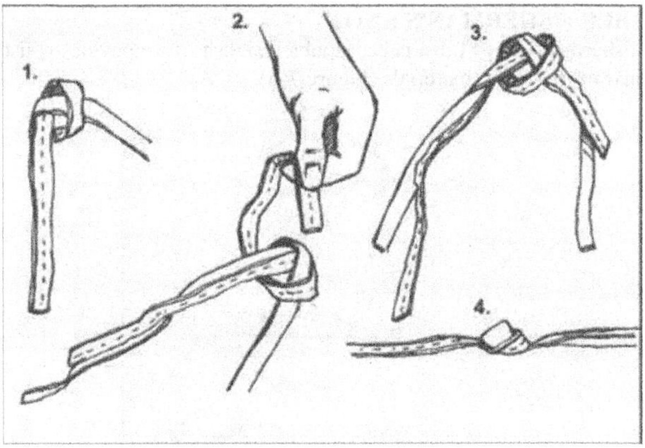

Figure 7-6. Water knot

a. **Tying the Knot.**

STEP 1. Tie an overhand knot in one of the ends.

STEP 2. Feed the other end back through the knot, following the path of the first rope in reverse.

b. **Checkpoints.**

(1) There are two overhand knots, one retracing the other.

(2) There is no slack in the knot, with the working ends coming out of the knot in opposite directions (at least 5 centimeters [2 inches] long).

Section II. ANCHOR KNOTS

7-7. BOWLINE KNOT

The bowline knot is used to tie a single fixed loop in the end of a rope (Figure 7-7). It does not slip under strain and is easily untied. This knot is always used when there is alternating tension. It can also be used to tie the end of a rope to an anchor.

Figure 7-7. Bowline knot.

a. **Tying the Knot.**

STEP 1. Bring the working end of the rope around the anchor, from right to left (as the rappeller faces the anchor).
STEP 2. Form an overhand loop in the standing part of the rope (on the rappeller's right) toward the anchor.
STEP 3. Reach through the loop and pull up a bight.
STEP 4. Place the working end of the rope (on the rappeller's left) through the bight, and bring it back onto itself. Dress the knot down.
STEP 5. Form an overhand knot with the tail from the bight.

b. **Checkpoints.**

(1) The loop is locked into place by a bight.

(2) The short portion of the bight is on the inside and on the loop around the anchor (or inside the fixed loop).

7-8. ROUND TURN AND TWO HALF HITCHES KNOT The round turn and two half hitches knot is used to tie the end of a rope to an anchor and must have constant tension (Figure 7-8).

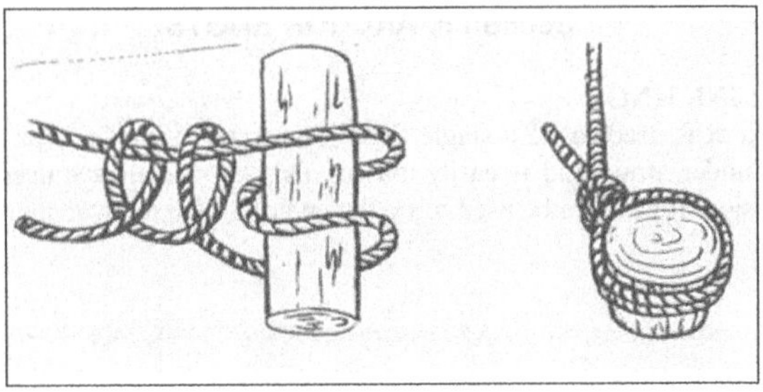

Figure 7-8. Round turn and two half hitches knot.

a. **Tying the Knot.**

STEP 1. Route the rope around the anchor from right to left and wrap down (must have two wraps in the rear of the anchor and one in the front). Run the loop around the object to provide 360-degree contact, distributing the load over the anchor.

STEP 2. Bring the working end of the rope left to right and over the standing part, forming a half hitch (first half hitch).

STEP 3. Repeat STEP 2 (last half hitch has a 15-centimeter [6-inch] pigtail).

STEP 4. Dress the knot down.

b. **Checkpoints.**

(1) A complete round turn should exist around the anchor with no crosses.

(2) Two half hitches should be held in place by a locking bar with no less than a 15-centimeter (6-inch) tail remaining.

(3) When the two half hitches are twisted to the left or the right, they will appear to for a figure eight.

(4) Looking at the top of the half hitches, the rope should appear to be one continuous rope.

7-9. FIGURE-EIGHT RETRACE KNOT

The figure-eight retrace knot is also called a rerouted figure eight. Although it produces the same result as a figure-eight loop, by tying the knot in a retrace it can be used to fasten the rope to trees or to places where the loop cannot be used (Figure 7-9).

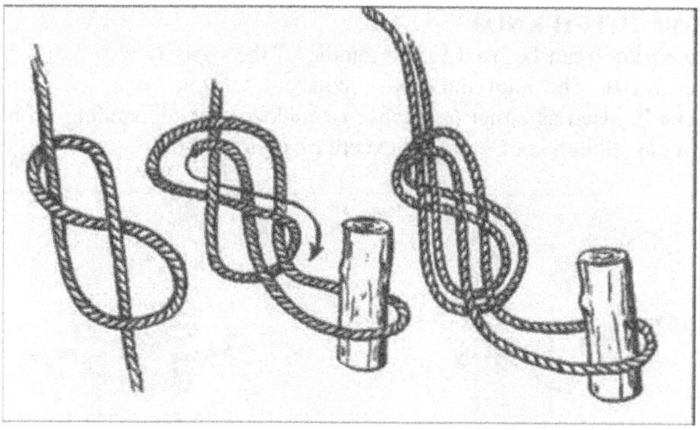

Figure 7-9. Figure-eight retrace.

a. **Tying the Knot.**

STEP 1. Use a length of rope long enough to go around the anchor, leaving enough rope to work with.

STEP 2. Tie a figure-eight knot in the standing part of the rope, leaving enough rope to go around the anchor. To tie a figure-eight knot:

- Form a loop in the rope.
- Wrap the working end around the standing part.
- Route the working end through the loop.
- The finished knot is dressed loosely.

STEP 3. Place the working end around the anchor point.

STEP 4. With the working end, insert the rope back through the loop of the knot in reverse.

STEP 5. Keep the original figure eight as the outside rope and retrace the knot around the wrap and back to the long sttanding part.

STEP 6. Safety off with an overhand knot.

b. **Checkpoints.**

(1) At least 15 centimeters (6 inches) of rope extends past the figure-eight knot.

(2) The original figure eight, tied with the standing end, remains as the outermost rope throughout the knot.

(3) The knot is dressed tightly.

7-10. CLOVE HITCH KNOT

The clove hitch knot can be used in the middle of the rope as well as at the end of the rope (Figure 7-10). The knot must have constant tension on it once tied to prevent slipping. It can be used as either an anchor or middle knot, depending on how it is tied. (Do not use a clove hitch knot on metal except on a picket-hold-fast.)

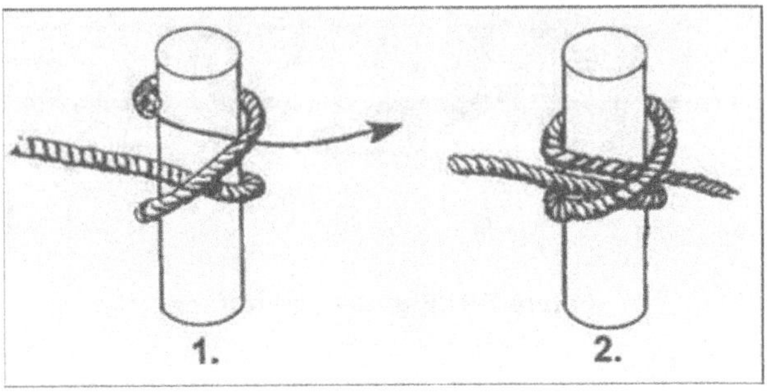

Figure 7-10. Clove hitch knot.

a. **Tying the Knot.**

(1) If there is access over the top of the anchor, the knot is tied as follows:

STEP 1. Hold rope in both hands, palms down with hands together. Slide the left hand to the left from 20 to 25 centimeters (8 to 10 inches).

STEP 2. Form a loop away from and back toward the right.

STEP 3. Slide the right hand from 20 to 25 centimeters (8 to 10 inches) to the right. Form a loop inward and back to the left hand.

STEP 4. Place the left loop on top of the right loop. Place both loops over the anchor and pull both ends of the rope in opposite directions to complete the knot.

(2) If the clove hitch knot must go around the anchor, tie the knot as follows (for instructional purposes, assume that the anchor is horizontal):

STEP 1. Place 76 centimeters (30 inches) of rope over the top of the anchor. Hold the standing end in the left hand. With the right hand, reach under the horizontal anchor, grasp the working end, and bring it inward.

STEP 2. Place the working end of the rope over the standing end (to form a loop). Hold the loop in the left hand. Place the working end over the anchor from 20 to 25 centimeters (8 to 10 inches) to the left of the loop.

STEP 3. With the right hand, reach down to the left-hand side of the loop under the anchor. Grasp the working end of the rope and bring it up and outward.

STEP 4. Dress down the knot.

b. **Checkpoints.**
(1) The knot has two round turns around the anchor with locking bar.
(2) The locking bar is facing 90 degrees from the direction of pull.
(3) The ends exit 180 degrees from each other.
(4) The knot has more than a 15-centimeter (6-inch) pigtail remaining.

SECTION III. MIDDLE ROPE KNOTS

7-11. WIREMAN'S KNOT

The wireman's knot forms a single, fixed loop in the middle of the rope (Figure 7-11).

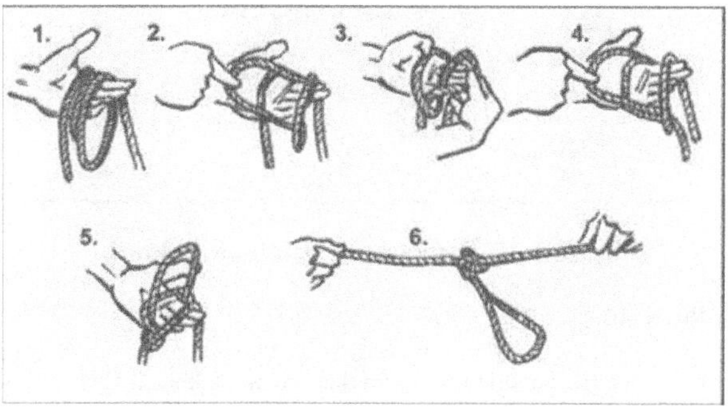

Figure 7-11. Wireman's knot.

a. **Tying the Knot.**

STEP 1. When tying this knot, face the anchor that the tie-off system will be tied to. Take up the slack from the anchor, and wrap two turns around the left hand (palm up) from left to right.
STEP 2. Take up a loop of 30 centimeters (12 inches) in the second round turn to create the fixed loop of the knot.
STEP 3. Name the wraps from the palm to the fingertips: heel, palm, and fingertip.
STEP 4. Grasp the palm wrap with the right thumb and forefinger, and place it over the heel wrap.
STEP 5. Grasp the heel wrap and place it over the fingertip wrap.
STEP 6. Grasp the fingertip wrap and place it over the palm wrap.
STEP 7. Grasp the palm wrap and pull up to form a fixed loop.
STEP 8. Dress the knot down by pulling on the fixed loop and the two working ends.
STEP 9. Pull the working ends apart to finish the knot.

b. **Checkpoints.**

(1) The completed knot should have four separate bights locking down on themselves with the fixed loop exiting from the top of the knot and laying toward the near side anchor point.

(2) Both ends should exit opposite each other without any bends.

7-12. DIRECTIONAL FIGURE-EIGHT KNOT

The directional figure-eight knot forms a single, fixed loop in the middle of the rope that lays back along the standing part of the rope (Figure 7-12).

Figure 7-12. Directional figure-eight knot.

a. **Tying the Knot.**

STEP 1. Face the far side anchor so that when the knot is tied, it lays inward.
STEP 2. Lay the rope from the far side anchor over the left palm. Make one wrap around the palm.
STEP 3. With the wrap thus formed, tie a figure-eight knot around the standing part that leads to the far side anchor.
STEP 4. Dress the knot down with the tail and the bight together.

b. **Checkpoints.**

(1) The loop should be large enough to accept a snaplink but no larger than a helmet-size loop.

(2) The tail and bight must be together.

(3) The figure eight is tied tightly.

(4) The bight in the knot faces back toward the near side.

7-13. BOWLINE-ON-A-BIGHT KNOT

The bowline-on-a-bight knot is used to form two fixed loops in the middle of a rope (Figure 7-13).

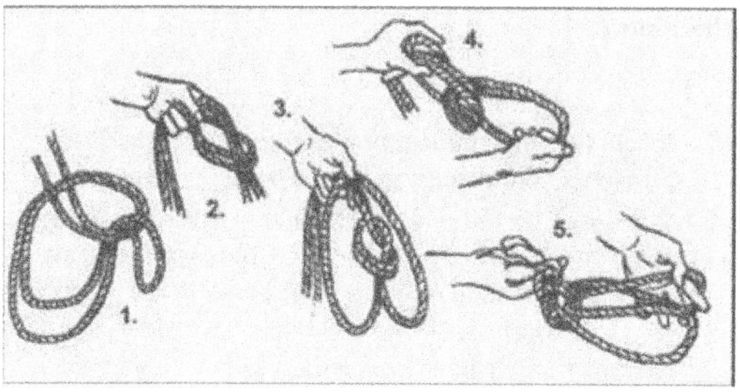

Figure 7-13. Bowline-on-a-bight knot.

a. **Tying the Knot.**

STEP 1. Form a bight in the rope about twice as long as the finished loops will be.
STEP 2. Tie an overhand knot on a bight.
STEP 3. Hold the overhand knot in the left hand so that the bight is running down and outward.
STEP 4. Grasp the bight with the right hand, fold it back over the overhand knot so that the overhand knot goes through the bight.
STEP 5. From the end (apex) of the bight, follow the bight back to where it forms the cross in the overhand knot. Grasp the two ropes that run down and outward and pull up, forming two loops.
STEP 6. Pull the two ropes out of the overhand knot and dress the knot down.
STEP 7. A final dress is required: grasp the ends of the two fixed loops and pull, spreading them apart to ensure the loops do not slip.

b. **Checkpoints.**

(1) There are two fixed loops that will not slip.
(2) There are no twists in the knot.
(3) A double loop is held in place by a bight.

7-14. FIGURE-EIGHT-ON-A-BIGHT KNOT

The figure-eight-on-a-bight knot is used to form two fixed loops in the middle of a rope . (Figure 7-14).

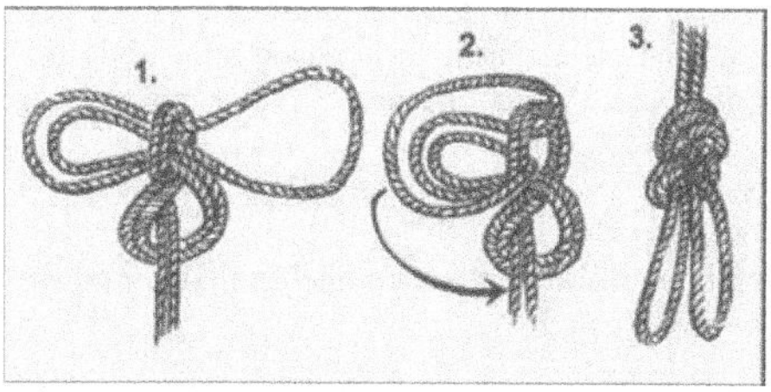

Figure 7-14. Figure-eight-on-a-bight knot.

a. **Tying the Knot.**

STEP 1. Using the doubled rope, form a 46-centimeter (18-inch) bight in the left hand with the running end facing to the left.

STEP 2. Grasp the bight with the right hand and make a 360-degree turn around the standing end in a counterclockwise direction.

STEP 3. With the working end, form another bight and place that bight through the loop just formed in the left hand.

STEP 4. Hold the bight with the left hand, and place the original bight (moving toward the left hand) over the knot.

STEP 5. Dress the knot down.

b. **Checkpoints.**

(1) There is a double figure eight with two loops.

(2) The knot will not slip.

7-15. OVERHAND LOOP KNOT

The overhand loop knot forms a single loop in the middle of a rope. It should not be used in a transport tightening system (Figure 7-15).

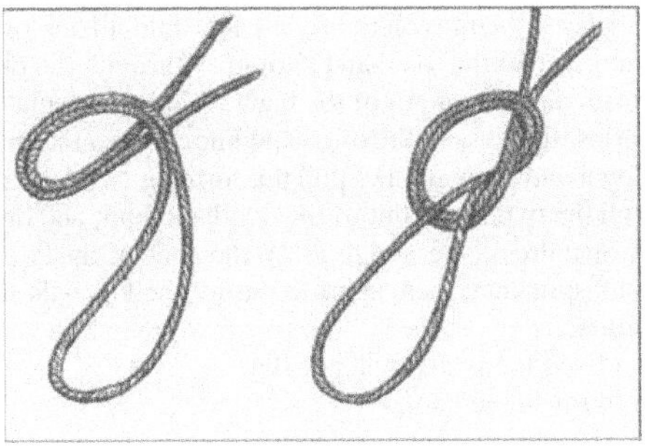

Figure 7-15. Overhand loop knot

a. **Tying the Knot.**

STEP 1. Form a bight in the rope about as long as the loop needed.

STEP 2. With the bight, form an overhand knot using the bight as the working end. Dress the knot tightly.

b. **Checkpoints.**

(1) The loop is the desired size.

(2) The ropes in the overhand knot are parallel and do not cross over each other.

(3) The knot is tightly dressed.

7-16. FIGURE-EIGHT LOOP KNOT

The figure-eight loop knot is a variation of the overhand loop and is easy to untie after it has held weight (Figure 7-16).

Figure 7-16. Figure eight loop knot.

a. **Tying the Knot.**

STEP 1. Form a bight in the rope about as large as the diameter of the desired loop.

STEP 2. With the bight as the working end, form a loop in rope (standing part).

STEP 3. Wrap the working end around the standing part 360 degrees and feed the working end through the loop. Dress the knot tightly.

b. **Checkpoints.**

(1) The loop is the desired size.

(2) The ropes in the loop are parallel and do not cross over each other.

(3) The knot is tightly dressed.

SECTION IV. SPECIAL KNOTS

7-17. SINGLE BUTTERFLY KNOT

The single butterfly knot is used to form a single, fixed loop in the middle of the rope without using the ends (Figure 7-17). The butterfly can be used for the middle man in a rope party as well as in a transport tightening system. The knot can be hard to untie when heavy weight has been placed on it for extended periods.

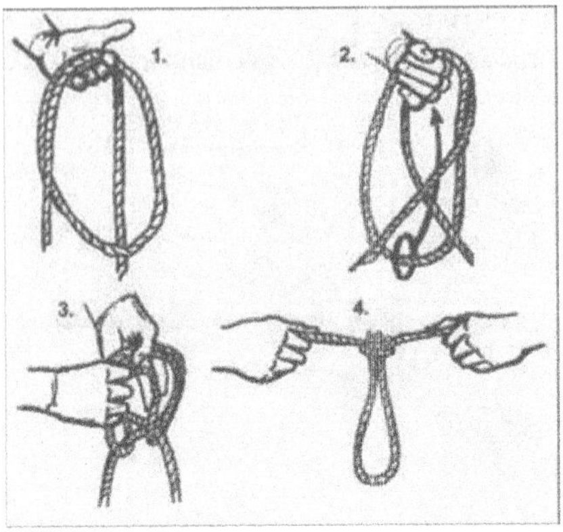

Figure 7-17. Single butterfly knot.

a. **Tying the Knot.**

STEP 1. Form a loop over the palm of the left hand.
STEP 2. With the right hand, cross the two long-standing ends; reach under and grasp the bottom of the loop formed in STEP 1.
STEP 3. Route the bottom of the loop over the crossed standing ends and back through the top of the loop.
STEP 4. Hold the newly formed loop and dress the knot down, removing all of the slack out of the knot.

b. **Checkpoints.**
(1) The wings of the knot are dressed down tightly and as close together as possible.
(2) The ropes between the wings are parallel with no crossovers.
(3) The loop is no larger than a helmet but large enough to accept a snaplink.
(4) All ropes in the knot are tightly dressed.

7-18. PRUSIK KNOT

The prusik knot is used to put a moveable rope on a fixed rope such as a prusik ascent or a tightening system. This knot can be tied as a middle, finger, or end knot.

a. **Tying the Knot.**

(1) *First method.* The prusik knot is tied at the middle or end of a rope; there are many ways to do this (Figure 7-18). The middle-of-the-rope prusik knot can be tied with a short rope to a long rope as follows:

STEP 1. Double the short rope, forming a bight, with the working ends even. Lay it over the long rope so that the closed end of the bight is 30 centimeters (12 inches) below the long rope and the remaining part of the rope (working ends) is the closest to the rappeller; spread the working end apart.

STEP 2. Reach down through the 30-centimeter (12-inch) bight. Pull up both of the working ends and lay them over the long rope. Repeat this process making sure that the working ends pass in the middle of the first two wraps. Now there are four wraps and a locking bar working across them on the long rope.

STEP 3. Dress the wraps and locking bar down to ensure they are tight and not twisted. Tie an overhand knot with both ropes to prevent the knot from slipping during periods of variable tension.

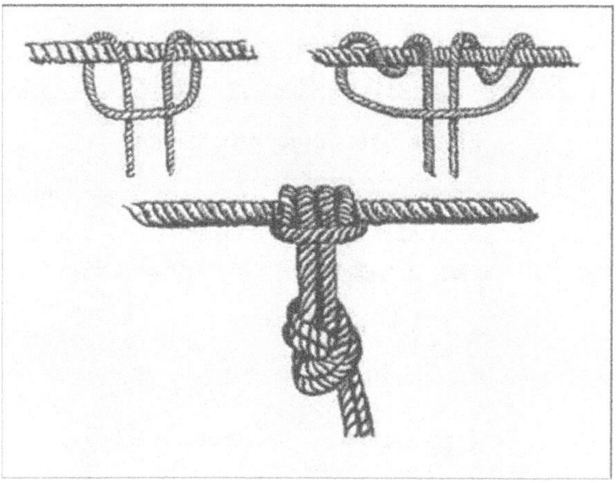

Figure 7-18. Middle-of-the-rope prusik knot.

(2) *Second method.* Another way of tying the prusik knot is the finger prusik (Figure 7-19).

STEP 1. Lay the rope across the thumb and index finger of the left hand.

STEP 2. Wrap the rope around the index finger twice (wrapping away from the rappeller). Wrap the rope around the thumb twice (wrapping towards the rappeller).

STEP 3. Place the thumb and index finger tip to tip. Push the wraps together so that the prusik is on the left thumb. Four wraps and a locking bar should be across them.

STEP 4. Pass the running end of the rope through the wraps of the prusik knot.

STEP 5. Dress the wraps and locking bar down to ensure they are tight and not twisted.

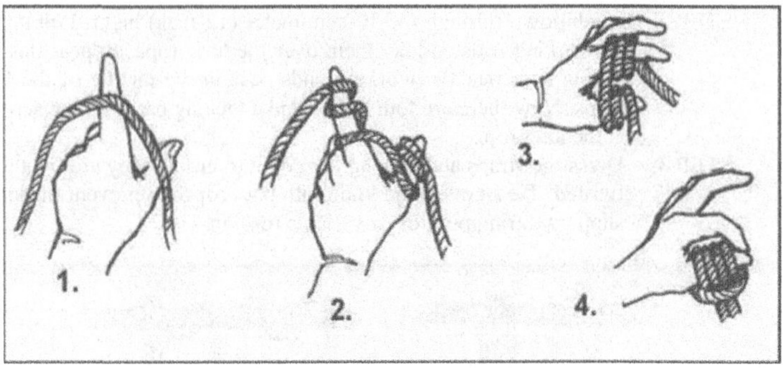

Figure 7-19. Finger prusik knot.

(3) *Third method.* Another way of tying a prusik knot is the end-of-the-rope knot (Figure 7-20).

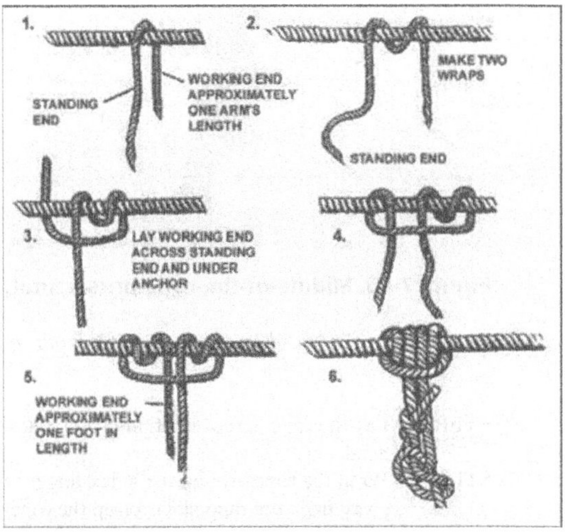

Figure 7-20. End-of-the-rope prusik knot.

STEP 1. Using an arm's length of rope, place it over the long rope.
STEP 2. Form a complete round turn in the rope.
STEP 3. Cross over the standing part of the short rope with the working end of the short rope.
STEP 4. Lay the working end under the long rope.
STEP 5. Form a complete round turn in the rope, working back toward the middle of the knot.

STEP 6. Four wraps and a locking bar should run across them on the long rope. Dress the wraps and locking bar down. Ensure they are tight, parallel, and not twisted.

STEP 7. Finish the knot with a bowline to ensure that the prusik knot will not slip out during periods of varying tension.

b. **Checkpoints.**

(1) Four wraps with a locking bar.

(2) The locking bar faces the rappeller.

(3) The knot is tight and dressed down with no ropes twisted or crossed.

(4) Other than a finger prusik, the knot should contain an overhand or bowline to prevent slipping.

7-19. BACHMAN KNOT

The bachman knot provides a means of using a makeshift mechanized ascender (Figure 7-21).

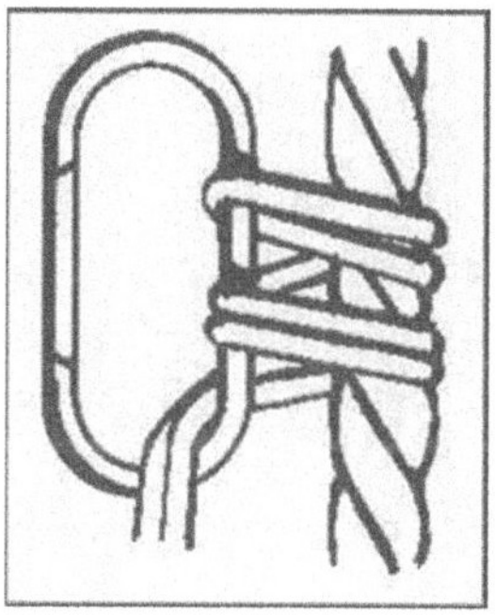

Figure 7-21. Bachman knot.

a. **Tying the Knot.**

STEP 1. Find the middle of a utility rope and insert it into a snaplink.

STEP 2. Place the snaplink and utility rope next to a long climbing rope.

STEP 3. With the two ropes parallel fromthe snaplink, make two or more wraps wraps around the climbing rope and through the inside portion of the snaplink.

Note: The rope can be tied into an etrier (stirrup) and used as a prusik-friction principal ascender.

b. **Checkpoints.**

(1) The bight of the climbing rope is at the top of the snaplink.

(2) The two ropes run parallel without twisting or crossing.

(3) Two or more wraps are made around the long climbing rope and through the inside portion of the snaplink.

7-20. BOWLINE-ON-A-COIL KNOT

The bowline-on-a-coil knot is an expedient tie-in used by rappellers in two- and three-man party climbs when a climbing harness is not available (Figure 7-22).

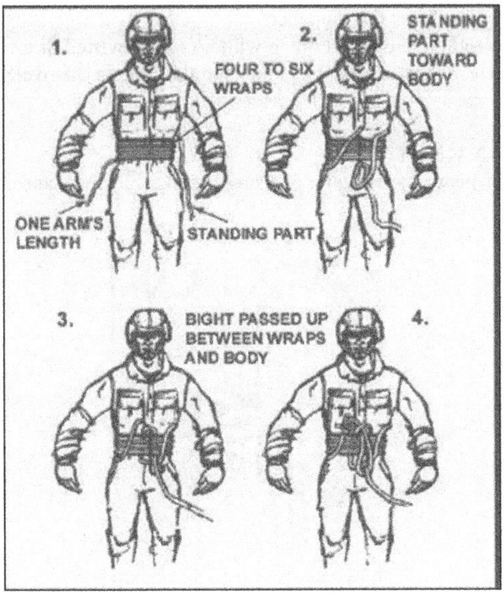

Figure 7-22. Bowline-on-a-coil knot.

Note: A safer tie in, when possible, is a swiss seat and figure eight around the shoulder, to avoid damaging the ribs if the climber falls.

a. **Tying the Knot.**

STEP 1. To tie a bowline-on-a-coil knot, start the same as tying a bowline around the waist. (The waist of the rappeller is the anchor point. The rappeller has about 40 centimeters [16 inches] of rope in the right hand. The standing end is coming from the remainder of the rope in his left hand.)

STEP 2. Draw slack from the standing end of the rope in the left hand, wrapping enough coils around the body to complete a minimum of four (but no more than six) coils.

STEP 3. With the rope in the left hand, make a loop inward with the standing end.

STEP 4. Slide the loop up between the horizontal wraps and the body, ensuring that the loop does not come uncrossed.

STEP 5. Bring a bight up through the loop. Bring the working end of the rope in the right hand through the bight and back onto itself.

OR

Insert the short working end to the right through the loop (just passed through the wraps). Pass the working end around the long standing end and back on itself, forming a bight.

STEP 6. Dress the knot down.

STEP 7. Safety off with an overhand knot against the knot formed on the top single coil. From 10 to 15 centimeters (4 to 6 inches) of rope should be left (standing part).

b. **Checkpoints.**

(1) There are a minimum of four (maximum of six) parallel wraps; the top and bottom ropes cross forward of the hips.

(2) The loop must be underneath all wraps.

(3) Other checkpoints are the same as the bowline.

7-21. THREE-LOOP BOWLINE KNOT

The three-loop bowline knot is used to form three fixed loops in the middle of a rope (Figure 7-23). It is used in a self-equalizing anchor system.

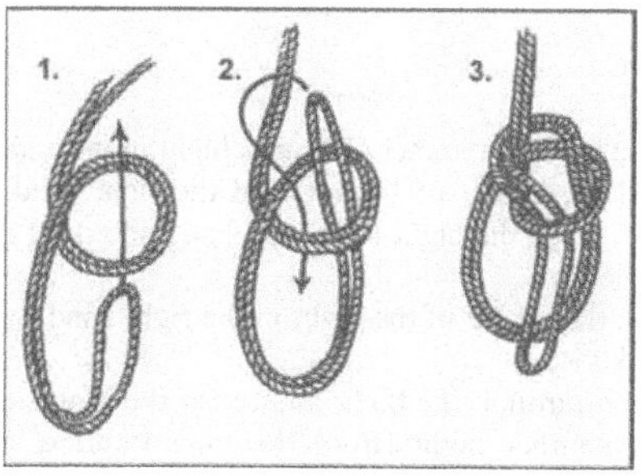

Figure 7-23. Three-loop bowline knot.

a. **Tying the Knot.**

STEP 1. Form about a 60 centimeter (24-inch) bight in the rope, laying it across the left hand, palm up, with the bight to the front.

STEP 2. With the right hand, form a loop with both ropes. Hold it in the left hand.

STEP 3. With the right hand, bring the long bight through the loop(formed in STEP 2) and around and around the two standing ends.

STEP 4. Follow the bight around the standing ends and back down into the loop.

STEP 5. Hold all three loops in the left hand. With the right hand, grasp both standing ropes and pull while dressing up the bowline.

b. **Checkpoints.**

(1) Checkpoints are the same as a regular bowline, except that it is tied with two ropes.

(2) An overhand knot must be tied if the third loop is not to be used as a secondary anchor.

7-22. FIGURE-EIGHT SLIP KNOT

The figure-eight slip knot forms an adjustable bight in a rope (Figure 7-24).

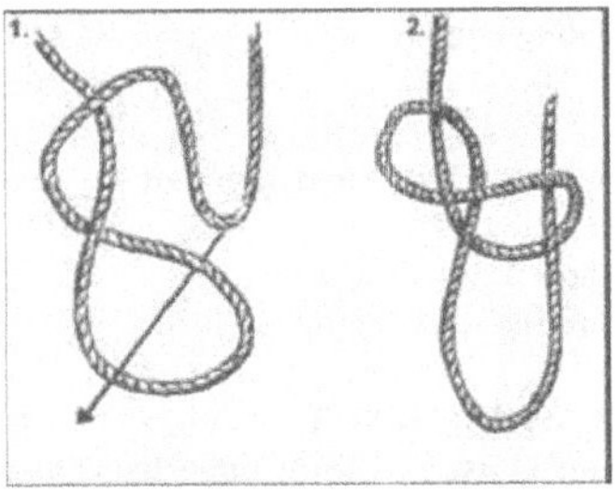

Figure 7-24. Figure-eight slip knot.

a. **Tying the Knot.**

STEP 1. Form a 30-centimeter (12-inch) bight in the end of the rope.

STEP 2. Hold the center of the bight in the right hand. Hold the two parallel ropes from the bight in the left hand about 30 centimeters (12 inches) up the rope.

STEP 3. With the center of the bight in the right hand, twist two complete turns clockwise.

STEP 4. Reach through the bight and grasp the long standing end of the rope. Pull another bight (from the long standing end) back through the original bight.

STEP 5. Pull down on the short working end of the rope and dress the knot down.

STEP 6. If the knot is to be used in a transport tightening system, take the working end of the rope and form a half hitch around the loop of the figure-eight knot.

b. **Checkpoints.**

(1) The knot is in the shape of a figure eight.

(2) The sliding portion of the rope is the long working end of the rope.

Section V. SPECIAL-PURPOSE KNOTS

7-23. KLEIMHIEST KNOT

The kleimhiest knot provides a moveable, easily adjustable, high-tension knot capable of holding extremely heavy loads while being pulled tight (Figure 7-25). This system requires a six-man pulling team when used in a transport tightening system.

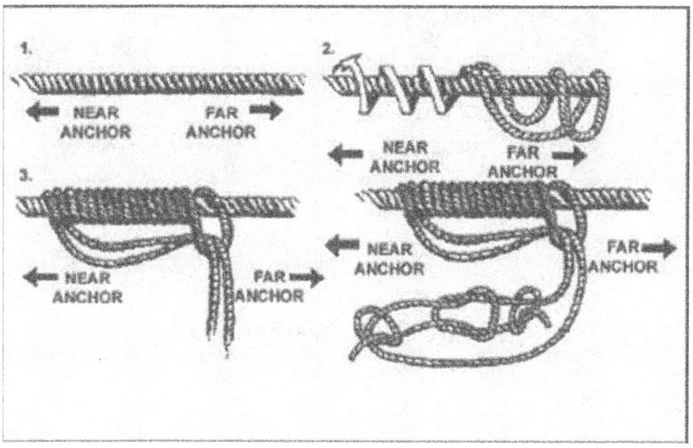

Figure 7-25. Kleimhiest knot.

a. **Tying the Knot.**

STEP 1. Using a utility rope, offset the ends by 30 centimeters (12inches). With the ends offset, find the center of the rope and form a bight. Lay the bight over a horizontal rope near the far side anchor.

STEP 2. Wrap the tails of the utility rope around the horizontal rope back toward the near side anchor. Wrap at least four complete turns.

STEP 3. Pass the remaining tails of the utility rope through the bight (see STEP 1).

STEP 4. Join the two ends of the tail with a joining knot (square knot with overhand safety knots).

STEP 5. Dress the knot down tightly so that all wraps are touching. Insert two snaplinks (opposing gates) into the large loop tied with the tails.

b. **Checkpoints.**

(1) The bight is closest to the far side anchor.
(2) All wraps are tight and touching.
(3) The ends of the utility rope are properly secured with the adjoining knot.

7-24. OVERHAND KNOT

The overhand knot (Figure 7-26) is used to make a knotted rope for a handline, to secure the ends of other knots, and to make stirrups in direct-aid climbing. This knot can also be used to temporarily whip the end of a rope. It is used with other knots.

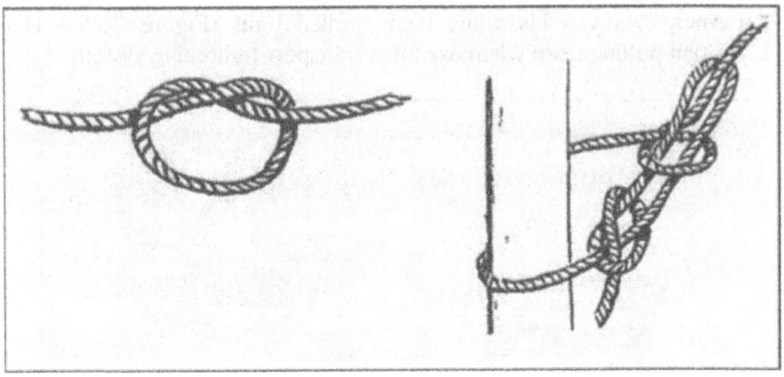

Figure 7-26. Overhand knot.

a. **Tying the Knot.**

STEP 1. Make a loop in the rope.
STEP 2. Pass the working end of the rope through the loop.
STEP 3. Dress down the knot.

b. **Checkpoints.**

(1) Running ends exit the loop in opposite directions.
(2) The knot is dressed down tightly.

7-25. FROST KNOT

The frost knot is used when working with flat webbing (Figure 7-27). It holds better than a standard overhand knot or square knot.

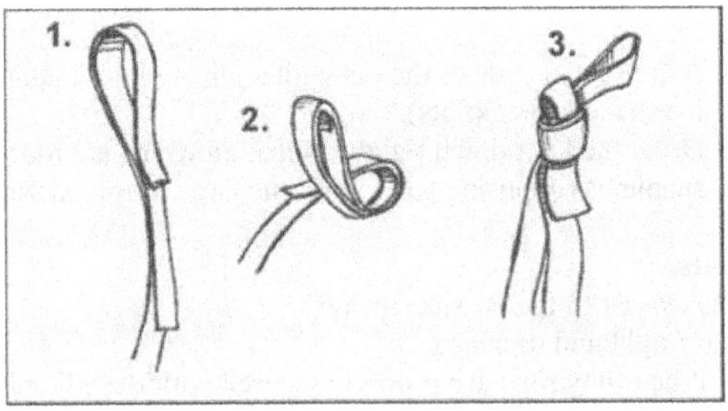

Figure 7-27. Frost knot.

a. **Tying the Knot.**

STEP 1. Lap one end (a bight) of webbing over the other (about 25 to 30 centimeters [10 to 12 inches]).
STEP 2. Tie an overhand knot with the newly formed triple-strand webbing; dress tightly.

b. **Checkpoints.**
(1) The tails of the webbing run in opposite directions.
(2) Three strands of webbing are formed into a tight overhand knot.
(3) A bight and tail exit the top of the overhand knot.

7-26. GIRTH HITCH KNOT

The girth hitch knot is used in tying a runner to a partly driven piton (Figure 7-28).

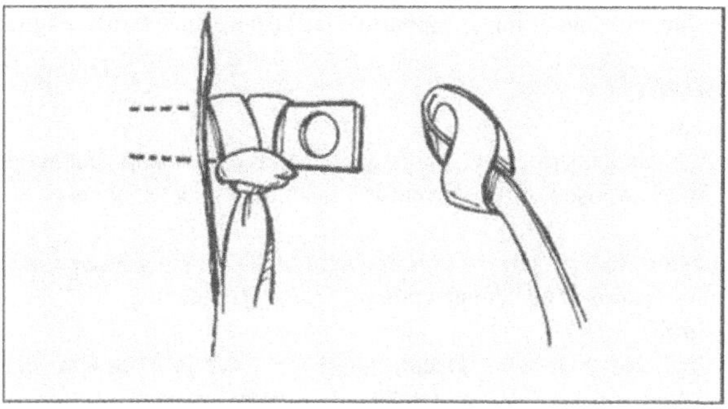

Figure 7-28. Girth hitch knot.

a. **Tying the Knot.**

STEP 1. Bring the standing ends back through the bight of a length of rope or webbing.
STEP 2. Dress the knot tightly.

b. **Checkpoints.**
(1) Two wraps exist with a locking bar running across the wraps.
(2) The knot is dressed tightly.

APPENDIX A

RISK ASSESSMENT

This appendix explains risk assessment for rappel operations. A risk assessement matrix is shown in Table A-1.

A-1. EFFECT

a. **Catastrophic.** Death or permanent total disability, system loss, major property damage.

b. **Critical.** Permanent partial disability, temporary total disability in excess of three months, major system damage, significant property damage.

c. **Moderate.** Minor injury, lost workday accident, compensable injury or illness, minor system damage, minor property damage.

d. **Negligible.** First aid or minor supportive medical treatment, minor system impairment.

A-2. PROBABILITY

a. **Frequent.**

(1) Individual soldier or item—Occurs often in career or equipment service life.

(2) All soldiers exposed or item inventory—Continuously experienced.

b. **Likely.**

(1) Individual soldier or item—Occurs several times in career or equipment service life.

(2) All soldier exposed or item inventory—Occurs frequently.

c. **Occasional.**

(1) Individual soldier or item—Occurs sometime in career or equipment service life.

(2) All soldiers exposed or item inventory—Occurs sporadically, or several times in inventory service life.

d. **Seldom.**

(1) Individual soldier or item—Possible to occur in career or equipment service life.

(2) All soldiers exposed or item inventory—Remote chance of occurrence; expected to occur sometime in inventory service life.

e. **Unlikely.**

(1) Individual soldier or item—Can assume will not occur in career or equipment service life.

(2) All soldiers exposed or item inventory—Possible, but improbable; occurs only very rarely.

A-3. RISK LEVELS

a. **Extremely High.** Loss of ability to accomplish mission.

b. **High.** Significantly degrades mission capabilities in terms of required mission standards.

c. **Medium.** Degrades mission capabilities in terms of required mission standards.

d. **Low.** Little or no impact on mission accomplishment.

RISK ASSESSMENT			HAZARD PROBABILITY				
			Frequent	Likely	Occasional	Seldom	Unlikely
			A	B	C	D	E
EFFECT	Catastrophic	I	*Extremely High*				
	Critical	II			*High*		
	Moderate	III		*Medium*			
	Negligible	IV				*Low*	

Table A-1. Risk assessment matrix.

APPENDIX B

BASIC EQUIPMENT FOR RAPPEL OPERATIONS

The equipment listed in this appendix is the standard basic items required to conduct ground and air rappelling operations.

B-1. GROUND OPERATIONS

For ground operations, the following equipment is needed:

a. Nylon rope, 120 feet (NSN 4020-00-931-8793).
b. Snaplinks (NSN 8465-00-360-0228).
c. Gloves (NSN 8415-00-268-7868).
d. Rope coiling log or standard deployment bag.
e. Rappel seat.
f. Rappel rings (used as expedient anchor points).

B-2. AIR OPERATIONS

For air operations, the following equipment is needed:

a. All items in paragraph B-1.
b. Donut ring (UH-1H operations only) (locally produced TSC item).
c. Safety floating ring.
d. Two 6-foot safety ropes (with snaplinks).
e. Eight 120-foot nylon ropes.
f. Monkey harness for static rappel master.
g. Sixteen snaplinks (UH-60 operations).
h. Twenty-two snaplinks (UH-1H operations).
i. FM communications.
j. Aircraft communications headset.

GLOSSARY

ACRONYMS AND ABBREVIATIONS

AFRM	assistant fast-rope master
AGL	above ground level
AMC	air mission commander
AR	Army regulation
ATP	aircrew training program
attn	attention
BDU	battle dress uniform
CG	commanding general
CGU	modified helicopter safety strap
chemlight	chemical light
cm	centimeter
DA	Department of the Army
D-bag	deployment bag
F	Fahrenheit
FM	frequency modulated; field manual
FRIES	fast-rope insertion/extraction system
FRM	fast-rope master
GSO	ground safety officer
HQ	headquarters
IAW	in accordance with
ICS	interagency communications system
intercomm	intercommunication
LBE	load-bearing equipment
LZ	landing zone
MEDEVAC	medical evacuation
mm	millimeter
mph	miles per hour
MSO	mountaineering safety officer
NCO	noncommissioned officer
NCOIC	noncommissioned officer in charge
NSN	national stock number
NVG	night vision goggles

NWTC	Northern Warfare Training Center
OIC	officer in charge
PIC	pilot in command
POI	program of instruction
POL	petroleum, oils, and lubricants
RATELO	radiotelephone operator
RSO	rappel safety officer
SFC	sergeant first class
SOAR	Special Operations Aviation Regiment
SOP	standing operating procedures
SPIES	special patrol infiltration/exfiltration system
STABO	a system for extracting personnel by helicopter (the combined first letters of the surnames of the five persons who designed the system)
SSO	STABO system operator
TAPO	Technology Applications Program Office
TC	training circular
TM	technical manual
TRADOC	Training Doctrine Command
TSC	training support center
UIAA	Union of International Alpine Association
UH	utility helicopter
US	United States
USAF	United States Air Force
USAIS	United States Army Infantry School
USASOC	United States Army Special Operations Command
VIP	very important person

REFERENCES

DOCUMENTS NEEDED

These documents must be available to the intended users of this publication.

DA Form 2028	Recommended Changes to Publications and Blank Forms. 1 February 1974.
DA Form 5752-R	Rope History and Usage. May 1989.
FM 57-38*	Pathfinder Operations. 9 April 1993.
FM 7-93*	Long Range Surveillance Unit Operations. 3 October 1995.
TC 90-6-1*	Military Mountaineering. 26 April 1989.

READINGS RECOMMENDED

These sources contain relevant supplemental information.

AR 95-3	Aviation: General Provisions, Training, and Standardization, A Resource Management. 27 September 1990.
FM 8-35	Evacuation of the Sick and Wounded. 22 December 1983
FM 57-220	Basic Parachuting Techniques and Training. 1 June 1990.
TM 10-1670-262-12	Operator's and Organizational Maintenance Manual (Including Repair Parts and Special Tools List) for Personnel STABO Extraction System and Cargo Personnel Lowering Anchor Device. 7 April 1988.
USASOC Reg 350-2	Training Airborne Operations. 1 June 1995.
USASOC Reg 350-6	Training Special Operations Forces Infiltration/Exfiltration Techniques. 12 May 1995.
	The Air Assault School Handbook.

*This source was also used to develop this publication.

INDEX

TC 21-24
10 SEPTEMBER 1997

By Order of the Secretary of the Army:

DENNIS J. REIMER
General, United States Army
Chief of Staff

Official:

JOEL B. HUDSON
Administrative Assistant to the
Secretary of the Army
03807

DISTRIBUTION:

Active Army, Army National Guard, and U.S. Army Reserve: To be distributed in accordance with the initial distribution number 115059, requirements for TC 21-24.

Made in the USA
Monee, IL
05 November 2026

67820240R00256